Physics Research and Technology

nova
science publishers
www.novapublishers.com

Physics Research and Technology

Spatial Homeostasis, Quantum Information Channel, and the Nature of Living Things Within the Framework of the Theory of Byuon
Yuriy Baurov, PhD
2022. ISBN: 978-1-68507-590-3 (eBook)

A Closer Look at Plasmonics
Christian Bosch (Editor)
2022. ISBN: 978-1-68507-709-9 (Hardcover)
2022. ISBN: 978-1-68507-725-9 (eBook)

An Introduction to Charge Carriers
Jai Singh, PhD (Editor)
2022. ISBN: 978-1-68507-456-2 (Hardcover)
2022. ISBN: 978-1-68507-531-6 (eBook)

Handbook of Research on Heat Transfer
Suvanjan Bhattacharyya, PhD (Editor)
Varun Goel, PhD (Editor)
2022. ISBN: 978-1-68507-459-3 (Hardcover)
2021. ISBN: 978-1-68507-542-2 (eBook)

Solar PV System: Design, Installation, Operation and Maintenance
L. Ashok Kumar, PhD
K. Mohana Sundaram
2021. ISBN: 978-1-68507-341-1 (Hardcover)
2021. ISBN: 978-1-68507-475-3 (eBook)

More information about this series can be found at
https://novapublishers.com/product-category/series/physics-research-and-technology/

Chandra Sekhar Rout
and Brahmananda Chakraborty
Editors

2D Metallic Transition Metal Dichalcogenides

Fundamentals and Applications

nova
science publishers
www.novapublishers.com

DOI: https://doi.org/10.52305/WPGU1394

Library of Congress Cataloging-in-Publication Data

ISBN: 978-1-68507-965-9

Published by Nova Science Publishers, Inc. † New York

My late parents Geeta Chakraborty and Tarani Chakraborty

-Brahmananda Chakraborty

Contents

Preface

Reports on graphene from 2004 have emerged and defined the birth of graphene analogous 2D materials. Understanding of different unusual physical properties and applications of graphene has provided the path to explore the vibrant research area on 2D layered materials during the past decade. Transition metal dichalcogenides (TMDs) are among the heavily studied 2D materials in recent years that differ from the semi metallic characters of graphene. They show novel, diverse and tunable layer-dependent properties and can be controlled by their crystalline structure, number of stacking layers and defect concentration. The 2D materials can be semiconductor, metal, semi-metal or superconductors. Also, the indirect band gap to direct band gap transition of the 2D material is possible by tuning the number of layers. By achieving the direct band gap of the material, it is possible to achieve enhanced photoluminescence, optoelectronic properties and controllable electrical properties suitable for fabrication of high-performance transistors. Similarly, hybridization with different 2D materials provides a very effective and powerful way not only to control the physiochemical properties of hybridized species but also to explore nanocomposites with novel functionalization.

In this book, we aim to focus on the metallic type 2D TMDs materials which have vast areas of applications. The book covers various aspects of 2D metallic materials including their fundamental properties, synthesis and characterization, recent theoretical investigations, applications in various fields such as energy storage and conversion, optical and optoelectronic devices, field emitters, catalysis, chemical and biosensors etc. Due to their sub-nanometre thickness, 2D materials have a high packing density which is suitable for the fabrication of thin film energy storage devices. Also, flexibility of 2D materials, good mechanical properties and high packing densities make them suitable for the development of thin and flexible devices. Curved and crumpled nanosheets generate porosity when deposited onto the current collectors in energy storage devices facilitating ion transport. Considering the versatile applications and economic aspect, a thorough revelation of the focus areas involving 2D materials is relevant and necessary. We strongly believe this book will provide most of the important properties and applications in a single reference.

Acknowledgements

CSR acknowledges Department of Science and Technology (DST)-SERB Early Career Research project (Grant No. ECR/2017/001850), DST-Nanomission (DST/NM/NT/2019/205(G), Karnataka Science and Technology Promotion Society (KSTePS/VGST-RGS-F/2018-19/GRD NO. 829/315). BC would like to thank Dr. Nandini Garg, Dr. T. Sakuntala, Dr. S.M. Yusuf and Dr. A. K Mohanty for their encouragement and support.

Chapter 1

Introduction to TMDs and Metallic TMDs

Chandra Sekhar Rout[1,*]
and Brahmananda Chakraborty[2,3]

[1]Centre for Nano and Material Science, Jain University, Jain Global Campus, Jakkasandra, Ramanagaram, Banglore, India
[2]High Pressure and Synchrotron Radiation Physics Division, Bhabha Atomic Research Centre, Trombay, Mumbai, India
[3]Homi Bhabha National Institute, Mumbai, India

Two dimensional (2D) layered materials have gained tremendous attention in past decades due to their novel and intriguing properties with potential applications in energy storage, conversion, optoelectronics, electrical devices, electrolysis, spintronics and other devices [1, 2]. Recently 2D transition metal dichalcogenides (TMDs) have revealed to be having interesting physicochemical properties due to their ultra-thin thickness, tunable bandgap, mechanical flexibility etc. [3, 4]. By employing these interesting properties, 2D TMDs reported to have multifunctional applications for energy devices, biosensors, field emitters and so on. 2D TMDs are usually represented by MX_2, where M is the transition metal (Mo, W, Ti, V, Nb, Hf, Ta etc.) and X represents either S, Se or Te [5]. In TMDs each MX_2 layer is sandwiched between two layers of chalcogen atoms with covalent bonds and the neighbouring MX_2 layers are separated by a van der Waals gap [6, 7].

Since different transition metals have varied number of d- electrons, its interaction and bonding with chalcogen atoms differ in the case of TMDs. This leads to the formation of a wide range of TMD materials with varied electronic properties with metallic, semiconducting, insulating or superconducting

* Corresponding Author's Email: r.chandrasekhar@jainuniversity.ac.in.

In: 2D Metallic Transition Metal Dichalcogenides
Editors: Chandra Sekhar Rout and Brahmananda Chakraborty
ISBN: 978-1-68507-965-9

nature [8, 9] Also, the physiochemical properties of TMDs are highly tunable by different methodologies including dimension tuning, doping, defect engineering, alloying, gating and heterojunction tuning etc. [10, 11].

At recent times, it is predicted both experimentally and theoretically that MX_2 based materials undergo stable structural phase transitions with 1T, 2H and 3R crystal structures with tetragonal, hexagonal and rhombohedral crystal symmetry respectively. It is predicted that the d- electron counts in TMDs can induce diverse Fermi-level locations when it interacts with different chalcogen atoms. Hence, the position of the Fermi-level significantly affects the properties of the TMDs [13, 14, 15].

Among TMDs, Mo and W based chalcogenides are known for their semiconducting nature with tunable layer-dependent indirect to direct band gap transitions [16]. On the other hand V, Nb, Ta based dichalcogenides are reported to show metallic properties with applications in several domains including superconductivities, charge density wave and magnetism etc. [17, 18]. These metallic TMDs are very fascinating compounds due to their unique properties of strongly competing effects of charge-density wave (CDW) formation and superconductivity. The overall electronic phase diagrams of their layered superconductors are strikingly similar to those found in high -T_C cuprates and the iron based pnictides [19, 20]. Further, the superconductivity and CDW transition temperature of these metallic TMDs can be tuned by changing the electron count through chemical substitution or intercalation and applying high pressure [21, 22, 23].

The goal of the book is to put a spotlight on emerging fundamental properties, synthesis approaches and emerging applications of these 2D metallic TMDs. The book consists of nine chapters. Various interesting properties of these TMDs includes the crystal and layer-dependent electronic properties, CDW, electrical properties, magnetic and optical properties. Structure and properties of metallic TMDs have been discussed in details in Chapter 2. Compared with much studied semiconducting TMDs, research about metallic TMDs is younger and thus their unique properties are still under the course of unfolding. In particular, Dirac semimetal and Weyl semimetal, whose electron energy is linearly connected to lattice momentum, exhibit many wonderful physical properties, e.g., the Fermi arcs in surface state, the magnetic unipole in momentum space, unique transport properties, etc. [24]. These arose as a highly thrilling material system for exotic devices e.g., the next generation ultrafast memories, spintronic devices, and quantum computing [25, 26]. Theoretical aspects and understanding of the relation between the structural changes of the TMDs and its electronic properties have been covered in a Chapter 3. DFT, being among the most popular and versatile available methods for computational quantum mechanical modelling tool for determining properties of a many-electron system, is ideal for investigating the electronic structure of metallic TMDs using functionals of spatially dependent electron density. Such investigations are almost beyond the scope of physical experiments. For instance, density of states, projection of atomic orbitals, binding energy and dielectric functions are shown to play a major role to find out the molecular adsorption characteristics targeted for sensing applications [27, 28]. Detailed synthesis approaches of these metallic TMDs by top-down, bottom-up methods and combined strategies have been extensively discoursed in Chapter 4. The methods which involve synthesis of TMDs from elementary (atom/molecule) level as starting materials, such as Chemical Vapour Deposition (CVD) or Molecular Beam Epitaxy (MBE) are collectively known as bottom-up approaches whereas the reverse in which bulk material is scalled down to form layered TMDs, e.g., Mechanical or liquid-phase exfoliation, ultrasonic exfoliation etc. are called top-down approaches [29]. Relative advantages and limitations of these approaches has been highlighted in detail through exclusive literature survey. In Chapters 5-8, the importance of metallic TMDs for such as chemical and biosensors, energy storage and conversion, catalysis and fuel cells, field emission respectively have been discussed at great depth. Detection of chemicals such as hazardous gas or liquid is vital for industrial and environmental safety. 2D layered materials in general and TMDs in particular are of paramount importance in this aspect for their tunability characteristics stemming from unfilled d orbital of the transition metal cation. A host of heterostructure, decoration, size and morphology alteration methodologies are in emergence for improving their sensing ability for targeted applications [30]. A sizable account of the state of the art findings about this theme has been furnished in Chapter 5. Similarly, catalytic suitability of metallic TMDs for usage as power bank in the form of battery or supercapacitor and efficient energy conversion through different evolution reactions such as OER, HER, photocatalysis, and usage as fuel cells have been illustrated through extensive literature survey in Chapters 6 and 7 respectively [31, 32, 33, 34]. TMDs and other 2D substances own more active sites, and strident edges in atomic scale along with point defects such as vacancies, favourable for fabricating cathodes with high current density even with a little turn-on voltages. Thus, expulsion of electrons from such materials are anticipated to be strikingly higher due to the drops in the dimensionality, quantum restraint consequence of electrons in 2D layers, and the incidence of sharp and random protuberant active edges [35]. Chapter

8 has thus been devoted for reviewing field emission applications of metallic TMDs. Finally, pertinent concluding remarks have been made in Chapter 9 for consolidation in the form of summary and future perspectives.

We strongly believe that this book will provide a thorough revelation of the four areas involving 2D metallic TMD materials. It will provide very useful information to the readers on the important properties, synthesis and applications of metallic TMDs in a single reference. This book will serve as a useful textbook for undergraduates, graduate students and researchers working in the 2D materials field. Moreover, the chapters are contributed by eminent editors and authors who have vast experience in these fields and would undoubtedly attract a major portion of the academic community working on 2D materials and its derivatives to read, consult, and refer this book.

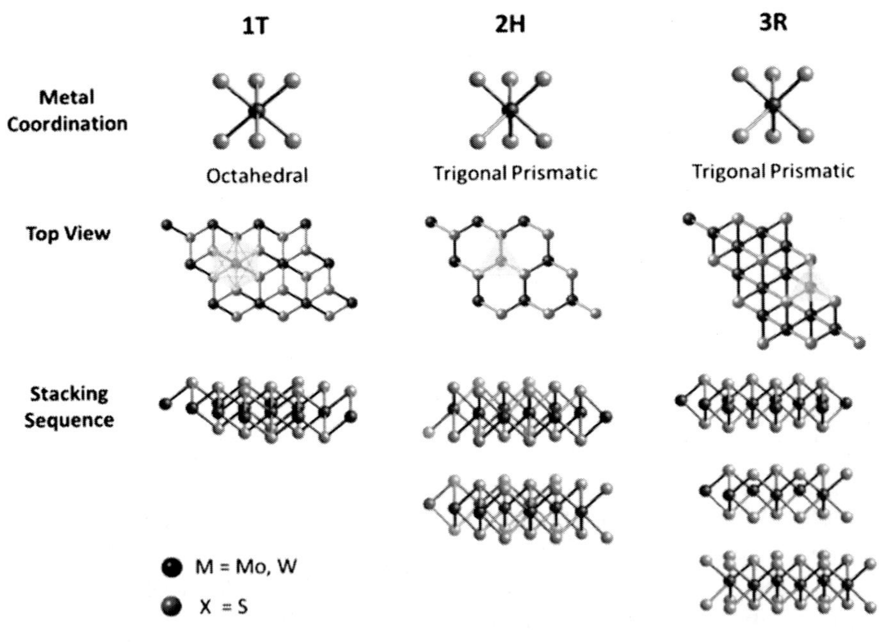

Figure 1.1. Metal coordination and stacking sequences of TMD structural unit cells. Metal coordination can be either octahedral or trigonal prismatic. The octahedral coordination allows stacking sequences, which yield a tetragonal symmetry (1T). Dissimilar stacking sequences of trigonal prismatic single layers can give rise to different symmetries: hexagonal symmetry (2H) and rhombohedral symmetry (3R). (Reprinted with permission from [12] Copyright (2017) Royal Society of Chemistry).

References

[1] Xu, M., Tao Liang, Minmin Shi, and Hongzheng Chen. "Graphene-Like Two-Dimensional Materials." *Chemical Reviews* 113, no. 5 (May 8, 2013): 3766-98. https://doi.org/10.1021/cr300263a.

[2] Zhou, J., Junhao Lin, Xiangwei Huang, Yao Zhou, Yu Chen, Juan Xia, Hong Wang, et al., "A Library of Atomically Thin Metal Chalcogenides." *Nature* 556, no. 7701 (April 2018): 355-59. https://doi.org/10.1038/s41586-018-0008-3.

[3] Chhowalla, M., Debdeep Jena, and Hua Zhang. "Two-Dimensional Semiconductors for Transistors." *Nature Reviews Materials* 1, no. 11 (2016): 1-15. https://doi.org/10.1038/natrevmats.2016.52.

[4] Chhowalla, M., Hyeon Suk Shin, Goki Eda, Lain-Jong Li, Kian Ping Loh, and Hua Zhang. "The Chemistry of Two-Dimensional Layered Transition Metal Dichalcogenide Nanosheets." *Nature Chemistry* 5, no. 4 (April 2013): 263-75. https://doi.org/10.1038/nchem.1589.

[5] Manzeli, S., Dmitry Ovchinnikov, Diego Pasquier, Oleg V. Yazyev, and Andras Kis. "2D transition metal dichalcogenides." *Nature Reviews Materials* 2, no. 8 (2017): 1-15. https://doi.org/10.1038/natrevmats.2017.33.

[6] Mak, K. F., Changgu Lee, James Hone, Jie Shan, and Tony F. Heinz. "Atomically Thin MoS 2 : A New Direct-Gap Semiconductor." *Physical Review Letters* 105, no. 13 (September 24, 2010): 136805. https://doi.org/10.1103/PhysRevLett.105.136805.

[7] Das, S., Joshua A. Robinson, Madan Dubey, Humberto Terrones, and Mauricio Terrones. "Beyond Graphene: Progress in Novel Two-Dimensional Materials and van Der Waals Solids." *Annual Review of Materials Research* 45, no. 1 (2015): 1-27. https://doi.org/10.1146/annurev-matsci-070214-021034.

[8] Yang, J., Weike Wang, Yan Liu, Haifeng Du, Wei Ning, Guolin Zheng, Chiming Jin, et al., "Thickness Dependence of the Charge-Density-Wave Transition Temperature in VSe2." *Applied Physics Letters* 105, no. 6 (2014): 14-18. https://doi.org/10.1063/1.4893027.

[9] Eichberger, M., Hanjo Schäfer, Marina Krumova, Markus Beyer, Jure Demsar, Helmuth Berger, Gustavo Moriena, Germán Sciaini, and R. J. Dwayne Miller. "Snapshots of Cooperative Atomic Motions in the Optical Suppression of Charge Density Waves." *Nature* 468, no. 7325 (December 2010): 799-802. https://doi.org/10.1038/nature09539.

[10] Li, L. J., E. C. T. O&amp, apos, Farrell, K. P. Loh, G. Eda, B. Ozyilmaz, and A. H. Castro Neto. "Controlling Many-Body States by the Electric-Field Effect in a Two-Dimensional Material." *Nature* 529, no. 7585 (January 14, 2016): 185-97.

[11] Wang, Q. H., Kourosh Kalantar-Zadeh, Andras Kis, Jonathan N. Coleman, and Michael S. Strano. "Electronics and Optoelectronics of Two-Dimensional Transition Metal Dichalcogenides." *Nature Nanotechnology* 7, no. 11 (2012): 699-712. https://doi.org/10.1038/nnano.2012.193.

[12] Toh, R. J., Zdeněk Sofer, Jan Luxa, David Sedmidubský, and Martin Pumera. "3R Phase of MoS2 and WS2 Outperforms the Corresponding 2H Phase for Hydrogen

Evolution." *Chemical Communications* 53, no. 21 (March 9, 2017): 3054-57. https://doi.org/10.1039/C6CC09952A.

[13] Jaramillo, T. F., Kristina P. Jørgensen, Jacob Bonde, Jane H. Nielsen, Sebastian Horch, and Ib Chorkendorff. "Identification of Active Edge Sites for Electrochemical H2 Evolution from MoS2 Nanocatalysts." *Science (New York, N.Y.)* 317, no. 5834 (July 6, 2007): 100-102. https://doi.org/10.1126/science.1141483.

[14] Lu, C. P., Guohong Li, Jinhai Mao, Li-Min Wang, and Eva Y. Andrei. "Bandgap, Mid-Gap States, and Gating Effects in MoS2." *Nano Letters* 14, no. 8 (August 13, 2014): 4628-33. https://doi.org/10.1021/nl501659n.

[15] Yang, H., Sung Wng Kim, Manish Chhowalla, and Young Hee Lee. "Structural and Quantum-State Phase Transition in van Der Waals Layered Materials." *Nature Physics* 13, no. 10 (October 4, 2017): 931-37. https://doi.org/10.1038/nphys4188.

[16] Zhang, Y. J., T. Oka, R. Suzuki, J. T. Ye, and Y. Iwasa. "Electrically Switchable Chiral Light-Emitting Transistor." *Science* 344, no. 6185 (May 16, 2014): 725-28. https://doi.org/10.1126/science.1251329.

[17] Pathan, M. A. K., Aakash Gupta, and Mihai E. Vaida. "Exploring the Growth and Oxidation of 2D-TaS2 on Cu(111)." *Nanotechnology* 32, no. 50 (September 2021): 505605. https://doi.org/10.1088/1361-6528/ac244e.

[18] Zhang, Z., Jingjing Niu, Pengfei Yang, Yue Gong, Qingqing Ji, Jianping Shi, Qiyi Fang, et al., "Van Der Waals Epitaxial Growth of 2D Metallic Vanadium Diselenide Single Crystals and Their Extra-High Electrical Conductivity." *Advanced Materials* 29, no. 37 (2017): 1702359. https://doi.org/10.1002/adma.201702359.

[19] Sahu, M. R., Pratap Raychaudhuri, and Anindya Das. "Andreev Reflection near the Dirac Point at Graphene - NbSe2 Junction." *Physical Review B* 94, no. 23 (December 30, 2016): 235451. https://doi.org/10.1103/PhysRevB.94.235451.

[20] Nakata, Y., Katsuaki Sugawara, Ryota Shimizu, Yoshinori Okada, Patrick Han, Taro Hitosugi, Keiji Ueno, Takafumi Sato, and Takashi Takahashi. "Monolayer 1T-NbSe2 as a Mott Insulator." *NPG Asia Materials* 8, no. 11 (November 2016): e321-e321. https://doi.org/10.1038/am.2016.157.

[21] Li, F., Kaixiong Tu, and Zhongfang Chen. "Versatile Electronic Properties of VSe2 Bulk, Few-Layers, Monolayer, Nanoribbons, and Nanotubes: A Computational Exploration." *The Journal of Physical Chemistry C* 118, no. 36 (September 11, 2014): 21264-74. https://doi.org/10.1021/jp507093t.

[22] Xing, Y., Kun Zhao, Pujia Shan, Feipeng Zheng, Yangwei Zhang, Hailong Fu, Yi Liu, et al., "Ising Superconductivity and Quantum Phase Transition in Macro-Size Monolayer NbSe2." *Nano Letters* 17, no. 11 (November 8, 2017): 6802-7. https://doi.org/10.1021/acs.nanolett.7b03026.

[23] Shi, J., Xina Wang, Shuai Zhang, Lingfeng Xiao, Yahuan Huan, Yue Gong, Zhepeng Zhang, et al., "Two-Dimensional Metallic Tantalum Disulfide as a Hydrogen Evolution Catalyst." *Nature Communications* 8, no. 1 (October 16, 2017): 958. https://doi.org/10.1038/s41467-017-01089-z.

[24] Yu, R., Quansheng Wu, Zhong Fang, and Hongming Weng. "From Nodal Chain Semimetal to Weyl Semimetal in HfC." *Physical Review Letters* 119, no. 3 (July 18, 2017): 036401. https://doi.org/10.1103/PhysRevLett.119.036401.

[25] He, L. P., X. C. Hong, J. K. Dong, J. Pan, Z. Zhang, J. Zhang, and S. Y. Li. "Quantum transport evidence for the three-dimensional Dirac semimetal phase in Cd 3 As 2." *Physical review letters* 113, no. 24 (2014): 246402. https://doi.org/10.1103/PhysRevLett.113.246402.

[26] Guo, P. J., Huan-Cheng Yang, Kai Liu, and Zhong-Yi Lu. "Type-II Dirac semimetals in the YPd 2 Sn class." *Physical Review B* 95, no. 15 (2017): 155112. https://doi.org/10.1103/PhysRevB.95.155112.

[27] Lee, E., Young Soo Yoon, and Dong-Joo Kim. "Two-Dimensional Transition Metal Dichalcogenides and Metal Oxide Hybrids for Gas Sensing." *ACS Sensors* 3, no. 10 (October 26, 2018): 2045-60. https://doi.org/10.1021/acssensors.8b01077.

[28] Joshi, N., Takeshi Hayasaka, Yumeng Liu, Huiliang Liu, Osvaldo N. Oliveira, and Liwei Lin. "A Review on Chemiresistive Room Temperature Gas Sensors Based on Metal Oxide Nanostructures, Graphene and 2D Transition Metal Dichalcogenides." *Mikrochimica Acta* 185, no. 4 (March 10, 2018): 213. https://doi.org/10.1007/s00604-018-2750-5.

[29] Zheng, B., Chao Ma, Dong Li, Jianyue Lan, Zhe Zhang, Xingxia Sun, Weihao Zheng, et al., "Band Alignment Engineering in Two-Dimensional Lateral Heterostructures." *Journal of the American Chemical Society* 140, no. 36 (September 12, 2018): 11193-97. https://doi.org/10.1021/jacs.8b07401.

[30] Late, D. J., Anha Bhat, and Chandra Sekhar Rout. "Chapter 2 - Fundamentals and Properties of 2D Materials in General and Sensing Applications." In *Fundamentals and Sensing Applications of 2D Materials*, edited by Morgan Hywel, Chandra Sekhar Rout, and Dattatray J. Late, 5-24. Woodhead Publishing Series in Electronic and Optical Materials. Woodhead Publishing, 2019. https://doi.org/10.1016/B978-0-08-102577-2.00002-6.

[31] Gao, Y. P., Xu Wu, Ke-Jing Huang, Ling-Li Xing, Ying-Ying Zhang, and Lu Liu. "Two-Dimensional Transition Metal Diseleniums for Energy Storage Application: A Review of Recent Developments." *CrystEngComm* 19, no. 3 (January 16, 2017): 404-18. https://doi.org/10.1039/C6CE02223E.

[32] Li, H., Xiaofan Jia, Qi Zhang, and Xun Wang. "Metallic Transition-Metal Dichalcogenide Nanocatalysts for Energy Conversion." *Chem* 4, no. 7 (July 12, 2018): 1510-37. https://doi.org/10.1016/j.chempr.2018.03.012.

[33] Sumesh, C. K., and Sebastian C. Peter. "Two-Dimensional Semiconductor Transition Metal Based Chalcogenide Based Heterostructures for Water Splitting Applications." *Dalton Transactions* 48, no. 34 (August 27, 2019): 12772-802. https://doi.org/10.1039/C9DT01581G.

[34] Khan, K., Ayesha Khan Tareen, Muhammad Aslam, Yupeng Zhang, Renheng Wang, Zhengbiao Ouyang, Zhongyi Gou, and Han Zhang. "Recent Advances in Two-Dimensional Materials and Their Nanocomposites in Sustainable Energy Conversion Applications." *Nanoscale* 11, no. 45 (November 21, 2019): 21622-78. https://doi.org/10.1039/C9NR05919A.

[35] Patra, A., Mahendra A. More, Dattatray J. Late, and Chandra Sekhar Rout. "Field Emission Applications of Graphene-Analogous Two-Dimensional Materials: Recent Developments and Future Perspectives." *Journal of Materials Chemistry C* 9, no. 34 (September 2, 2021): 11059-78. https://doi.org/10.1039/D1TC02054D.

Chapter 2

Structure and Properties of Metallic TMDs

Gopal Sanyal[1] and Brahmananda Chakraborty[2,3,*]
[1]Mechanical Metallurgy Division, Bhabha Atomic Research Centre, Trombay, Mumbai, India
[2]High Pressure and Synchrotron Radiation Physics Division, Bhabha Atomic Research Centre, Trombay, Mumbai, India
[3]Homi Bhabha National Institute, Mumbai, India

Abstract

Transition metal dichalcogenides (TMDs) are incrusted materials that contain transition metals and group VIA chalcogens in 1:2 share. Such layered materials have been expansively explored for quite a few decades. These may be insulators, semiconductors, or metallics disclosing all sorts of solid-state characteristics since a magnetic matrix slanted to superconducting traits. Metallic TMDs are unique for many interesting features such as charge density wave (CDW), magneto-resistance, superconductivity, topological insulator, and Weyl semimetal. These characteristics are imperative for applications in various arenas like energy storage devices and catalysts for energy transformation. This review is to present the readers a state-of-the-art concept regarding development of metallic TMDs owing to their above exclusive properties and electronic structure.

Keywords: TMD, band structure, electronics, sensing, magnetism, super-conductivity

---

* Corresponding Author's Email: brahma@barc.gov.in.

In: 2D Metallic Transition Metal Dichalcogenides
Editors: Chandra Sekhar Rout and Brahmananda Chakraborty
ISBN: 978-1-68507-965-9

2.1. Introduction

Two-dimensional (2D) van der Waals (vdW) layered materials, that possess a robust covalent attachment inside the sheet and feeble vdW interface between the coats, expose themselves as partaking exceptional layer-dependent characters prominently different from bulk substances. For instance, graphene is known as one of the vividly cultivated classes [1] for their amazing optical and electrical properties, starting since 2004. Subsequently, countless 2D materials have been revealed for a decade. Boron nitride has been well-thought-out as a pattern to advance graphene's carrier mobility while compared with usual insulator substrates e.g., SiO_2 [2]. Further 2D materials, like black phosphorus [3], borophene [4], silicone [5], and stanene [6], also come under the 2D category. Moreover, there are many compounds with transition metals and chalcogens to shape a huge class of layered transition metal dichalcogenides (TMDs). Reentering of the first TMDs has been done since the semiconducting types e.g., MoS_2, that has long been recognized as a lubricant alike graphene. Fascinating physical characteristics of a host of semiconducting TMDs have previously been conferred somewhere else [7]. Unlike semiconducting TMDs, metallic TMDs (Figure 2.1) possess a limited density of states at Fermi level and their study domains somewhat differ from semiconductors. Thus, it is valuable to revisit a few phenomena that are renowned in bulk substances e.g., periodic lattice distortion (PLD) or charge density wave (CDW), magnetism, and superconductivity in metallic 2D vdW constituents. Recent developments in metallic TMDs have supplementarily expanded this particular study field. For example, Weyl semimetals and phase manipulation involved in contact resistance are quite new domains [8]. From synthesis viewpoint, some progresses have been done with chemical vapor deposition (CVD) process. Monolayer metallic TMDs are effortlessly synthesizable with CVD, that is somewhat hard to attain by other means like flux or CVT approaches. Fresh studies on metallic TMDs have hardly been steered copious in contrast with the semiconducting counterparts. Numerous chemical and physical phenomena are yet being investigated for metallic TMDs.

This review composes nine sections counting Introduction. In the following part (section 2.2), we present the basic H and T edifices of metallic TMDs and their protracted assemblies by bearing in mind the stacking sequences. We precisely familiarize the metallic chalcogenides that have different stoichiometric ratios e.g., 1:1 or 2:3 or ternary (quaternary), though, such materials are out of space and omitted in this focused review. We deliver them in tabular form at the end of the section. It is critical to understand thoroughly the electronic band structure of this kind of materials. Theoretical calculation, in combination with angle-resolved photoemission spectroscopy (ARPES), is capable to present the band structure more clearly [9].

Figure 2.1. Kinds of metals associated in layered dichalcogenides. Blue color designates transition metals for at present developed metallic TMDs; yellow, green, and red small rectangles are S, Se, and Te compounds. White rectangles for chalcogen are unreported or nonexistent phase in metallic TMDs.

Section 2.3 discourses metallic TMDs' band structure. We highlight the electronic structures of chalcogens and transition metals to relate the band structures of demonstrative metallic TMDs such as $TiSe_2$, TiS_2, and VSe_2 with hypothetical computations and experiments with ARPES. We additionally acquaint with the electronic band structures of bulk $NbSe_2$ and TaS_2, having both H and T phases. Dependence of their properties on number of layers will be conferred in the last portion of section 2.3. The density functional theory (DFT) is the furthermost adaptable and general method for calculations of band structure [10]. The main impression of DFT is to substitute the wave function basic set with the electron charge density, which is termed as the mean field approximation [11]. By such practice, a large quantity of electrons may be considered as electron density, that is a three-dimensional space function having three degrees of freedom. Nevertheless, this tactic originates problems in reckoning exchange (e.g., the interface sourced from the blurry attribute of electrons) and connection (e.g., the energy error while thinking electron clouds to be an autonomous object) of electrons. It is the inspiration for the advance of assortments of several exchange-correlation functions since the local density approximation, generalized gradient approximation to hybrid functionals [12]. Here, the band structure utilizing different functionals for one metallic TMD will also be demonstrated.

The presence of CDW phases supplementary to the superconductivity changeover happens in several metallic TMDs, and that turns this kind of layered things striking to the condensed matter scientists. This subject is the major attention of section 2.4. The spectacle of PLD conversions along with their machinery are labelled by an innate depiction of one-dimensional Peierls distortion. It is important to note that the CDW transformation yields a subordinate symmetry construction, that destroys the decadence of normal phase's electronic band. This is analogous to chemistry's Jahn−Teller effect, in which molecules exhibit a propensity to garble their edifice into lower symmetry. Significantly, one energy stage will be fragmented to form various states, in which electrons can inhabit a lower energy arrangement in contrast to the one of greater symmetry constructions. In disparity, the Peierls model merely ponders one band model deprived of any degeneracy. It will be discussed how to outspread Peierls model for CDW transformation in actual things. Clarifications will be furnished to show that entirely exposing a band gap to uphold the PLD phase's low energy state is not compulsory. The phenomenon of phonon relaxing and its relevance to superconductivity will also be revisited. The association with Kohn anomaly, that also exposes the phonon unstiffening for alteration of the electron screening, is scrutinized too. We would also confer portrayal approaches to determine PLD transition of phase.

Lately, the advent of topological substances and 2D ferromagnetism fetches special treatment to metallic TMDs, that is discoursed in two subsequent sections. The section 2.5 has been dedicated to point out restricted optical properties of metallic TMDs. In element, we familiarize ferromagnetism in VS_2 and VSe_2 and converse the impact of doping, strain, and imperfections on magnetic characteristics in section 2.6. On the other hand, precisely, we will narrate the perception and band structure of Weyl semimetals, that is emergent in 1T phase $MoTe_2$ and T_d phase WTe_2 and their blend creation in section 2.7. Change in characteristics while foreign atoms are introduced through doping, intercalation or formation of heterojunction either with other TMD or carbon-based nanomaterials has been briefed in section 2.8.

The final section (2.9) abridges perspectives and issues of phase transformation, physical properties and devices. The likelihood of phase transformation through the alloy and residual matters on Weyl semimetals and superconductivity will be conversed also. In addition to that, ways of controlling the properties of metallic TMDs towards targeted applications will also be covered.

2.2. Crystal Structures

2.2.1. Organizational Categories

Chalcogen atoms are extremely responsive to metals, generating many metal chalcogenide blends with a number of stoichiometric proportions. Amid these complexes, TMDs partake layered constructions through a 1:2 share among metals and chalcogen atoms [13]. Elemental assembly of TMDs comprises a layer of transition metal (M) crammed between a couple of chalcogen atoms' (X) sheets. M-X jots generate robust covalent bonds inside the film. Every sheet is loaded plumb through a feeble vdW connections, and the incrusted substance can classically be exfoliated into discrete films. The basic edifices of H and T phases of TMDs are introduced in the following.

2.2.1.1. H and T Phases

TMDs, whichever semiconducting or metallic phase, contain two tetrahedrons (Figure 2.2). Each tetrahedron designates the context of H and T phases by ball-and-stick replicas; blue and yellow balls specify M and X atoms, respectively. Bottom and top tetrahedrons are settled evenly from metal surface (center) to shape a trigonal prismatic construction (H-phase) (Figure 2.2a). The other chief component, an octahedral or T phase, revealed in Figure 2.2b, is built by spinning the top (or bottom) tetrahedron by 180°. The plan view of H phase (Figure 2.2a, inset) displays a hexagonal edifice. Nevertheless, in realism, two indistinguishable chalcogen sheets are separated at top and bottom with reverence to the metal film. For T phase, X atoms in upper sheet are projected amid X atoms in the lower sheet (Figure 2.2b, inset).

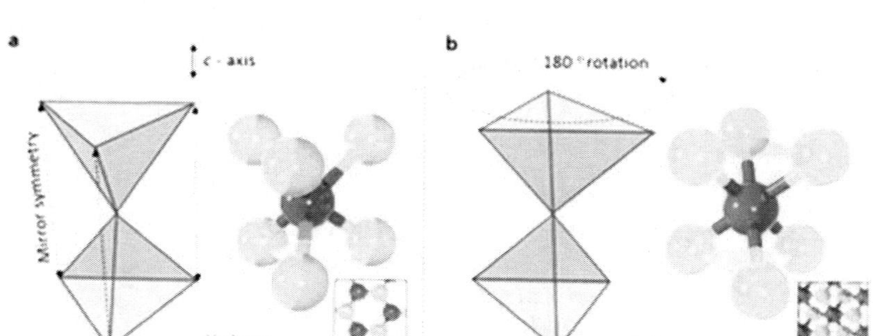

Figure 2.2. Two chief phases for 1:2 (metal:chalcogen) ratio metallic TMDs: (a) H phase, (b) T phase.

2.2.1.2. Extended Structures

Piling order influences protuberant physical attributes of matters e.g., electronic band structures, optical properties and phonon vibrations [14]. Figures with the phase terminology e.g., "2"H, "1"T, "3"R, and "4"H designates the heaping arrangement of TMDs. For instance, 2H-MoS_2 specifies the alliance of two sheets with AB-sequencing, whereas 3R-NbS_2 pronounces the retelling of three coats as one cluster, i.e., ABC stacking. Plan and lateral sights of the protracted assemblies are revealed in Figure 2.3. 2H edifice is intricate in the hexagonal space group (P63/mmc); the plan view displays a hexagonal trellis by flashing two X atoms and one M atom. Though mainstream 2H assemblies in TMDs show semiconducting characteristics, very scarce 2H phases are acknowledged to be metallic TMDs counting 2H-$NbSe_2$, 2H-NbS_2, 2H-$TaSe_2$, and 2H-TaS_2 [15]. On the other hand, 1T phase is essentially the prime assembly belonging to hexagonal space group ($P\bar{3}m1$) in metallic TMDs. The construction molded in that case is an octahedron instead of a trigonal prismatic for the spinning of a tetrahedron.

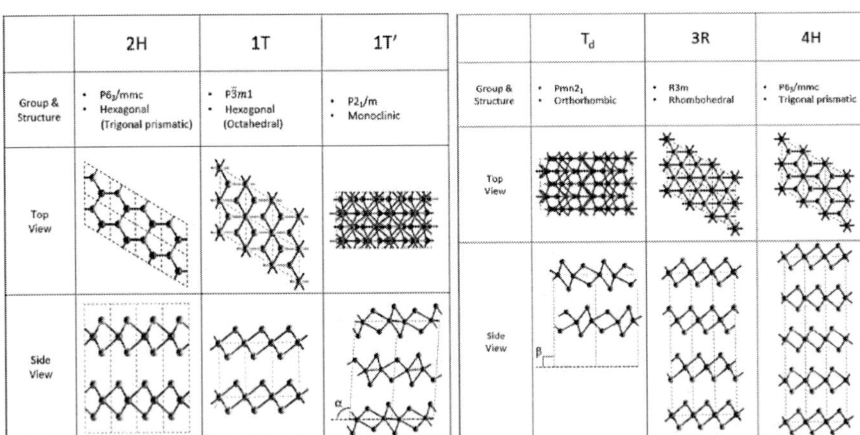

Figure 2.3. The phases with stacking sequences in metallic TMDs.

Dimerization of M atoms persuades the alteration of 1T assembly to shape 1T′ phase (P21/ m). Particularly, it induces shift of X atoms in an out-of-plane course and a 3- to 2-fold regularity transition. 1T′-$MoTe_2$ is a typical such example in metallic TMDs [16]. The T_d construction, observed in WTe_2 [17], is comparable to the 1T′ edifice. Dissimilarity among 1T and T_d is the c-axis angle ($\alpha \neq 90$, $\beta = 90$), as specified in Figure 2.3. 3R-phase (R3m) embraces three films in the unit cell. For instance, TaS_2, $TaSe_2$, NbS_2, and $NbSe_2$, which look as if in both T and H phases too, possess 3R phases [15]. In 4H phase, two distinct heaping sequences occur. The 4H construction in Figure 2.3 is the $4H_b$ kind, which can be labelled as a blinking phase of T−H−T−H layers, though the unit cell of the $4H_a$ edifice is made of H phase coatings only [18-19].

2.2.1.3. Allied Compounds

In tallying to the archetypal 1:2 ratio stoichiometry for metallic TMDs, there are extra kinds of transition metal chalcogenides (TMCs) and complexes. For instance, T phase of VS_2 and TiS_2 or their selenide amalgams display metallic comportment [20-21]. Alloy formed through partial substitution of M or X atom may lead to formation of $Ti_xV_{1-x}S_2$ and $Ti_xV_{1-x}Se_2$, or TiS_xSe_{1-x} or VS_xSe_{1-x} that uphold a 1:2 M-X ratio though the topology does not vary by alloying [22]. These atomic conversations, whether it is termed doping or alloying, could persuade its inflection on phononic and electrical responses. Phosphor-chalcogenides or transition metal phosphorus trichalcogenides (C12/m1, monoclinic), characteristically shortened as MPX_3 (M, metal; P, phosphorus; X, chalcogens), is a novel category of encrusted compounds for spintronics [23-25]. There is existence of substances with 1:1 stoichiometry between M and X atoms. For illustration, FeSe possesses a tetragonal structure (P4/nmm) [26]. Since the Fe plane, X atoms are interchangeably situated in a z-orientation. Fe places can be substituted or nobbled by Cu atoms. These compounds have fascinating physical properties e.g., high-temperature superconductivity overhead 100 K on strontium titanate ($SrTiO_3$) substrate [27]. Metal chalcogenides having 2(M):3(X) generates a quintuple deposit. Bi_2Te_3, Bi_2Se_3, and Sb_2Te_3 are typical substances for such quintuple layer prearrangement and recognized as superior thermoelectric resources. These compounds also disclose features of topological insulators [28]. As there are plentiful blends of common metal chalcogenides, the current review encompasses only a 1:2 ratio incrusted metallic TMDs.

A list of metallic TMDs is given in Table 2.1. The first and second pillars display kinds of M and X, respectively. The M column comprises transition metal classes e.g., V, Ti, W, Nb, Zr, Mo, Pd, Ta, Pt, Ir, Hf, and Ni, which joined with chalcogens such as S, Se, and Te. The third pilaster defines the agreeing phases for each metal-chalcogen grouping. H, T, and R are essentially the shortenings of hexagonal, tetragonal, and rhombohedral assemblies, correspondingly.

Table 2.1. Various existent metallic TMDs and their phases

Metal	Chalcogen	Phase	Metal	Chalcogen	Phase
Ti	S	T	Hf	Te	T
	Se	T	Ta	S	H, T, R
	Te	T		Se	H, T, R
V	S	T		Te	T′
	Se	T	Ir	Te	T
	Te	T	Pt	S	T
Zr	S	T		Se	T
	Se	T		Te	T
Nb	S	H, T, R	Mo	S Te	T (mix) T_d, T′
	Se	H, T, R	W	Te	T_d, T′
	Te	T	Pd	Te	T

2.3. Electronic Structures

Electrons in atoms generate discrete energy stages whereas in molecules they create a sequence of new electronic vigor levels via orbitals' hybridization, gathered as bonding (engaged) and antibonding (untenanted) levels [29]. Solids contain a vast number of atoms, that creates an assortment of isolated energy echelons with minor energy disparities between themselves. These energy heights can be assembled into valence and conduction bands, analogous to engaged and untenanted oomph echelons in molecules, correspondingly. In case, the atoms are organized in an order, a crystal is shaped. Then, the electronic construction can characteristically be labelled in the momentum or reciprocal, space (e.g., k-space) [30]. If there is no overlay between valence and conduction bands that happens when the sum of electrons in the primitive cell is even, a band gap occurs among these two bands, founding an insulator or semiconductor. The gap amid the two bands is termed as the energy band gap. The Fermi stage is positioned in principle at the bandgap's central area. If quantity of electrons in primitive cell is odd, the solid turns into metallic since an energy band is partly occupied. Alternative condition is when there is a minor overlay in k-space amongst valence and conduction band, so that the notion of valence and conduction band is not vibrant any further. In that situation, irrespective of whether total quantity of electrons in primitive cell is even or odd, the resources have metallic features with a huge amount of free electrons at Fermi superficial. If the overlain energy area is small, such things are named semimetal with a little electron and hole densities at Fermi layer. These substances are intriguing, with two dissimilar kinds of transporters: hole- and electron- like carriers, that generate a hefty magnetoresistance [31-32]. The condition turns out to be more interesting while there is a node amid valence and conduction bands. Weyl semimetals and topological insulators fall into this group, which are deliberated in section 2.6 [8, 28]. As crystals are designed by a convincingly well-ordered atomic arrangement, their electronic band structures are sturdily linked to their crystal constructions (e.g., bond angles, lengths and coordination) and the electronic arrangements of the component elements [33]. Crystal assembly of the solid is also fixed by electronic edifices of the participating elements. Though the forecast for presence of complexes (e.g., chemical conformation) and their constructions is still exciting from hypothetical aspect, consideration of the elements' electronic energy is a critical footstep to form an innate image of the crystals' attributes. Chalcogen atoms consume six valence electrons in *s* and *p* orbitals (s^2p^4), though transition metals partake an extensive array of electron arrangements with the farthest electrons mostly positioned at *d* and *s* orbitals. The electron features close to Fermi level of TMDs are obtained by exchanges of charge amid electrons in the *p*, *s*, and *d* orbitals of every type. Energy stage of the *s* orbital in chalcogen atoms is comparatively profound in valence band and hardly subsidize much to the band structure by Femi level [34]. Group IVB transition metals e.g., Ti, Zr, and Hf partake four electrons at external shell, i.e., two for each in *d* and *s* orbitals with a d^2s^2 arrangement. Most happening segment of this group with chalcogens in TMDs is the 1T phase. Source of such a 1T phase is hardly known at the moment. Alteration in the band structure of MX_2 of group IV can be unthinkingly elucidated by bearing in mind the comparative energy difference between chalcogens and metals. Since total sum of valence electrons in the unit cell is 10, i.e., an even number, which is expected to show semiconducting characteristics. For instance, ZrS_2, $ZrSe_2$, HfS_2, $HfSe_2$ and TiS_2 are semiconductors [35]. Energy range of the *d* orbital upsurges from Ti to Zr to Hf, that results in a greater energy disparity from energy level of *p* orbital of chalcogens. So, the bandgap amid valence and conduction bands upsurges, resulting in a bigger bandgap in HfS_2 ($HfSe_2$) comparing with ZrS_2 ($ZrSe_2$) [36]. Likewise, the band gap reduces when S is substituted by Se. Ultimately, semimetals are obtained in Te compounds [35]. $TiSe_2$ is apparently analogous with a semiconductor. Nevertheless, energy stages of Se *p* orbital and Ti *d* orbital are approximately same, leading to an overlay in energy to generate a semimetallic phase. Band structure of TiX_2 has been reviewed in more details later. The TMDs shaped since group VB transition metals are constantly metallic for occurrence of odd quantum of electrons in their detours [37]. VX_2 is steady with a 1T phase, whereas in cooperation NbX_2 and TaX_2 possess 2H phase. 1T phases are chemically robust at only elevated temperature [37]. Rather, at ambient temperature, they transmute to distorted phases. An inferior symmetry edifice fitting to space group 12 of MTe_2 is also detected in this group. Group VIB transition metals possess six valence electrons. The robust electron arrangement of both Mo and Cr is d^5s^1, that is the half-filled arrangement. In distinction, the steadiest form in W is d^4s^2. Though, it is worth mentioning that d^5s^1 arrangement has an energy alike d^4s^2 configuration in W [33]. CrX_2 is essentially a metastable phase with a 1T construction, that hardly appear in the equipoise phase diagram of Cr-X [38]. On the other hand, MoX_2 and WX_2 stake a mutual 2H steady phase with semiconductor properties at ambient temperature, exclusive of WTe_2, having a T_d structure [39]. The $MoTe_2$ exposes a T′ (a partisan form of T_d) phase at elevated temperature [40]. The T′ phase can transmute to T_d phase and contrariwise by governing the Te vacancy, pressure, or temperature [41].

2.3.1. Typical Illustrations

2.3.1.1. 1T-TiS_2 and 1T-$TiSe_2$

Elementary band structure of solid $TiSe_2$ and TiS_2 were not defined for an extended time [33]. It is so, as the pragmatic band structures can't be simply attained since the entire k-space; furthermore, the replication technique for band structure computation generally undervalues actual bandgap [42]. The band structures of bulk $TiSe_2$ and TiS_2 have lately been determined with MBJ potential, B3LYP hybrid functional, and a mature ARPES practice. Figure 2.4 displays computed and practical band structures of 1T-$TiSe_2$ and 1T-TiS_2 [43-44]. In TiS_2, no intersection between valence and conduction band is there. Consequently, TiS_2 is indeed a semiconductor having a minor indirect band gap of 0.53 eV, where the supreme (least) of valence (conduction) band is situated at the Γ point (M point). The theoretical and practical data are pretty matching, even though a slighter gap (0.4 eV) is foretold since the conjectural result, as revealed in Figure 2.4a, b [45]. A minor electron pocket in conduction band appears at the M point for the n-doping action of the S vacancies, that are generated throughout the crystal growth, resulting in six Fermi curves at M points, as revealed in Figure 2.4c. Compared to TiS_2, a trivial overlay amongst conduction and bands valence amidst Γ, A. and L points arises in $TiSe_2$ (Figures 2.4d–f). As a result, $TiSe_2$ performs as semimetal with an electron compact nearabout the L point and a hole concise at the Γ point [44].

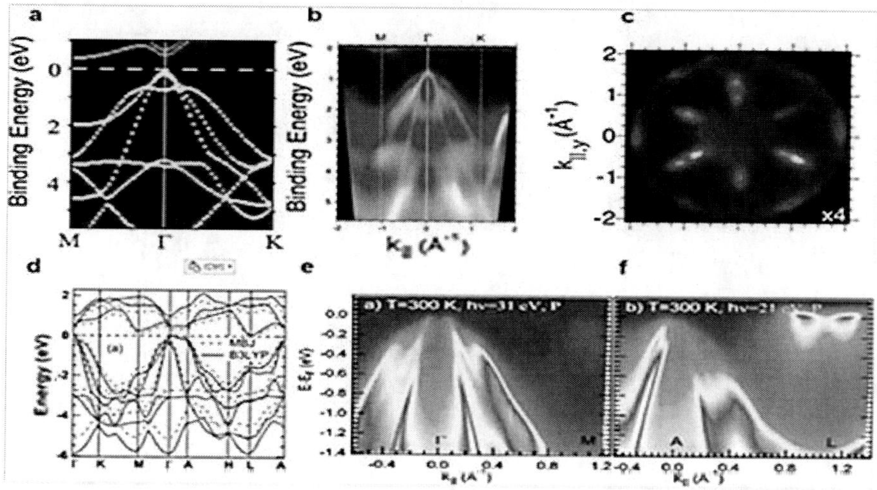

Figure 2.4. (a) Computed and (b) pragmatic ARPES of TiS_2 and (c) the Fermi surface. Reproduced with permission [45]. Copyright 2015 Institute of Physics Publishing. (d) Premeditated and (e, f) investigational ARPES of $TiSe_2$. (Reproduced with permission [44]. Copyright 2017 Elsevier).

2.3.1.2. 1T-VSe₂

Vanadium fits in group VB at the periodic table. There is an extra electron in unit cell of 1T-VSe_2 than in 1T-$TiSe_2$. Thus, it is noticeable that 1T-VSe_2 is metallic having an incompletely occupied band [46]. Figure 2.5 illustrates Fermi surface and band structure of VSe_2 quantified by ARPES at 10 K with soft X-ray [46]. It is evident that Fermi level passes the band lengthwise the Γ–K course (Figure 2.5e), representing metallic origin of VSe_2. Sideways the z track (Γ–A), the band dispersal is lesser than that in the planer courses (Γ–M and Γ–K). There exist two virtual flat bands with minor scattering, one is about Fermi stage and other is nearly 2 eV underneath the Fermi layer. VSe_2's Fermi interface is a locked plane with six twists. Parts f and g of Figure 2.5 display computed and investigational Fermi superficial in several faces. The hypothetical recreation settles fine with the trial data.

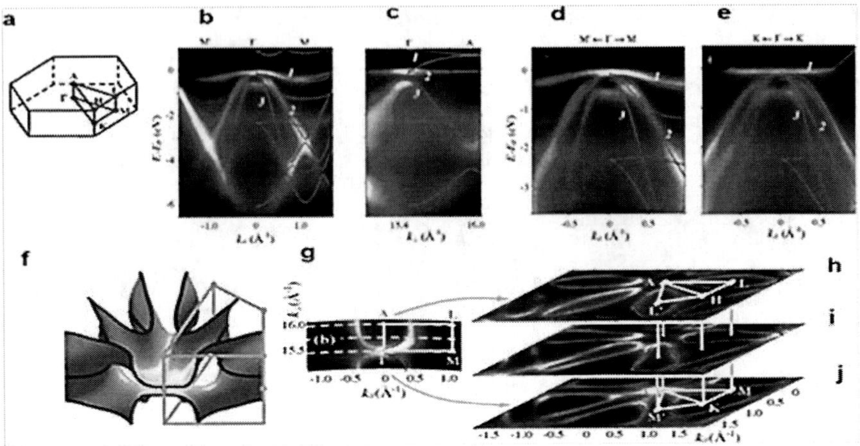

Figure 2.5. (a) Hexagonal Brilluoin zone (b, c) low and (d, e) high resolution ARPES band structure of VSe_2. (f–j) The Fermi surface slices of nominated BZ faces. (Reproduced with permission [46]. Copyright 2012 American Physical Society).

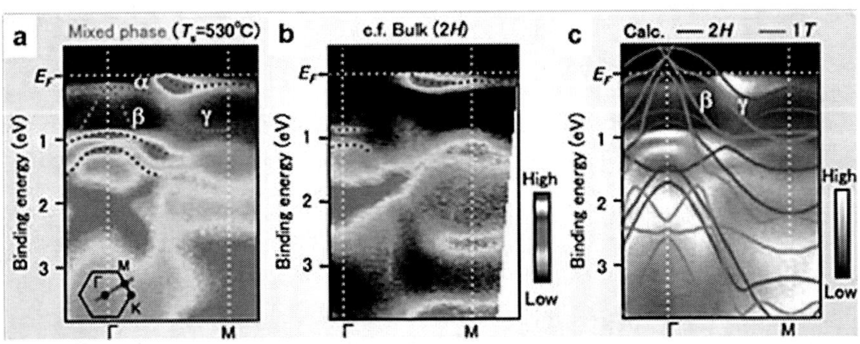

Figure 2.6. Band structure of $2H\text{-}NbSe_2$ and $1T\text{-}NbSe_2$ to display the Mott–Hubbard band in the Mott insulator phase (1T). (Reprinted with permission [47]. Copyright 2016 Macmillan Publishers Ltd).

2.3.1.3. 1T- and 2H-NbSe₂

While Nb has its place at group VB alike V, $NbSe_2$ possesses two crystal edifices, 2H and 1T phases [33]. Both are metallic for the odd quantity of valence electrons in Nb. Yet, the $NbSe_2$ 1T phase is not the utmost robust phase at ambiance [37]. There is indeed a conversion of phase from 1T to its distorted one that is steady at room temperature. Hence, to find the band structure of 1T phase, calculations need to be done at elevated temperatures. The band structure discussed above is not of 1T phase but of its variants with distortion. Figure 2.6 is to show the band structure of a combination of 1T and 2H phases at 40 K, in which the partisan phase of 1T construction has been generated [47]. The assorted specimen of 1T and 2H phases is created by annealing the co-evaporation film of Se and Nb atoms at 530°C on double-layer graphene/6H-SiC (0001). Comparing with bulk 2H phase (Figure 2.6b), many extra bands come into appearance, signified as the band structure of 1T phase with the red curve (Figure 2.6c). Remarkably, the plain band (α band) hardly agrees well with DFT calculation for 1T phase, representing the Mott–Hubbard band in the Mott insulator phase [47]. This infers that the partisan phase discloses Mott insulator in 1T phase edifice.

2.3.1.4. 1T-, 2H- and 4H-TaS_2

In accumulation to the organizational distinctions amid chemical complexes with the identical conformation, a motivating characteristic of TMDs is the creation of various amassing orders lengthwise the c- course [33]. TaS_2 exhibits both 2H and 1T assemblies. 4H polytype is a blend of 2H and 1T phases, that has four films, two sheets each of the 2H and 1T assembly [19, 33]. So, band structure and characteristics of 4H phase are a concoction of such of 2H and 1T phases [19, 48]. Fermi layer intersects about the two (one) strips in 2H (1T) phase with two (one) sheets in unit cell. For 4H construction, there are four bands, two since 2H and two from1T. The two kinds of bands cross along with the M–K course. Interface amongst these bands results in new characteristics of 4H phase, that are absent in 2H and 1T phases. Nevertheless, 4H phase up till now stays intangible.

2.3.2. Layer Reliance of the Band Structure

Even if communication among sheets in TMDs is the feeble van der Waals might, this collaboration yet possesses a vital part to hypothesis the band structures and characteristics of such substances [49]. A palpable quantity of indication is that the band dispersal lengthways Γ to A of numerous bands is yet not insignificant [49, 50]. As a result, band structures of TMDs are powerfully reliant on the count of sheets. Figure 2.7 displays band structure of $TiSe_2$ from single-layer to six-layer samples of the usual and CDW stages [51]. By the few layers, the A (L) and Γ (M) points are identical. The imbrication amid bottom conduction and top valence bands at M and Γ points upsurges as the count of sheets rises for both CDW and normal phases. Therefore, a huge band overlay is evidently exposed among the L and A points in wholesale. Interplanar communication turns into more overriding in group VIIIB (Ni, Pt, Pd). Metallic character of the bulk can transmute to semiconducting properties in the limited sheet specimen [52]. Though bilayer PdS_2 is a semimetal, monolayer is in fact a semiconductor having a bandgap of 1.1 eV. This wonder is instigated by the robust vdW interlayer interface, that turns the conduction band more spread, overlayed with the valence band, suggesting metallic or semimetallic conduct. The marvel happens in extra kinds of layered things such as antimonene (Sb) and GeP_3 and is anticipated to disclose in other sheet substances having a sturdy vdW communication [53]. This intriguing miracle might cause designing of a metal–semiconductor connection devoid of any bodily borders [52].

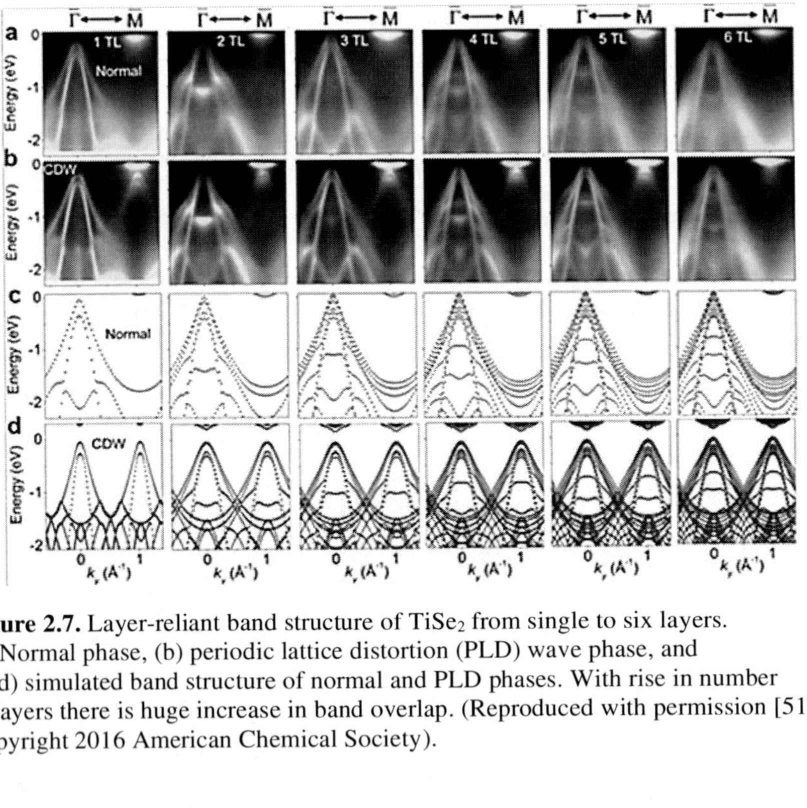

Figure 2.7. Layer-reliant band structure of $TiSe_2$ from single to six layers. (a) Normal phase, (b) periodic lattice distortion (PLD) wave phase, and (c, d) simulated band structure of normal and PLD phases. With rise in number of layers there is huge increase in band overlap. (Reproduced with permission [51]. Copyright 2016 American Chemical Society).

2.4. Periodic Lattice Distortion or Charge Density Wave

2.4.1. The Phenomenon

In several metals, the state having the rudimentary cell construction (termed the standard phase) is not stable while the temperature is dropped [37, 54].

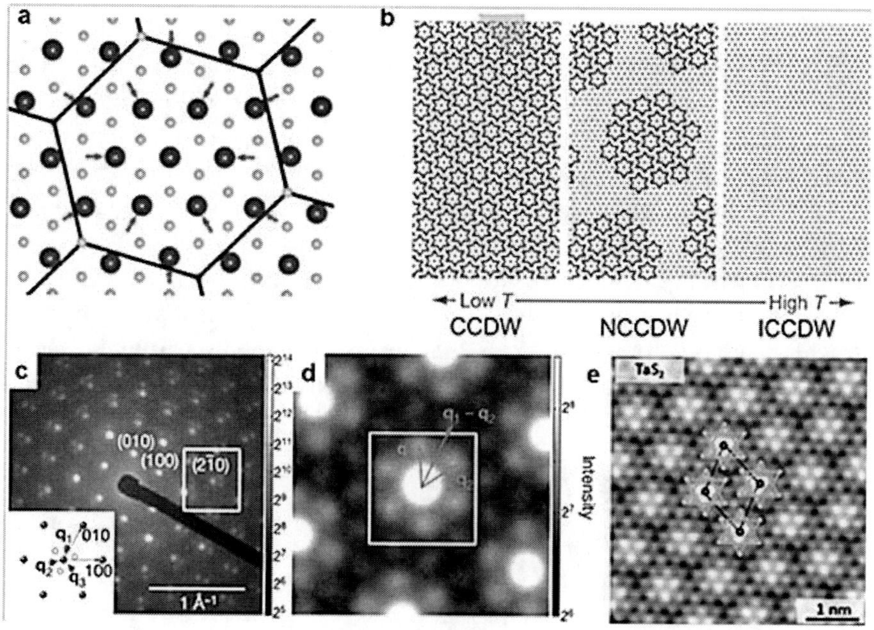

Figure 2.8. Periodic lattice distortion in 1T-TaS_2. (a) The schematic of the relative movement of the corresponding PLD phase equated to the normal phase. Reprinted with permission [37]. Copyright 2015 Macmillan Publishers Ltd. (b) Graphic of development of different transition phase in 1T-TaS_2. Reproduced with permission [55] Copyright 2015 American Association for the Advancement of Science. (c, d) The TEM image of the C-PLD phase, that clearly discloses the star form of the C-PLD phase. (Reprinted by permission [56]. Copyright 2010 Macmillan Publishers Ltd. (e) STM image of the C-PLD phase. Reproduced with permission [57]. Copyright 2016 Cornell University Library).

Their assemblies garble to produce new organizational segments, recognized as the charge density wave (CDW) phase [or periodic lattice distortion (PLD) phase]. The term CDW does not obviously disclose bodily instinct of partisan state since the wave shape of charge density initiating after the episodic prearrangement of atoms stays even in usual phase. Hence, in this review, the name PLD is preferred to CDW in places. The marvel of PLD happens in numerous 2D incrusted metallic resources [37]. CDW changeovers happen in an extensive temperature range, depending on the elemental systems. In addition to that, one substance may have potential for numerous distinct transformations. Construction of the CDW stage may be commensurate (C) or incommensurate (IC) with the usual edifice. Figure 2.8a displays partisan assembly of TaS_2 at a low-slung temperature. The extent of the unit cell in the

C-CDW phase is $\sqrt{13} \times \sqrt{13}$ of the usual phase, as exposed in Figure 2.8a [37]. In TaS_2, three transformations are there: since the standard phase to the IC at 543 K, since the IC to the almost (nearly) commensurate (NC) phase at 353 K, and after the NC to the C phase at 190 K [37]. The graphic consistent with each stage changeover is abridged in Figure 2.8b [55]. Atoms' location in CDW phase is somewhat partisan from the initial location, that may be exposed from scanning tunneling microscope (STM) and TEM findings (Figure 2.8c–e) [56-57]. It's important to note that the largeness of the falsification is very minor (~0.1 Å) while equated with that of the usual structural phase transformation, that causes a totally different crystal assembly. The CDW evolution is at all times escorted by a demulcent phonon,

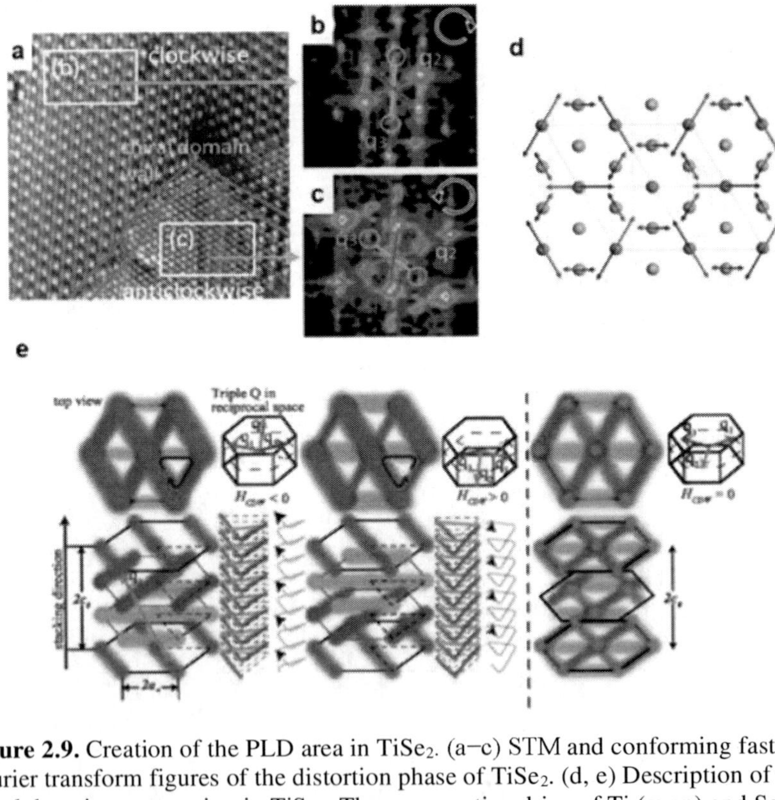

Figure 2.9. Creation of the PLD area in $TiSe_2$. (a–c) STM and conforming fast-Fourier transform figures of the distortion phase of $TiSe_2$. (d, e) Description of the chiral domain construction in $TiSe_2$. The comparative drive of Ti (cyan) and Se (yellow) associated to the normal phase. The red (blue) arrows designate the distortion of the top (bottom) layer. The movement of the bottom and top layers is transposable, which can occur in the same sheet but in a different zone in real space, leading to the formation of chiral domains. (Reproduced with permission [58-59]. Copyright 2010 and 2015 American Physical Society, respectively).

named 'lattice unsteadiness', which is conversed more in subsequent parts [37]. Figure 2.9 demonstrates an extra instance of CDW transformation in $TiSe_2$ [58-60]. There is just a transformation at 200 K after ordinary stage to the partisan one with 2×2×2 assembly. Grippingly, $TiSe_2$ has areas with varied chiralities (clockwise and counterclockwise) in CDW [59]. The creation of these various domain assemblies may be unspoken by depending on slanted structure of Se on bottom and top layers in the 2×2×2 framework. Lengthways the in-plane partisan structure, there is comparative drive of Se atoms on upper and lower layers, manifested with red and blue arrows, correspondingly [58]. It can be realized that courses of top and bottom coats are conflicting, first is clockwise and second is counterclockwise. The character of upper and lower layers is interchangeable, causing fields having distinct chiralities in actual space (as exposed in Figure 2.9d) [59].

2.4.2. Mechanism for PLD Transformation

The uncertainty of a one-dimensional metal at short temperature was formerly hypothesized by Rudolf Peierls [61-62]. For its simple and intuitive character, the Peierls archetypal is classically utilized to clarify CDW transformation in several references and is a reasonably well initial point beforehand heading to the primer of a byzantine microscopic concept. However, this modest theory is not appropriate in various cases when considering real 2D things. Figure 2.10a displays band structure of one electron model of a 1D periodic potential having a matrix spacing 'a' in first Brillouin zone [54, 61-62]. In this postulate, electronic band is half-filled, which designates that the Fermi level crossing the posse at $\pm\pi$. Nevertheless, this metallic band is hardly the least energy construction. If the assembly is a little distorted, creating an episodic potential with 2a, the band structure will develop a gap at the location $\pm\pi$ at the Fermi layer, as revealed in Figure 2.10b. Since the engaged phases, which trace underneath the Fermi level, possess lesser energy linked to those of usual state, the slanted edifice is steadier than the normal one. At elevated temperatures, the electrons beneath Fermi layer will be thermally enthusiastic to the vacant states (more exactly, electrons are reorganized to phases both above and below the Fermi stage, ensuing Fermi−Dirac dispersal), which has more energies than vacant states in the standard phase [54]. Resultantly, gap in energy in distorted phase relative to usual phase is lowered once temperature is amplified, resulting in a transformation to ordinary phase at elevated temperatures. It is interesting to note that the alteration takes place at 2a and not at 3a or 4a, or anywhere else. If it is assumed that the falsification happens at 3a, the gap will happen at $\pm\pi$ position, situated underneath the Fermi stage. Both higher and lower energy statuses are engaged, and hereafter, no energy improvement is attained then. That is why it is supposed that CDW changeover is at all times escorted by a metal–insulator conversion, with the breach at Fermi level [61-62]. Though, this condition is just appropriate for a one-band prototype. In actual substances, normal phase's electronic band is debauched. A minor misrepresentation will discriminate the decadence of this band, and energy alteration cannot be identical to one band. In numerous situations, entire energy circulation of electrons to this band will reduce the energy as exposed in the right board of Figure 2.10c lacking any development of a band gap at Fermi stage. Figure 2.11 depicts two distinct situations for CDW phases of 1T-TaS_2 and, $TiSe_2$ where the energy gap is created at and beneath the Fermi layer, respectively [63-64]. For $TiSe_2$, the CDW phase possess 2×2×2 supercell, and their Brillouin zone (red) is slighter than that of the standard stage (blue) (Figure 2.11a). Normal phase's M point is folded to PLD phase's Γ point. As a value, electron band of usual phase at M point seems at Γ point in CDW phase, which is evidently exposed in Figure 2.11b. Electronic band at Γ and M points are inferior in CDW phase linked to usual stage, representing a discount of electronic energy. An analogous tendency is again detected at L point [44, 65]. Essentially, no opening is shaped at Fermi layer. Numerous comparable instances may be obtained in other things like 2H-TaS_2 and 2H-$NbSe_2$ [58, 65]. The hypothesized excitonic machinery in $TiSe_2$ is observed to be not applicable for the circumstance as all variations in the band structure happens under the Fermi stage [66]. The time-resolved X-ray spectroscopy also rejects the excitonic machinery [67]. Figure 2.11c displays the investigational (left) and computed (right) band structures of the C-PLD phase of 1T-TaS_2. In comparison to the $TiSe_2$ case, a gap is created in the C-PLD state of 1T-TaS_2 lengthwise all planer courses, though there exists yet an electronic band crossing the Fermi layer through the Γ−A course [64]. Moreover, 1T-TaS_2 is a sturdy electron-correlated scheme and Mott transformation is hypothesized to elucidate PLD construction. Nevertheless, a relaxing of the phonon means does divulge in DFT calculation without deliberation of the stout association effects, signifying important aspect of the disparity of electronic energy subsequent to the lattice alteration, termed electron−phonon coupling [68]. Overall lashing machinery to minimalize electronic energy in together circumstances is comparable to Jahn−Teller effect in chemistry [69].

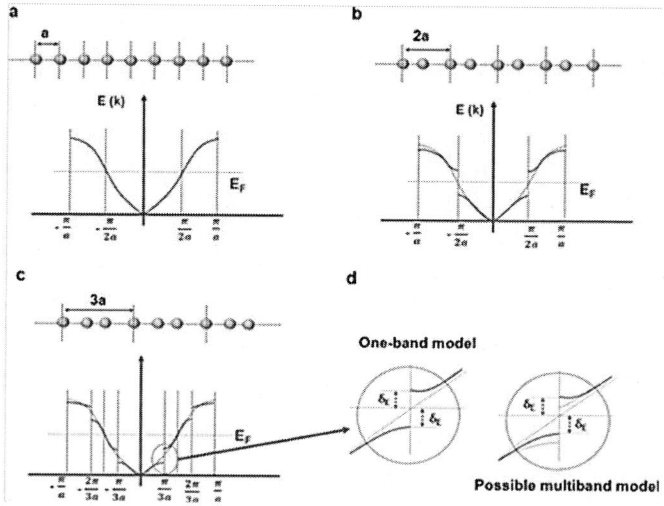

Figure 2.10. (a–c) 1D Peierls distortion (d) Extended model for multiband model that typically happens in real materials. (Reprinted with permission [54]. Copyright 1996 American Institute of Physics).

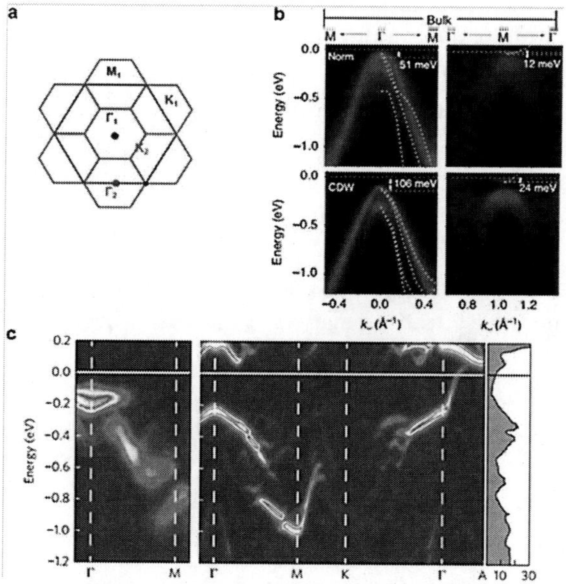

Figure 2.11. (a, b) Band structure of $TiSe_2$' distortion phase. The Γ point's band of the normal phase is pleated into the M point in the distortion phase. No band gap is created at the Fermi level. The energy level at the M and Γ points declines. Reprinted with permission [63]. Copyright 2015 Macmillan Publishers Ltd. (c) Investigational ARPES (left) and DFT calculation (right) band structures of 1T-TaS_2. (Reprinted with permission [64]. Copyright 2015 Macmillan Publishers Ltd).

It is value citing the thought of Kohn anomaly organized with Fermi surface nesting and the way these relate to Peierls distortion. Kohn anomaly is essentially a wonder, in which the phonon gamut is relaxed because of lessening of electron screening at a definite k-point [70]. For 1D model, the salving of phonon happens at q = 2kF. The unstiffening of phonon for Kohn anomaly is unswerving with volatility of the one-electron postulate in Peierls distortion that creates PLD phase's 2a supercell [71]. This phenomenon leads to the idea of Fermi surface nesting, in which electron screening abruptly switches in q space. The bodily measure directly linked to Fermi surface nesting is the plain electronic susceptibility, articulated by [65]

$$X_0(q) \sim 2 \sum_{nn'} \sum_{k} \frac{f^0(E^0_{n'k-q}) - f^0(E^0_{nk})}{E^0_{nk} - E^0_{n'k-q}}$$

This magnitude is only reliant on temperature and Fermi interface. As stated overhead, CDW may happen starved of a wide gap at Fermi stage in actual things. In those situations, idea of Fermi surface nesting has an entirely virtual sense and thus may not be useful to signify the CDW transformation. Even for the gap opening, there exist many countless distinctive characteristics amid PLD transformation and Kohn anomaly [65]. One demonstrative distinction is that the energy advantage for the Peierls distortion is underwritten since energy straddling crossways an extensive span of k- space, whereas Kohn anomaly abruptly happens at a singly definitive k- point. Thus, it is strongly believed that thought of Fermi surface nesting ought to be utilized prudently for elucidating CDW changeover, as explained underneath.

For studying phase transformation, a macroscopical philosophy founded on free energy must be taken into account. The concept of electronic free energy for the clarification of the PLD phase change has been advanced for quite a few years [37]. In this tactic, energy of the partisan phase may be articulated by

$$F = F_0 + \Delta F$$

where F_0 is energy of usual phase and ΔF is energy alteration for distortion. This ΔF is reliant on q vector and phonon mode λ, denoted with

$$\Delta F_{q\lambda} = -\frac{1}{2}\chi(q\lambda)|Q_{q\lambda}|^2 + \frac{1}{2}D_2(q\lambda)|Q_{q\lambda}|^2$$

The former term is connected to electronic change and the later factor linked the elastic distorted energy for falsification. Parameter $\chi(q\lambda)$ is termed the generalized electronic susceptibility:

$$X_0(q) = 2\sum_{nn'}\sum_{k}\frac{f^0(E^0_{n'k-q}) - f^0(E^0_{nk})}{E^0_{nk} - E^0_{n'k-q}}|V_1^\lambda(nk, n'k-q)|^2$$

where V_1^λ is a term related with k and q vector electron− phonon coupling strength. If V_1^λ is not hooked on k and q vectors, the generalized electronic susceptibility is proportionate to simple electronic susceptibility:

$$X_0(q) \sim 2\sum_{nn'}\sum_{k}\frac{f^0(E^0_{n'k-q}) - f^0(E^0_{nk})}{E^0_{nk} - E^0_{n'k-q}}$$

Obviously, the idea of bare electronic susceptibility may only be practical to learn CDW transformation where V_1^λ is not reliant on k and q. This condition is not simply fulfilled in actual resources. However, it may happen relaxed for 1D substances, in which the nesting notion has been altered to deficient nesting concept [72].

CDW transformation at all times goes together with the salving of at best one phonon mode at a definite q point [37]. This feature can be quantified with inelastic scattering, that can straightly detect the phonon band of things [73-74]. Figure 2.12a demonstrates the pragmatic phonon scattering of 2H- $NbSe_2$, that evidently discloses the relaxing of the acoustic phonon mode while the temperature is dropped to transformation temperature, 33 K [73]. Phonon energy is totally zero at this temperature, representing CDW transformation. At a still less temperature, the phonon mode reappears, which essentially is the new PLD phase's phonon mode. PLD changeover of $TiSe_2$ may be sensed by X-ray thermal diffuse scattering (Figure 2.12b) [75]. The extent intensity at L-point abruptly enlarged while the temperature spreads to 190 K, demonstrating the conversion from CDW phase to standard phase. This phonon salving may also be exposed in hypothetical DFT calculation [58]. It's worth noting that the electronic temperature unfolding the heat-associated excitement of electrons may be involved in overall free energy computation. This issue is recognized as the smearing factor, that is usually utilized for merging of the numerical matter in DFT. (Figure 2.12c, d) The phonon gamut of the usual phase vividly alters with the spreading value. It's evidently revealed in $TiSe_2$, 2H-$NbSe_2$, and further CDW substances.

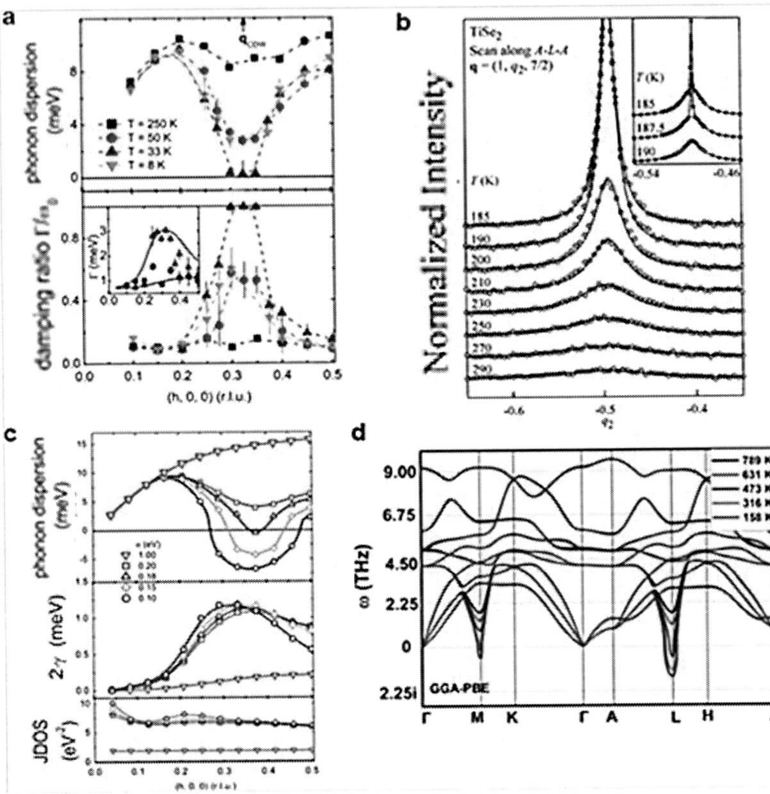

Figure 2.12. (a) Development of phonon dispersion with temperature in 2H-$NbSe_2$. A salving of the acoustic phonon mode is evidently revealed. Reproduced with permission [73]. Copyright 2011 American Physical Society. (b) The evolution of the X-ray thermal diffuse scattering intensity of $TiSe_2$ with temperature along the A−L−A line with fine and coarse scans (inset), that displays a strong Bragg peak below 190 K. (Reproduced with permission [74]. Copyright 2001 American Physical Society). (c, d) The replicated phonon dispersion of (c) 2H-$NbSe_2$. (Reproduced with permission [74]. Copyright 2011 American Physical Society). (d) $TiSe_2$. (Reproduced with permission [58] Copyright 2015 American Physical Society).

2.4.3. Characterization of CDW

In accumulation to the straight assembly portrayals such as TEM or STM and XRD, the PLD transformation may also be sensed by secondary approaches e.g., Raman scattering and electronic transport [75]. While there is no mutual feature midst distinct substances, the representative of the resistance along with temperature (R−T graph) demonstrates an anomalous metallic comportment [76]. Figure 2.13 displays R−T arcs of TaS_2 and $TiSe_2$ [77-78]. Although $TiSe_2$ exhibits a slickly variation of resistance, transformation in TaS_2 is quite abrupt along with a gigantic hysteresis. For $TiSe_2$, it's required to find the differentiation of R−T graph to fix the transformation temperature, that is about 200 K. The changeover of TaS_2 is visibly revealed in R−T graph at 353 as well as 230 K, steady with the STM and XRD results. The difference in behavior amid TaS_2 and $TiSe_2$ can be clarified by several appearances in band structure of the CDW phase, at Fermi level as stated in past. The hysteresis is classically characterized for a metal− insulator transformation.

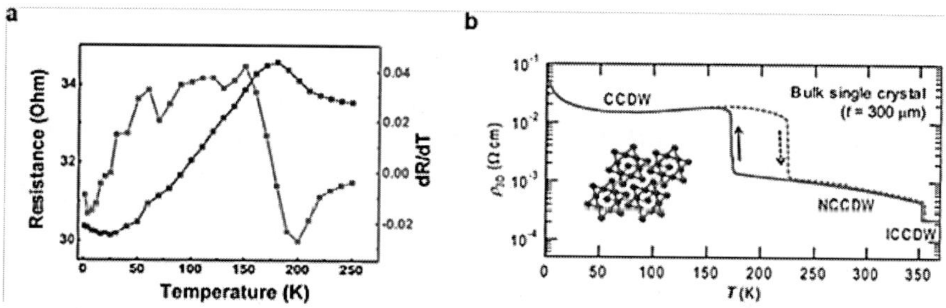

Figure 2.13. Evolution of resistance with temperature in (a) $TiSe_2$. Reproduced with permission [77]. Copyright 2017 American Chemical Society. (b) 1T-TaS_2. (Reprinted by permission [78]. Copyright 2014 Macmillan Publishers Ltd).

While the changeover to CDW state happens, the alteration decreases regularity of the substances, leading to new dynamic Raman modes. Through tracing the advent of these summits, the phase transformation may be fixed [76-77, 79]. Figure 2.14 displays development of Raman spectra of 1T- TaS_2 and 2H-$NbSe_2$ with temperature fluctuations [76, 79]. Entrance of new summits near about 30 and 200 cm^{-1} in 2H-$NbSe_2$ are evidently detected while chilling. Conforming strengths of the relaxed peak are composed in Figure 2.14b with different specimen depths. The transition temperature may be inferred since on the variation in the crowning strength. The T_c is visibly incremented once the depth of the sample is reduced. (Figure 2.14c) This technique is somewhat complex in experimentations since a strong strength of Raman peak is essential to decrease fault throughout extrapolation. A better approach founded on the peak location is more precise resolve of T_c, as verified in the events of TaS_2 and $TiSe_2$ [76-77]. In that method, a few Raman peaks even in usual phase display a random disparity. T_c of TaS_2 for C-PLD changeover was found to be 21 and 140 K by increasing and decreasing temperature, correspondingly. It is consistent with the transport quantification (Figure 2.14d-e).

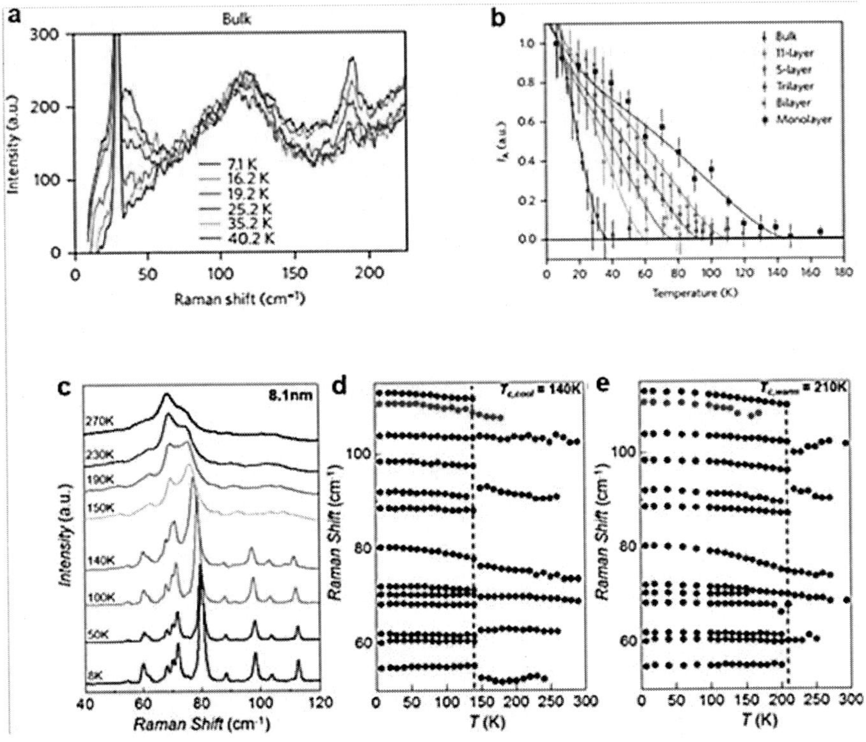

Figure 2.14. (a, b) Evolution of Raman spectroscopy in 2H-$NbSe_2$. (c–e) Development of Raman spectroscopy in 1T-TaS_2. (Reprinted by permission [76]. Copyright 2015 Macmillan Publishers Ltd. Reproduced with permission [79]. Copyright 2016 American Physical Society).

2.4.4. Tailoring CDW Transition for Superconductivity Phase Generation

Along with the motivating physics behindhand the PLD transformation, this phenomenon occurs in numerous schemes and can be adjusted to exhibit superconductivity. Usual "superconductivity dome" form, which occurs in all kinds of superconductors, does disclose in things displaying CDW conversion.

For just about 2D metallic schemes, it is extensively putative that the Bardeen–Cooper–Schrieffer (BCS) superconducting philosophy encompasses the science behindhand that [80]. Both PLD transitions and BCS superconductivity are connected to phonons, and their communications with electrons [81]. Hence, research of the phonon spectra of the 2D metallic schemes comprises rich physics linked to PLD and superconductivity via phonon–electron connection. There exist two tactics for fine-tuning BCS superconducting phase and PLD transition: to regulate the carrier density using pressure or by nobbling [82-83]. Figure 2.15a displays magnetic susceptibility of Cu-doped $TiSe_2$ that shows a fall neighboring CDW transformation [84]. This changeover happens at a lower temperature with growing Cu amount. Resistivity too discloses the identical drift up to 0.6% Cu. Superconductivity at 0.6% Cu, looks as if with a drop in resistance at 2 K. Critical temperature, T_c, rises more with growing dopant density and arrives with 8% Cu the supreme transformation temperature at 4 K. These performances of superconductivity and PLD also appears by smearing pressure, as revealed in Figure 2.15c [85]. Figure 2.15d abridges the cupola form of $TiSe_2$ by changing the pressure and dopant. Nobbling has a stouter influence in contrast to that of pressure in this circumstance. It is stimulating to look at the way the phonon alters while the CDW is repressed. Figure 2.16 depicts development of Raman spectrum, quantified at 3.5 K, under pressure [86]. The E_g (70 cm^{-1}) and A_{1g} (110–140 cm^{-1}) modes of PLD phase are stifled with growing pressure, representing a tender of CDW conversion. Even though the replication info of the phonon spectrum of $TiSe_2$ by density functional perturbation theory also demonstrates the salving means at L and M points, the actual magnitude is upheld, suggesting constancy of the usual phase at squat temperatures [58]. Analogous influences of pressure are practical also in 2H-$TaSe_2$ and 2H-TaS_2, though, the condition is not same in 2H-$NbSe_2$. Some grade of association is detected amid superconducting phase and PLD state but not as robust as in $TiSe_2$ [85]. As exposed in Figure 2.17, CDW evolution is sturdily adjusted whereas superconductivity changeover is not affected. This may be clarified by bearing in mind the connection of such two-phase transformations with phonon. Whereas CDW is powerfully attached with the longitudinal aural manner, superconductivity phase is stalwartly joined equally with optical phonon modes and longitudinal acoustic mode [87]. This is the dissimilarity in contrast with $TiSe_2$, where mutually the quantum phases are attached with identical acoustic phonon [88].

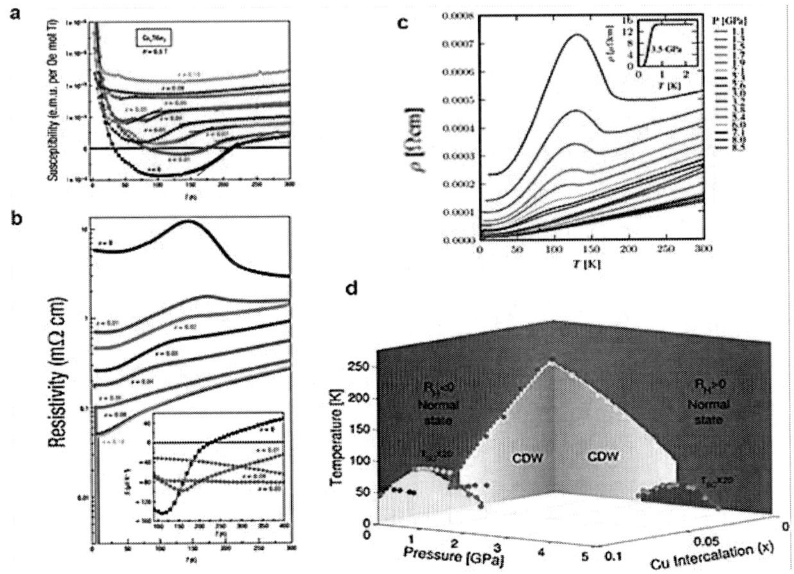

Figure 2.15. (a, b) Resistivity and magnetic susceptibility of Cu-doped $TiSe_2$ at different temperature. Reprinted with permission [84]. Copyright 2006 Macmillan Publishers Ltd. (c) Resistivity of $TiSe_2$ under different temperatures and pressures. (d) Superconductor dome shape of $TiSe_2$. (Reproduced with permission [85]. Copyright 2009 American Physical Society).

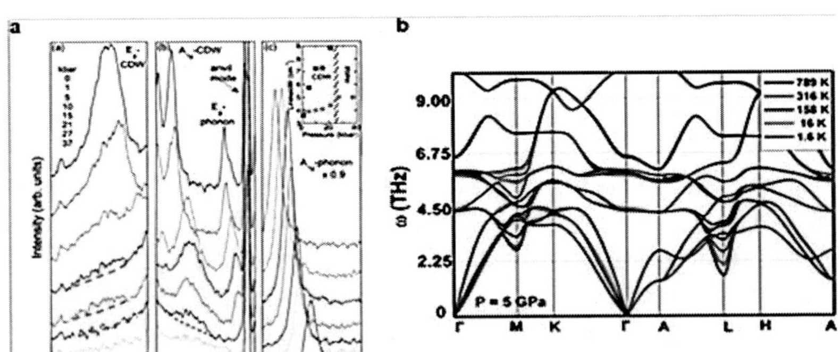

Figure 2.16. (a) Progress of the Raman spectrum of $TiSe_2$ at 4 K at different pressures. Reproduced with permission [86]. Copyright 2003 American Physical Society. (b) Computed phonon dispersion of $TiSe_2$ at 5 GPa. The relaxed phonon mode is still reserved but upholds real at low electronic temperature, representing the suppression of the distortion phase. (Reproduced with permission [58]. Copyright 2015 American Physical Society).

A notable characteristic of metallic TMDs is that the sum of sheets can be regulated by a modest scotch tape technique [89-90]. The thickness-reliant traits are also intriguing for research of the quantum state at 2D boundary. This tactic has been studied for quite a long time, exhibiting that the superconducting transformation temperature is lowered while the depth of 2H-$NbSe_2$ reduces [91]. Nonetheless, CDW phase is alleviated at greater temperature while the sample depth is reduced, as revealed in Figure 2.17a. This tendency is reversed in 1T-TaS_2, i.e., their CDW phase has a propensity to be repressed in the shrill samples with a decrease of the changeover temperature [79, 89]. The transporter meditations of metallic phase may be regulated by peripheral nobbling approaches with scarce layer samples e.g., common oxide gate structure and superficial nobbling by molecules. Figure 2.18a displays phase illustration of 1T-TaS_2 with various gate bias, that regulates the transporter compactness of the sample [89]. At little temperatures, a vibrant transformation since C-PLD to NC-PLD and to SC stage can be perceived with growing the gate bias. Superconducting phase fades when a greater bias is functional, again creating a superconducting cupola form. The alike tendency is detected in $TiSe_2$ (Figure 2.18b) [90]. This method is quite beneficial for examining the carrier-dependent characteristics compared to usual bulk tactic, in which the vigorous interaction creating the new attaching phase is not easy to eliminate.

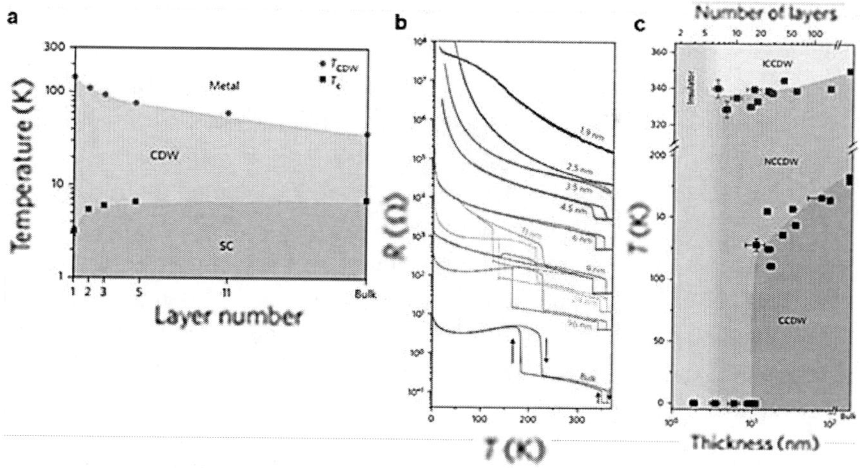

Figure 2.17. Influence of thickness on the PLD and superconducting phase transitions in (a) 2H-$NbSe_2$ and (b, c) 1T-TaS_2. (Reprinted by permission [87-88]. Copyright 2015 Macmillan Publishers Ltd).

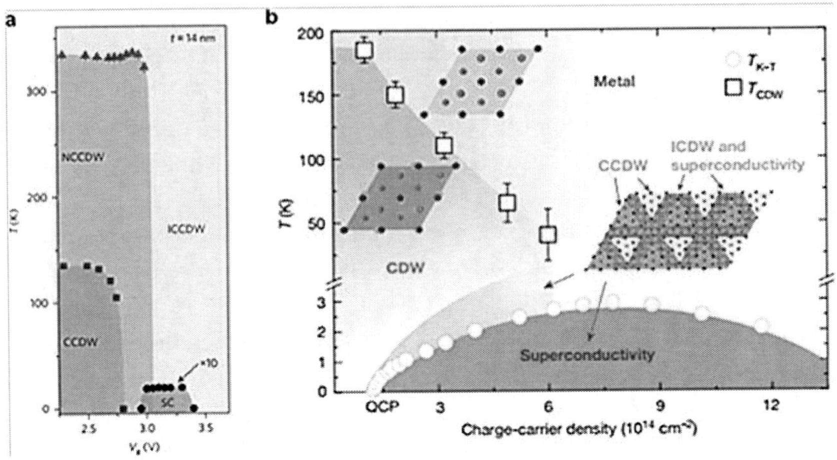

Figure 2.18. Gate-mediated phase transformations in (a) 1T-TaS_2 and (b) $TiSe_2$. (Reprinted with permission [89-90]. Copyright 2015 Macmillan Publishers Ltd).

2.5. Optical Properties

Unlike vividly studied semiconducting TMDs, metallic TMDs do not have well-defined optical band gap. Study about their optical properties is thus restricted to observation and determination of dielectric function, valley polarization, absorption coefficient, reflectance, refractive index, optical conductivity, photoluminescence, transmittance etc. Reshak et al. calculated anisotropic frequency-dependent optical properties of TiX_2 [e.g., real and imaginary parts of dielectric function $\varepsilon(\omega)$, reflectivity $R(\omega)$, absorption coefficient $I(\omega)$, refractive index $n(\omega)$, and extinction coefficient $k(\omega)$] and noticed excellent agreement with the accessible experimental findings (Figure 2.19) [92-95]. The optical properties exhibited three major structures that may be ascribed to transformations between the X p states and the Ti d states. Recently, Dey calculated the similar parameters under influence of electric field and biaxial strain for TiX_2 [96]. As per his findings, optical, conductivity, refractive index, dielectric function, absorption coefficient, and reflectance are highly tunable under influence of such external factors (Figure 2.20).

In yet another elegant work, Ghobadi et al. experimentally demonstrated for VSe_2 nanosheet that the surface defects persuaded during the pulsed laser ablation (PLA) process are the source of photoluminescence [97]. Their findings proved that photoluminescent VSe_2 nanosheets can be a promising semimetal for light emitting diode and photovoltaic applications.

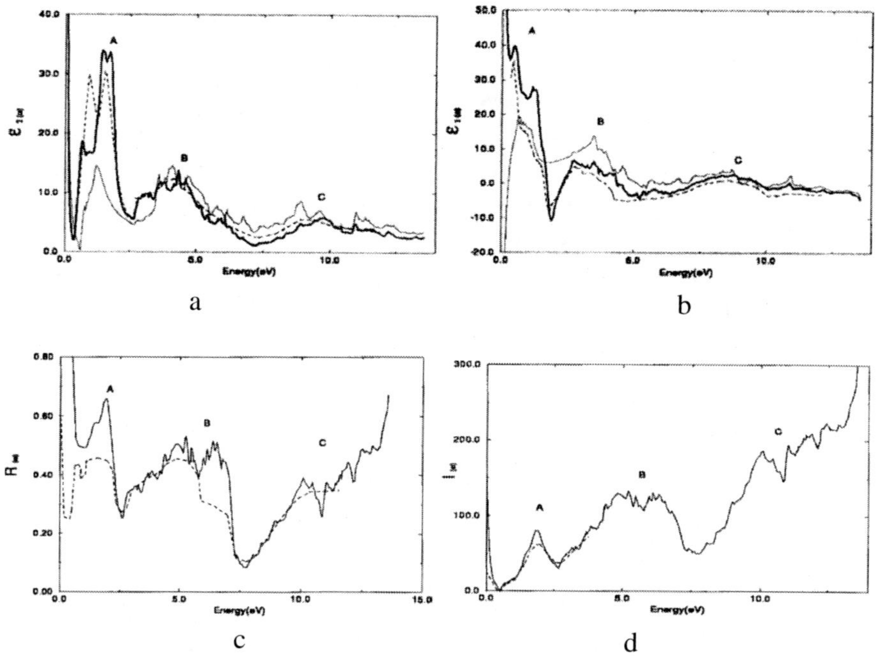

Figure 2.19. Calculated (a) $\varepsilon_2^{\parallel}(\omega)$ (light line) and $\varepsilon_2^{\perp}(\omega)$, (dark line) and (b) $\varepsilon_1^{\parallel}(\omega)$ (light line) and $\varepsilon_1^{\perp}(\omega)$, (dark line) along with the experimental data [93] for $\vec{E} \perp c$, (c) reflectivity spectrum along with experimental data [94] and (d) absorption coefficient along with experimental data [95] for TiS_2. (Reprinted with permission [92]. Copyright American Physical Society).

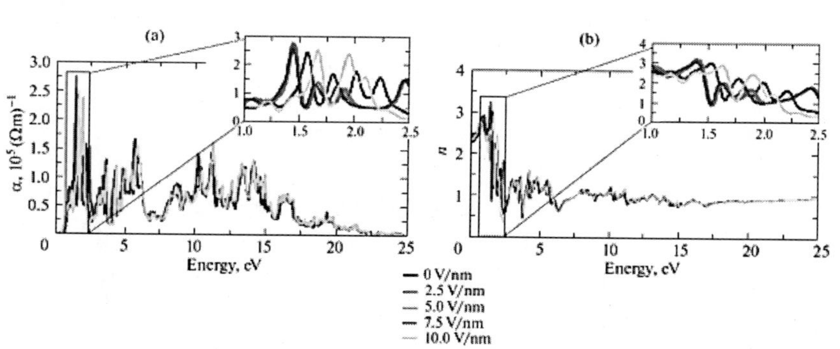

Figure 2.20. (a) Optical conductivity and (b) refractive index of TiS_2 under the effect of electric field. The peak regions are shown in the insets. (Reprinted with permission [96]. Copyright Springer).

PL spectroscopy results (Figure 2.21) concludes that quantum confinement in lateral dimension of VSe_2 nanosheet during sonication may be a possible explanation behind such PL observation. Reshak et al. [98] calculated optical properties such as anisotropic frequency dependent dielectric function, reflectivity and the absorption coefficient of $ZrTe_2$ and found them to be in great agreement with experimental findings [99-100]. Zhou et al. [101] exhibited by transient transmission spectra measurements that $NbSe_2$ being metallic does not have size-dependent optical properties (non-linear optical response) unlike MoS_2 and WS_2.

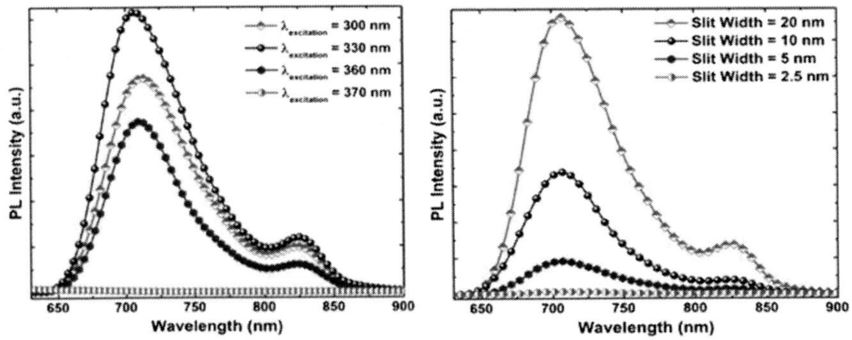

a b

Figure 2.21. (a) The change of the emission spectra of VSe_2 nanosheet sample under different excitation wavelengths, (b) The dependence of PL on the excitation light intensity at an excitation wavelength of 330 nm with different slit width values. (Reprinted with permission [97]. Copyright. The Optical Society).

2.6. Magnetism in Metallic TMDs

Absenteeism of 2D magnetic substance has been forecast after the 2D isotropic Heisenberg model by Mermin−Wagner theorem [102]. However, magnetically ordered phases e.g., ferromagnetism (FM) and antiferromagnetism (AFM) have lately been exposed in single layer 2D materials e.g., metallic FM $GdAu_2$, semiconducting AFM MPS_3 (M is Ni, Fe, Mn) and semiconducting FM CrI_3 family [103-105]. 2D magnetic resources possess a host of rewards. Maximum van der Waals magnets possess an intrinsic magnetocrystalline anisotropy due to the condensed crystal symmetry, that is a hopeful intrant for forthcoming spintronic substances. A list of metallic TMDs in Table 2.2 delivers info on magnetic attributes e.g., kind of magnetic ordering, magnetic moment, and Curie (Neel) temperature. It is important to note that a few of TMDs are not incrusted structures, even though they are in a system of MX_2 e.g., FeS_2 and CoS_2. Amongst the TMD things itemized in Table 2.1, just monolayer vanadium dichalcogenides (VX_2 where X = S, Se, and Te) are magnetic layered materials. Though chromium dichalcogenides (CrX_2) are also encrusted structures, they are metastable and chiefly occur as a steady system of naturally occurring Cr_2X_3.

Table 2.2. List of magnetic metallic TMDs with basic magnetic properties[a]

Materials	Space group	Experimental or theory	Structure	Magnetic ordering	T_c or N_c (magnetic moment)
VS_2	$P\bar{3}m1$	experimental	thin flake	FM	RT (0.51 μB/V atom)
VSe_2	$P\bar{3}m1$	theory	monolayer	FM	514 K (0.68 μB/V atom)
VTe_2	$P\bar{3}m1$	experimental	bulk	AFM	410 K (–)
CrS_2	$P\bar{3}m1$	theory	bulk	FM	-
$CrSe_2$	$P\bar{3}m1$	theory	bulk	AFM	~550 K (–)
$CrTe_2$	$P\bar{3}m1$	theory	bulk	FM	-
CoS_2	$Pa\bar{3}$	experimental	bulk	FM	120 K (0.84 μB/Co atom)
$CoSe_2$	$Pa\bar{3}$	experimental	bulk	FM	50 K (8.4 μB/g)
FeS_2	$P\bar{3}m1$	theory	monolayer	FM	144 K (–)
$FeTe_2$	Pnn2	experimental	thin flake	FM	RT (–)

[a]T_c (TN) indicates Curie (Neel) temperature.

2.6.1. Ferromagnetism

Vanadium dichalcogenides (VX_2) is often forecast as long-range FM well-organized substances by theoretical computations even at single layer depth [106]. The density of states (DOS) and atomic site projected DOS of VX_2 (VSe_2 and VS_2) are shown in Figure 2.22a. The distinction among upward and downward spin phases of VX_2 designates single layer VX_2 to be ferromagnetic. The *d* stages of V atoms are extremely delocalized and mongrelized with *p* stages of S and Se atoms, representing covalent bonding amid the X and V atoms. Ferromagnetism is chiefly underwritten by V atoms, while X atoms subsidize feebly to the overall magnetic moment. Figure 2.22 b–d show pragmatic findings from superslim VS_2 flakes [107]. The magnetization against magnetic field graphs for thin VS_2 and bulk $VS_2.NH_3$ are revealed in Figure 2.22b. Both the bulk and thin VS_2 exhibit paramagnetic conduct in large magnetic field area, whereas only slim VS_2 displays hysteresis in stumpy field domain that is one of the foremost hints for ferromagnetism (enlarged chart in the insertion of Figure 2.22b). Figure 2.22c demonstrates a vibrant inundation of magnetization upon deducting signal from paramagnetic background. Tendency of lessening in the drenched magnetization (M_s) and coercivity (H_c) with growing temperature additionally affirms the FM stage in shrill VS_2. Temperature reliance of magnetization in case of reedy VS_2 nanosheets are revealed in Figures 2.22d,e. The variance among field cooling and zero-field cooling up to 330 K at 100 Oe specifies presence of a magnetic phase beneath 330 K, as displayed in Figure 2.19d. Under huge magnetic field (8000 Oe) (Figure 2.19e), magnetic moment quickly boosted in short temperature area (underneath 50 K) while compared to tall temperature overhead 50 K since paramagnetic and FM indications are overlapped. A fitted graph summing the $T^{3/2}$ law Curie– Weiss law for FM and the paramagnetic state is exposed in Figure 2.22e, correspondingly, as labelled with the succeeding equation:

$$M(T) = M_{s0}\left(1 - AT^{\frac{3}{2}}\right) + CH/(T - \theta_P)$$

where M_{s0} is the saturated magnetization at 0 K, θ_p is the paramagnetic Curie temperature, C is the Curie constant, and A is a structure-associated coefficient [108]. The mined magnitude of θ_p is negative mark (−6.6 K), representing cohabitation of AFM along with FM states [107]. Reedy VS_2 also possesses a negative magnetoresistance of 6.5% at ambient temperature with a magnetic field in line with direction of electrical current. The vibrant hysteresis in the magnetoresistance suggests the FM behavior of VS_2 nanosheets (Figure 2.22f).

While some investigational findings exhibit the magnetic characteristics of VSe_2, their cause is somewhat obstruse. Magnetic comportment could be persuaded by defect formation all through the creation procedure or substrate effect. Yet, monolayer VSe_2 development by the MBE was lately established on the 2D base (e.g., MoS_2 and graphite), that displays the magnetic hysteresis [109]. This reflection eradicates all opacities of chemical blending method. It fuses yet again the supremacy of the hypothetical prophecy in 2D materials. Analogous to VSe_2, 1T-TaS_2 is foretold as a magnetic semiconductor in CDW state, which discloses that valence and conduction bands possess dissimilar spin phases [110]. Figure 2.23 displays band structure of monolayer and bulk

TaS_2 in usual and CDW stages. In the standard state, the band structure of single layer TaS_2 is alike the planer band structure of the wholesale TaS_2. No remarkable alteration is noticed, as revealed in Figure 2.23a, b. Still, the spin-polarized band looks as if in the single layer of PLD phase, while majority favors the no-spin stage in Figure 2.23c, d. This guess has not been established by experimental effort, as of now.

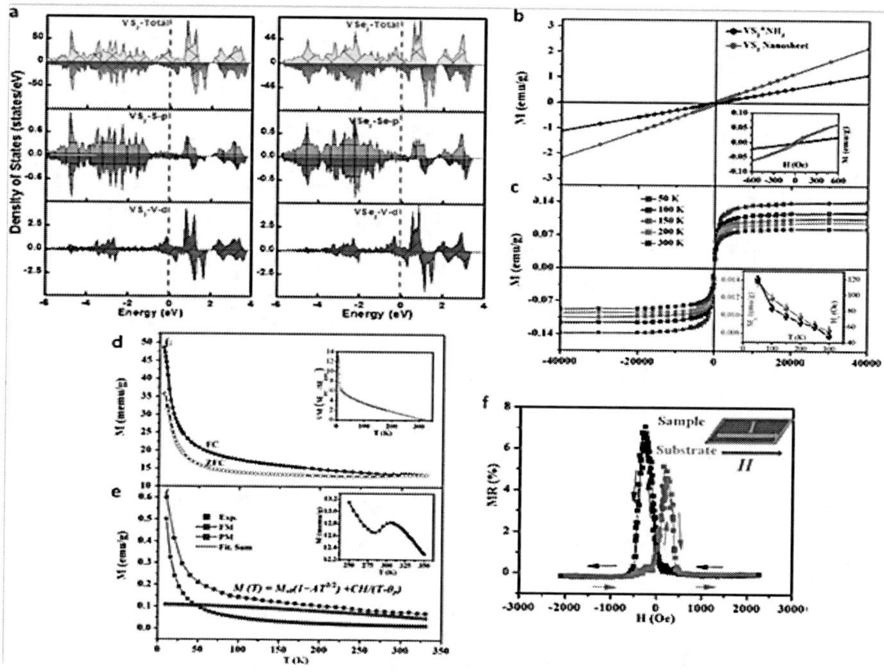

Figure 2.22. (a) Density of states for single layer VS_2 and VSe_2. The magnetism mainly initiates since the V atoms. Reproduced with permission [97]. Copyright 2012 American Chemical Society. (b) Magnetization of thin and bulk VS_2 ($VS_2.NH_3$) at room temperature, which displays a hysteresis in the inset. (c) Paramagnetic signal-subtracted curves for thin VS_2 at various temperatures. The inset demonstrates the temperature dependence of M_s and H_c. (d) Field cooling and zero field cooling curves for thin VS_2. (e) Temperature dependency of magnetization of the thin VS_2 fitted by totaling the FM and paramagnetic behavior. The inset displays zoomed curves from 250 to 350 K. (f) Magnetoresistance of thin VS_2 as a function of the magnetic field at room temperature. (Reproduced with permission [98]. Copyright 2013 Royal Society of Chemistry).

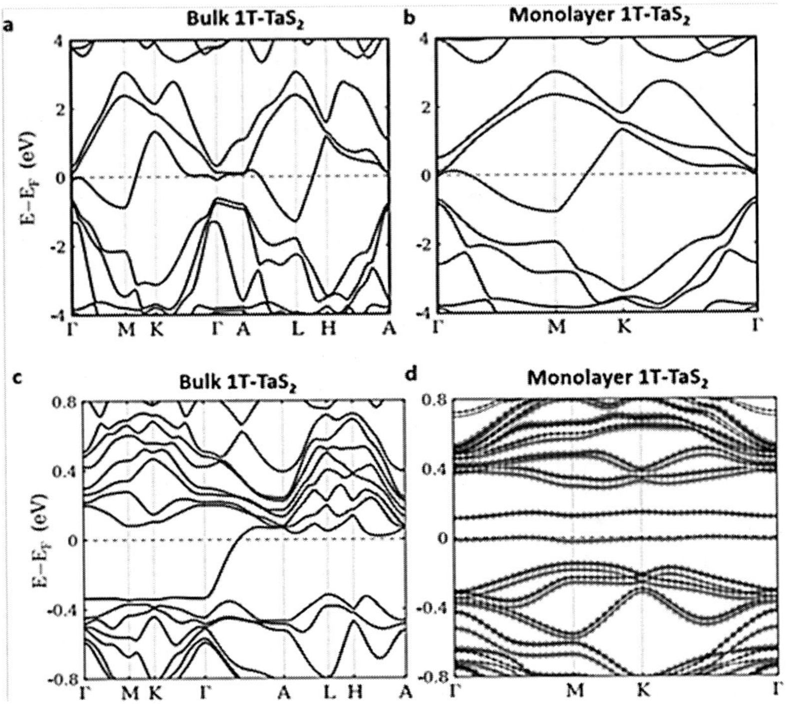

Figure 2.23. Band structure of (a) bulk and (b) monolayer TaS_2 in the normal phase and (c, d) PLD phase. (Reproduced with permission [100]. Copyright 2014 American Physical Society).

2.6.2. Strain Effect on Magnetism

Magnetism may be effectively controlled by strain in metallic TMDs [111]. After 2D materials turn out to be ultrathin, they can be easily elongated, being flexible. Figure 2.24 displays influence of planer strain on magnetic moments per V atom (upper panel) and per X atom (lower panel) for single layer 1T-VX_2 since DFT computations [106]. Tensile strain augments both magnetic moments, while compressive strain slakes the magnetic moments. This inflection of magnetic moment with strain may be realized since the mutual contest of covalent and ionic bonding communications amongst X and V atoms. To exemplify, when tensile strain is functional, the separation between X and V atoms become enlarged, leading to lessening of covalent bonding effect but improvement of ionic bonding dealings. Therefore, the heightened ionic bonding collaboration could cause an upsurge in populace of unpaired electrons on V atom. Strain influences not just magnetic moment but spin ordering also. FM spin ordering stage in VS_2 single layer is sturdier than AFM phase as the functional strain is moderated from compressive to tensile strain (Figure 2.24a-b); a parallel trend is found in CrX_2 scheme too [106, 112]. This marvel of magnetic ordering may be elucidated well by uniting the through-bond and through-space spin polarization models [113-115]. FM phase in single layer VS_2 might initiate since through-bond spin polarization, in which an 'a' atom may persuade contrary spin on a head-to-head 'b' atom which is unswervingly attached to one more 'a' atom. For instance, a V atom's up-spin tempts down-spin to S atom. As a significance, down-spin S atom once more encourages up-spin to an extra V atom, causing long-range magnetic ordering. Since V atoms' magnetic moment is superior to that of S atoms, overall magnetic moment is finite and causes long-range FM collation. One more magnetic ordering machinery is the through-space model, in which an 'a' atom persuades to converse spin to an 'a' atom deprived of an arbitration by a 'b' atom. Afterwards, the through-bond and through-space communications lead to FM and AFM phase in magnetic substances, correspondingly.

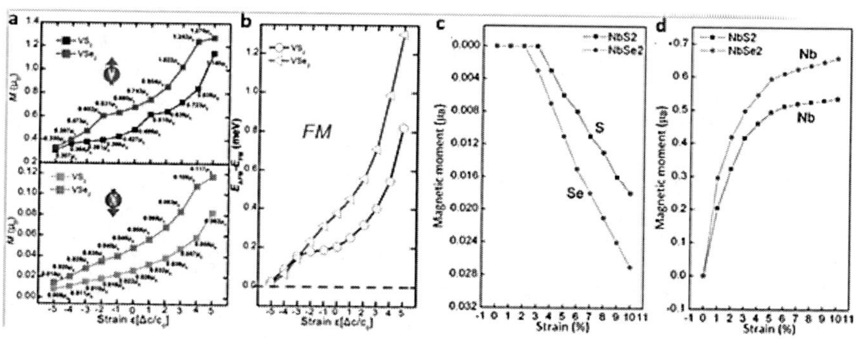

Figure 2.24. (a) Computed magnetic moment per V atom and X (S and Se) atom as a function of functional strain for monolayer VX_2. (b) Energy difference between the FM and AFM state of monolayer VX_2 as a function of applied strain. Reproduced with permission [96]. Copyright 2012 American Chemical Society. (c, d) Magnetic moments of S (Se) (c) and Nb atoms (d) for monolayer $NbS(Se)_2$ as a function of practical strain. (Reproduced with permission [108]. Copyright 2012 American Chemical Society).

One more thought-provoking marvel is that strain even persuades magnetism to the otherwise nonmagnetic substances [116]. NbX_2 is a renowned superconducting thing, with no magnetic ordering [117]. While tensile strain is functional to single layer NbX_2, a limited amount of magnetic moment is formed, which improves more with greater strain. As conversed, tensile strain may lessen covalent bonding and augment ionic bonding forte amid X and Nb atoms. Consequently, the heightened ionic configuration amid atoms leads to unpaired electrons and accordingly the magnetic moment may upsurge [118]. Figure 2.24c, d display magnetic moments of NbX_2 (X = S and Se) on X and Nb atoms by strain change [118].

2.6.3. Effect of Foreign Atoms on Magnetism

Nobbling or doping has been utilized for adjusting material characteristics such as magnetism, mechanical strength, and carrier density [119]. There are three chief methods for nobbling of layered TMDs: substitution by dopants, surface hydrogenation and vacancy creation. Chalcogen vacancies possess a remarkable effect on magnetic attributes in layered TMDs. The organizational graphic and associated iso-surfaces of the spin density for four kinds of S vacancies (V_{no}, V_{1S}, V_{2S}, and V_{3S}) are characterized for single layer VS_2 in Figure 2.25a [120]. Spin density of V atoms next to S vacant site is reallocated. Magnetic moments of V atoms neighboring the S vacant site are progressively amplified with further S vacancies (1.17, 1.67, 2.32, and 2.74 μB for V_{no}, V_{1S}, V_{2S}, and V_{3S}, correspondingly). Residual electrons of the V atom next to the S vacancy wane the covalent bond amid V and S atoms for improving magnetic moment of V atom. S vacancy may also alter the magnetic ordering. For pristine VS_2 (no S vacancy), FM ordering is steady although AFM collation is constant with attendance of S vacancies. Hydrogenation of 2D substances may alter their magnetism and persuade FM since nonmagnetic things [121-122]. Hydrogenation of single layer VX_2 is also assessed by DFT computations. Figure 2.25b demonstrates illustrative assemblies of unadulterated single layer VX_2 and its one- (1H) and two-site (2H) hydrogen functionalization [123]. The rudimentary magnetic data is abridged in Figure 2.25c. Two intriguing characteristics are obtainable: (i) the exchange energy (E_{ex}), where $E_{ex} = (E_{AFM} - E_{FM})$, of single layer VX_2 upsurges as the conformation is changed from S to Te, representing a greater Curie temperature; (ii) the magnetic moment and spin ordering of VX_2 are remarkably influenced by hydrogenation. For 1H, VX_2 mislays its magnetic characteristics and additional hydrogenation to 2H leads to AFM order in VX_2. Hydrogen acts as an n-dopant to VX_2, causing redistribution of the electronic structure and carrier density, that extremely affects magnetism. The replacement of the transition metals is one of the capable approaches to persuade magnetism to nonmagnetic things [124]. $CrSe_2$ has inherent AFM, and its magnetic collation can be altered by Ti or V atom swap [125]. Figure 2.25d displays the converse magnetization of $Cr_{1-x}V_xSe_2$; the elevated temperature performance trails a Curie law. The emblem of the Curie constant is altered from plus to minus, which designates that AFM is produced as density of replaced V atoms rises. No visible variation from Ti atom replacement was detected under 20%, while above 20% Ti swap, the complexes exhibited a FM comportment (Figure 2.25e). Ti replacement may lengthen the bond length of $Cr_{1-x}Te_xSe_2$ alloys and errands ferromagnetic dealings, that is analogous to the situation of $Cr_xTi_yS_2$ [126]. Moreover, chalcogen alloying in the Cr−Se−Te system displays a tailorable magnetic ordering. For instance, AFM is steady in the Se-based complex, while FM is more ideal in Te-based complex [127].

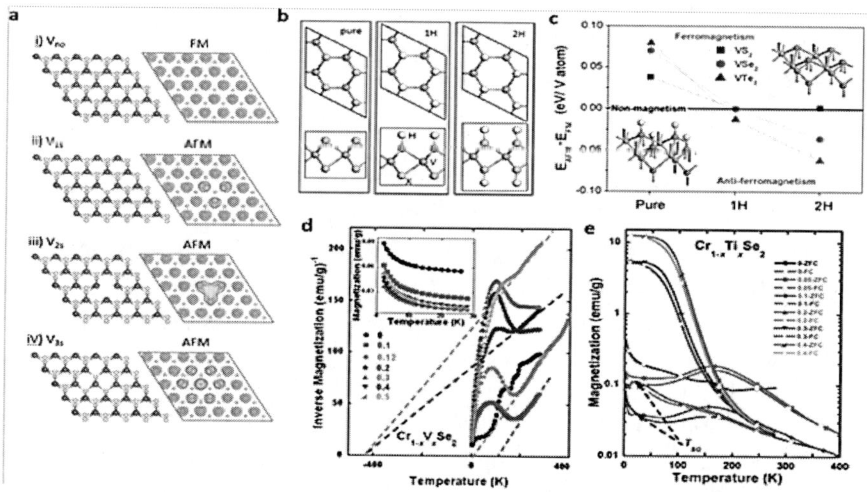

Figure 2.25. (a) Atomic structures and computed isosurfaces of spin density for monolayer VS_2 with sulfur vacancies (V_{no}, V_{1s}, V_{2s}, and V_{3s}). (Reproduced with permission [110]. Copyright 2016 Elsevier). (b) Atomic structures of pure, one side fully hydrogen passivated (1H) and both sides fully hydrogen passivated (2H) monolayer VX_2. (c) Calculated energy difference between FM and AFM states of monolayer VX_2 in terms of hydrogenation. (Reproduced with permission [113]. Copyright 2014 American Chemical Society). (d) Inverse field cooling magnetization curves of $Cr_{1-x}V_xSe_2$ samples at a magnetic field of over 0.1 T. The inset displays Curie−Weiss fitting in the low temperature region. (e) FC and ZFC curves of $Cr_{1-x}Ti_xSe_2$ samples at a magnetic field of over 0.1 T. (Reproduced with permission [115]. Copyright 2013 American Physical Society).

2.7. Weyl Semimetal

In circumstance where the two bands are crisscrossed, it is called band inversion [128]. The term "band inversion" originates since the energy difference amongst the two energy bands shifting from plus to minus in momentum space. This wonder can usually occur in solids' band structure. In that case, the intriguing physical idea for topological band structure has to be well thought-out, particularly when these bands are upturned for spin–orbit coupling (SOC) phenomenon [129-130]. Figure 2.26 displays different conduct in instance of two bands decussating each other. Such decadences (e.g., two conditions having the same energy at a convinced k-point) in quantum mechanics, ought not to be permitted because of the dealings amid the two bands [130-131]. This singularity is analogous to that exhibits while carrying two alike atoms nearby together; their original energy stages will fragment into two dissimilar energy heights. The ingress of the band–band dealings is the development of an energy gap. That is the circumstance of topological insulator, as revealed in Figure 2.26a [8]. Yet, here are quite a few special situations, in which, two band-crossing points persist steady, as revealed in Figure 2.26b [132]. These points are termed Weyl or Dirac points, contingent on their gradation of decadence, due to the type of the rectilinear conduct of these bands about the intersecting points [133]. It is interesting to observe that linear band dispersal in topological insulator invents since the surface state, while the 3D Weyl or Dirac points are truthfully bulk states. In divergence to the locked curve of Fermi surfaces of the surface conditions in topological insulators, the Fermi surfaces of Weyl semimetals are sort of arc, that joins a couple of two Weyl points [134]. Two kinds of Weyl semimetal are there, founded on the form of Weyl cone. Type I is illustrative for Fermi surface dwindling to Weyl points. In distinction, type II of Weyl semimetals has slanting Weyl cones, leading to a Fermi surface comprising of both hole and electron carriers, as revealed in Figure 2.26c, d [135, 136]. Type II Weyl semimetals may be observed as overturned band structures of indirect semiconductors. The probing procedures for things manifesting Weyl semimetals are yet strange. Nonetheless, there is a great likelihood that Weyl semimetal can stay in substances with no upturn or wrecked time reversal symmetry (e.g., magnetic materials) [130, 132]. This finding is consequent since the hypothetical model founded on two-band scheme. The band is not degenerate in such circumstances and situations for upholding Weyl cone may be attained more simply. This implies that not all magnetic materials or noninversion symmetry will at all times possess Weyl cone. Here is yet another significant situation. Though the Weyl points occur, it would be situated at or quite close to Fermi level and here ought to be no other phase since other bands contributive to DOS at Fermi face [132], excluding the bodily characteristics of Weyl points are not easy to discriminate and be quantified. While a slender band gap semiconductor deprived of inversion symmetry is projected as a fine thing to reconnoiter Weyl semimetals, there is no vibrant recommendation for probing things nourishing all overhead situations [132].

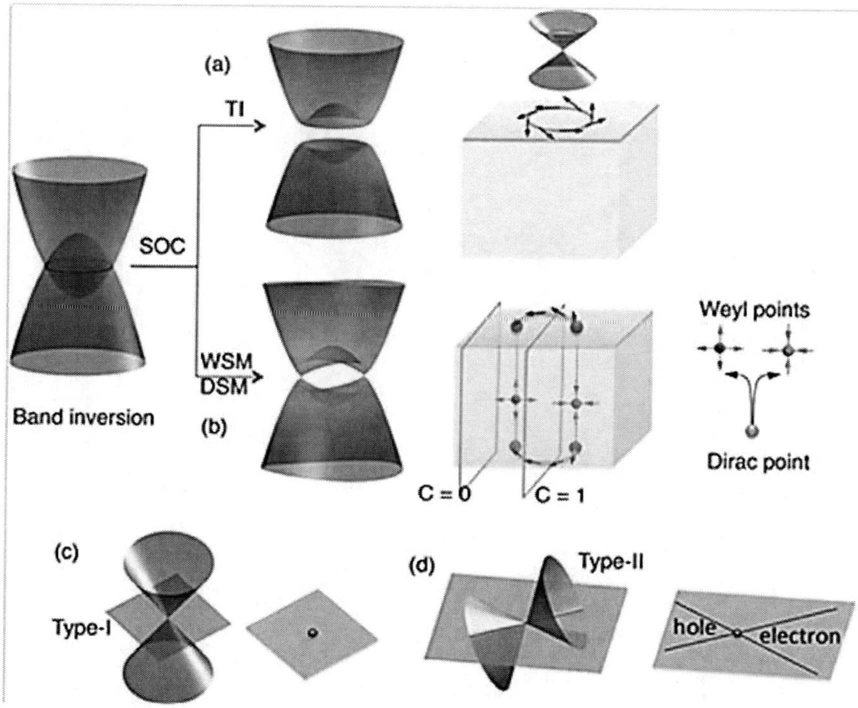

Figure 2.26. Graphic band structure of topological insulators (a) and Weyl semimetal (b) and their topological surface states. Fermi surface of type I (c) and type II (d) Weyl semimetals with dot and electron–hole pocket shapes, respectively. (Reproduced with permission [8]. Copyright 2017 Annual Reviews).

The presence of Fermi arcs and Weyl cones have been verified from two separate modules of substance: type I, TaAs category (e.g., TaAs, NbP, TaP, NbAs) and type II, layered transition dichalcogenides, $MoTe_2$. The WTe_2, an allied compound of $MoTe_2$, is foretold to be type II Weyl semimetal, yet the uncovering of the Weyl features is extra thought-provoking for the brief remoteness amongst the two Weyl points [137]. Here, the focus is on the band structure of $1T_d$ WTe_2 and $MoTe_2$.

Figure 2.27 demonstrates band structure of $1T_d$-WTe_2 through (b, c) and short of SOC (a) since the Σ to Γ points [135]. The SOC is selfsame sturdy in WTe_2, designated by a distinguished feature in the band structure without and with SOC. The minor gap of intersecting points is completely unlocked once SOC is involved. Fascinatingly, Weyl points look as if in the $k_z = 0$ plane lengthways the K−K′ course, as exposed in Figure 2.27c. Nevertheless, these dots are around 0.052 and 0.058 eV overhead the Fermi energy. Besides, the aloofness amongst the two Weyls dot is too trivial (about 0.7% of the BZ breadth), causing limitations in recognition of these dots for the restraint of accuracy in the ARPES quantities [137]. To resolve this problem, the searching course is being moved near $MoTe_2$ and $(W_{1-x}Mo_x)Te_2$ amalgam [138].

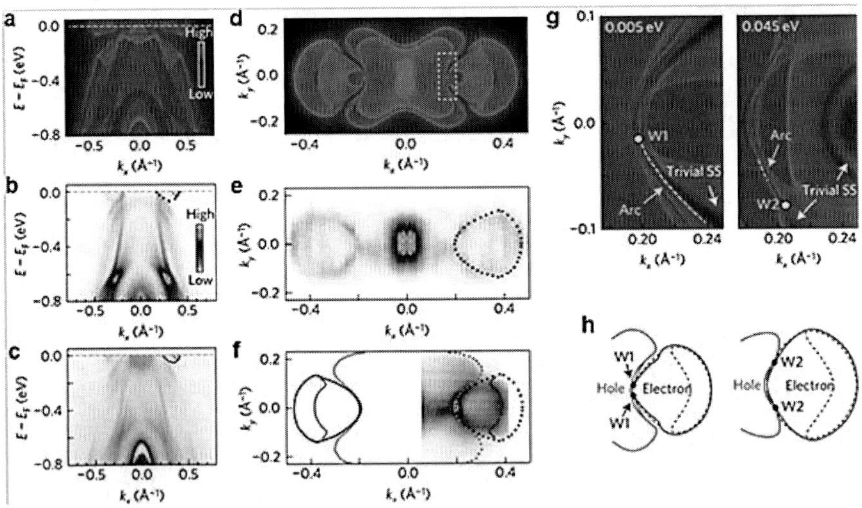

Figure 2.27. (a, d, g, h) Hypothetical and (b, c, d, f) investigational band structure of $1T_d$-$MoTe_2$. A decent agreement between the theoretical and experimental consequences is observed. (Reprinted with permission [125]. Copyright 2016 Macmillan Publishers Ltd.).

Figure 2.28 displays the experimental and theoretical band structures of $MoTe_2$ [135]. The computed band structure counting the superficial state is revealed in Figure 2.28a. The wholesale band structure is characterized by the incessant spectrum for robust dispersal laterally the k_z axis, though a shrill line designates the shallow state. The investigational ARPES is evidently steady with the computed band structure, excluding that the band is *p*-doped with Fermi stage is moved down by 20 meV. To generate distinction amid the surface and bulk phases, band structures in Figure 2.28b, c is calculated by *p*- and *s*-polarized light, correspondingly. Surface state (shown as a black arc) and bulk electron pocket (earmarked as blue) are plainly viewed by means of p-polarized light. Spectral function at Fermi face (e.g., DOS at Fermi plane) discloses a bowtie and bell form for the hole and electron pocket, separately, as also detected in ARPES outcomes (Figure 2.25e, f). It is perceived that the Weyl points of $MoTe_2$ are positioned at 5 and 45 meV overhead the Fermi level since hypothetical computation as exposed in Figure 2.28g.

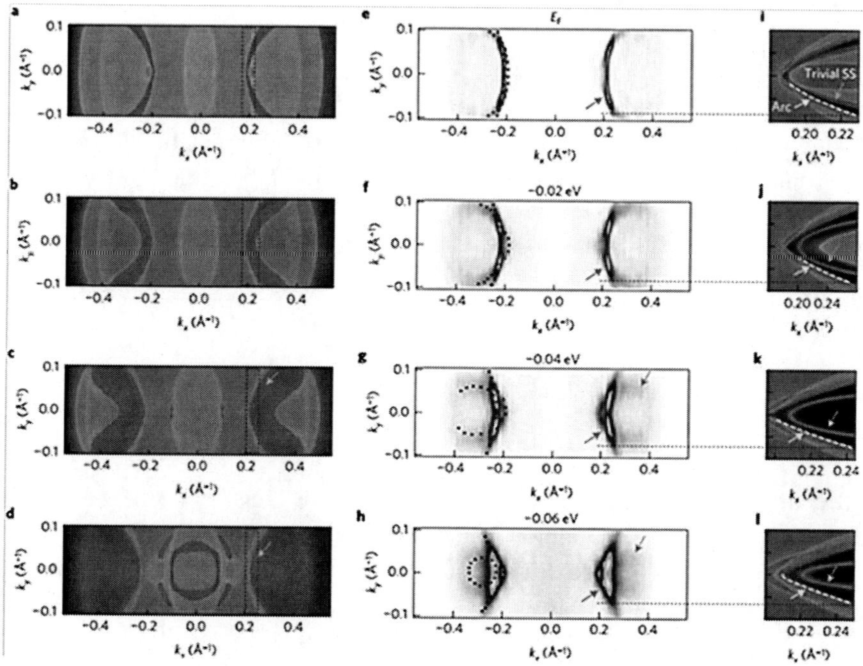

Figure 2.28. (a–d, i–l) Topological Fermi arcs in $MoTe_2$ from simulation and (e–h) ARPES measurements at different energy levels. (Reprinted with permission [125]. Copyright 2016 Macmillan Publishers Ltd.).

A pertinent characteristic of indication for Weyl semimetal is the presence of Fermi arc (e.g., the topologic face status). Whereas the arc may clearly be exposed in theoretic computation, it is round about to verify through experimentations. In the situation of $MoTe_2$, the face phases are embraced among the hole and electron pockets, that is not easy to be spotted. A secondary method is to utilize hypothetical computation as a standard to view the extent of steadiness amid experimental and theoretical consequences, even by the iso-energetic plane. A motivating characteristic of Weyl semimetal is the huge unsaturated magnetoresistance. Both types of Weyl semimetals disclose this property [139, 140]. This phenomenon initiates since the synchronicity of hole and electron carriers, that is thought to be the chief aspect causing hefty magnetoresistance, analogous to the instance of Bi crystal [141].

2.8. Combination with Foreign Atoms

It is well known that properties of 2D materials can be controlled by introducing foreign substance [142] for metallic TMDs also, it can be achieved through three broad methods: doping [143], intercalation [144] and heterojunction formation [145]. Effect of atom substitution on magnetic properties in metallic TMDs has previously been discussed in detail in section 2.6.3. In this section, other notable aspects of such phenomenon as revealed by researchers will be discussed. Most of the studies are theoretical in nature. However, there do exist a few classic experimental efforts.

2.8.1. Doping

Influence of Te doping in titanium dichalcogenides has been studied by Mali et al. through computation (Perdew-Burke-Ernzerhof, Perdew-Wang generalized gradient approximation and von Barth-Hedin) and comparing with experimental Compton profile results [146]. As per their findings, doping of Te at S site disappears the narrow band gap of TiS_2 resulting in metallic character of TiSTe. Zhao et al. theoretically studied optical properties of ZrS_2 monolayer by n- (F, Cl, Br and I) and p- (N, P and As) type doping [147]. The dopants are combined by substituting S atom. A few calculated band structures have been compared in Figure 2.29. The formation energy upsurges with growing impurity atom size for each doped case and it is comparatively easier to integrate impurity atoms under Zr-rich experimental condition. Change in optical properties in terms of absorption coefficient $\alpha(\omega)$, refractive index $n(\omega)$, reflectivity $R(\omega)$ and extinction coefficient $\kappa(\omega)$ have also been studied. Pristine or doped ZrS_2 monolayer is transparent for all light frequencies. Absorption effect of ultraviolet light improves with the rise of incident light energy and the absorption edge happens red shift phenomenon in each doped system. Particularly, the optical properties of the system are much affected by As-doping.

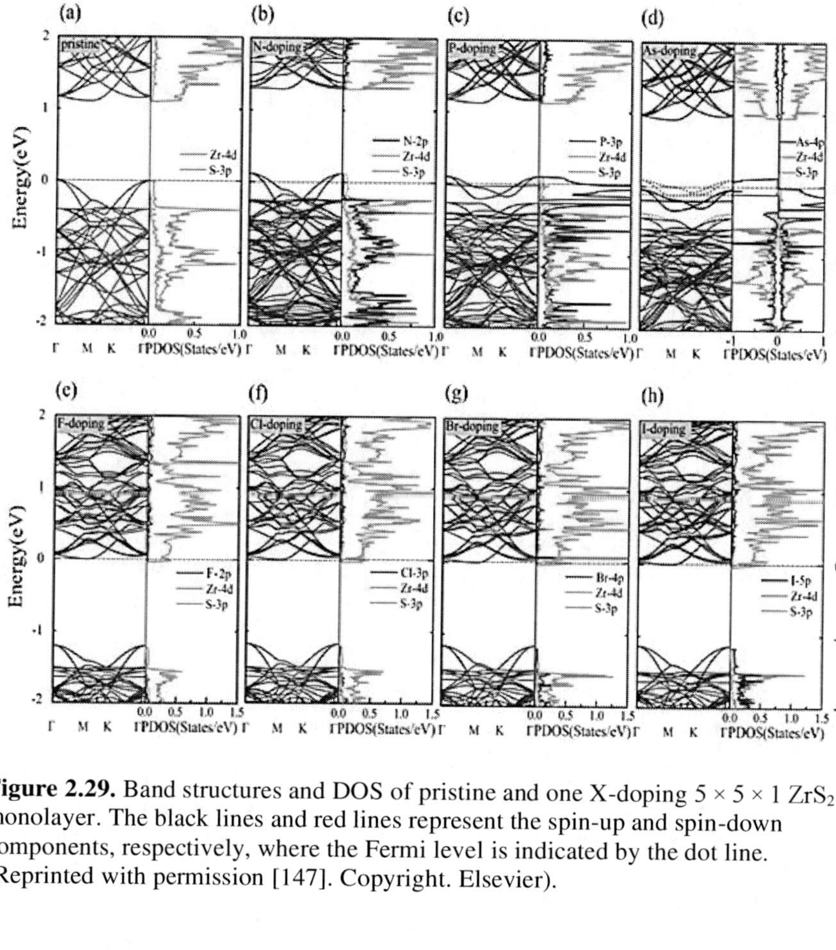

Figure 2.29. Band structures and DOS of pristine and one X-doping 5 × 5 × 1 ZrS_2 monolayer. The black lines and red lines represent the spin-up and spin-down components, respectively, where the Fermi level is indicated by the dot line. (Reprinted with permission [147]. Copyright. Elsevier).

2.8.2. Intercalation

Lithium intercalation of exfoliated VX_2 has been reported to exhibit quite different properties while compared with bulk counterpart [148]. Both bulk and exfoliated VS_2, VSe_2, and VTe_2 possess inherent electrochemical activity, which can limit their usage for electrocatalysts and as electrode materials. HER performance study of bulk and exfoliated VX_2s demonstrates that the presence of lithium vanadates and vanadium oxides generated during the exfoliation process can highly affect the HER capability. Sanchez-Perez et al. synthesized and characterized Fe-intercalated $ZrSe_2$ thin films on quartz substrates utilizing a low-pressure chemical vapor deposition of a single-source precursor $[Fe(\eta_5\text{-}C_5H_4Se)2Zr(\eta_5\text{-}C_5H_5)_2]_2$ [149]. Powder X-ray diffraction of the film scraping and succeeding Rietveld enhancement of data exhibited successful preparation of the $Fe_{0.14}ZrSe_2$ phase, along with minor phases of FeSe and ZrO_2. through intercalation, a slight optical band gap enhancement ($E_{g(direct)opt} = 1.72$ eV) is sensed in contrast with that of the host substance (Figure 2.30).

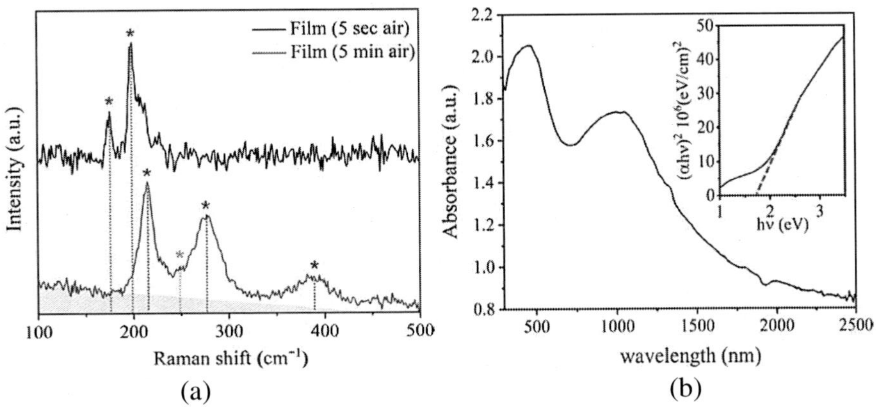

Figure 2.30. (a) Raman spectra of the LPCVD grown $FexZrSe_2$ films after minimal (ca. 5 s) and 5 min exposure to air. Bands corresponding to the contribution of $FexZrSe_2$ (purple), iron oxide (α-Fe_2O_3, green), selenium (Se, orange), and amorphous zirconia (ZrO_2, red) are detected in the spectra and (b) Absorbance (black line) of the thin film after minimal exposure to atmospheric conditions showing a $FexZrSe_2$ absorption band centered at ~1000 nm and a second absorption band centered at ~450 nm due to the formation of α-Fe_2O_3 52 upon exposure of the thin film to air prior to/during measurement. Graphical representation of band gap calculation using a Tauc plot is depicted in the inset. (Reprinted with permission [149]. Copyright. American Chemical Society).

2.8.3. Heterojunction

Popov et al. studied structural, electronic and magnetic properties of graphene/VX_2 heterostructures using a DFT+U tactic. The research with vertical junctions comprising of graphene and VX_2 monolayers confirmed that interlayer interactions resulted in the generation of robust spin polarization of both VX_2 and graphene fragments while conserving the linear dispersal of graphene-originated bands [150]. It was discovered that the addition of Mo atoms between the sheets led to n-doping of graphene with a discerning transition of graphene bands maintaining the spin-down Dirac cone unbroken. (Figure 2.31) Vu et al. recently developed a one-step synthesis process for generating $NbSe_2$/Nb-doped-WSe_2 metal-doped-semiconductor (M/d-S) van der Waals heterostructures (vdWHs) using a niobium molar ratio-controlled solution-phase precursor [151]. This resulted in tuning the hole carrier concentration through control in molar ratio in precursor solution and a corresponding reduction in contact resistance and gain in mobility and on-off ratio. The results are promising for developing upcoming high performance integrated circuits.

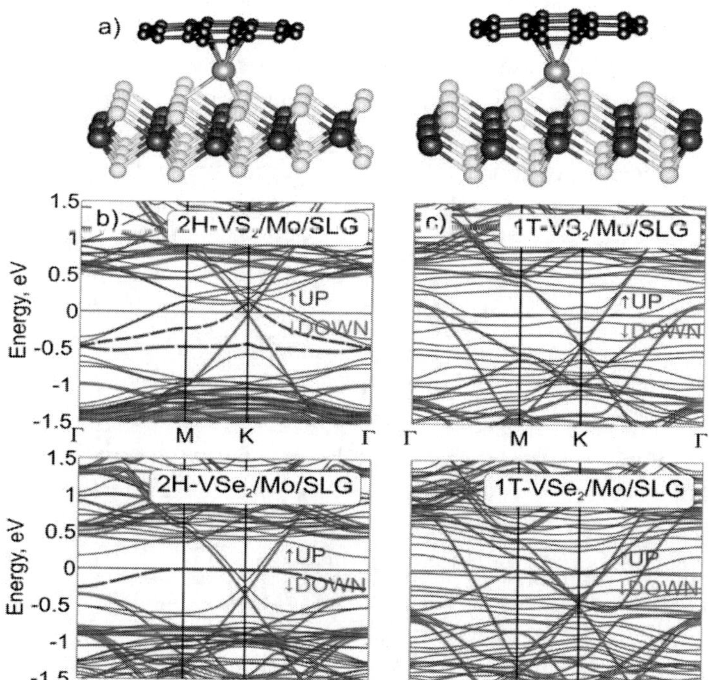

Figure 2.31. Intercalation of Mo atoms between VX_2 and graphene. The atomic structures (a) and electronic bands of 2H (b) and 1T (c) graphene/Mo/2H-VS_2 and graphene/Mo/2H-VSe_2 phases. In (a) carbon, vanadium, sulfur/selenium and molybdenum atoms are marked by brown, red, yellow and violet colors, respectively. In (b) and (c) SLG denotes single layer graphene. The dashed lines in (b) indicate the Mo-originated impurity bands. The Fermi energy is shifted to zero and marked by the horizontal red line. (Reprinted with permission [150]. Copyright. Royal Society of Chemistry).

Conclusion

Metallic TMDs possess a host of unique properties which turn them suitable candidate material for many sophisticated applications in the field of electronics in general. Contrasting semiconducting counterparts, mono- or few-layer metallic films are yet in the premature stage. They can easily be oxidized at ambient situations or during chemical processing. Challenges are thus prevalent in device applications, and one of the major issues is the attainment of ohmic contact. Frequency reliant conductivity in precise at small frequency bound is fascinating for RF devices. Whereas $1T_d$ WTe_2 and $MoTe_2$ are evidenced to be the Weyl semimetal, the Weyl points are positioned overhead the Fermi energy. This averts to directly portray the characteristics of the Weyl points by the electrical transport quantities. So, a good stratagem of nobbling ought to be settled to transfer the Fermi energy arrive to the Weyl points. This effort should include both first-principle computations and practical quantifications. Exploring for rich modules of 2D substances for Weyl semimetal is yet another open research area.

Numerous efforts have been expended for probing high temperature superconductivity in the metallic TMDs. Most metallic TMDs trail the BCS theory. This needs a high energy phonon, that is sturdily attached with electrons to generate electron pairs. Though, the creation of the CDW state designates the relaxed phonons i.e., phonons with a rather little energy. It results in low T_c. Yet, the metallic TMDs are however a podium for confirming the endorsement of the theory useful for the superconductivity. Moreover, the innate of the atomic depth may cause more recompences. One archetypal instance is the juxtaposition effect, where the interface amongst the 2D things with the substratum will sturdily influence their characteristics. Whether this connecter effect can augment or abolish the superconductivity and magnetism of the atomic shrill layer is yet strange.

Controlling of phase is critical in optimizing properties of metallic TMDs for targeted applications. This can be done in various means as per the situation and intended application. Strain is a good option in modulating phase transition. However, the contact between TMDs and substrate is a vital matter that should be well-thought-out as it regulates whether or not the strain can be completely transmitted to TMDs. Both the thermal treatment and interlayer coupling may not be good adoptions for activating phase transition. Thermal treatment will unavoidably introduce vacancies or defects into TMDs, causing irreversible phase transition. Moreover, for many TMDs, high temperature treatment will abolish the structure of the materials before prompting phase transition. Interlayer coupling can successfully modulate the (opto) electronic properties of TMDs or TMD heterostructures. Though, this strongly rests on on the layer numbers of TMDs which is very tough to precisely control under current fabrication capability. TMD doping with electrostatic gating is a comparatively better choice, as it evades destroying material structure or presenting impurities originating from other doping methods, such as surface chemical modification and alkali ion intercalation. Moreover, electrostatic gating can tweak the doping concentration with external gate voltage and therefore augment the controllability of the phase transition. Yet, the doping ability of electrostatic gating is still restricted to $\approx 10^{13}–10^{14}$e cm^{-2}, that is lower than the necessities for phase transition in most of the metallic TMDs. Furthermore, the doping depth is also limited to only nanometer scale. In nutshell, there is hugely unexplored potential for exploiting benefits from metallic TMDs in numerous fine-scale electronic applications and scope of research in this field is very high in attaining such goals.

References

[1] Geim, Andre Konstantin. "Graphene: status and prospects." *Science* 324, no. 5934 (2009): 1530-1534. https://doi.org/10.1126/science.1158877.

[2] Wang, Jinying, Ruiqi Zhao, Zhongfan Liu, and Zhirong Liu. "Widely tunable carrier mobility of boron nitride-embedded graphene." *Small* 9, no. 8 (2013): 1373-1378. https://doi.org/10.1002/smll.201202978.

[3] Li, Likai, Yijun Yu, Guo Jun Ye, Qingqin Ge, Xuedong Ou, Hua Wu, Donglai Feng, Xian Hui Chen, and Yuanbo Zhang. "Black phosphorus field-effect transistors." *Nature nanotechnology* 9, no. 5 (2014): 372-377. https://doi.org/10.1038/nnano.2014.35.

[4] Wang, Zhi-Qiang, Tie-Yu Lü, Hui-Qiong Wang, Yuan Ping Feng, and Jin-Cheng Zheng. "Review of borophene and its potential applications." *Frontiers of Physics* 14, no. 3 (2019): 1-20. https://doi.org/10.1007/s11467-019-0884-5.

[5] Vogt, Patrick, Paola De Padova, Claudio Quaresima, Jose Avila, Emmanouil Frantzeskakis, Maria Carmen Asensio, Andrea Resta, Bénédicte Ealet, and Guy Le Lay. "Silicene: compelling experimental evidence for graphenelike two-dimensional silicon." *Physical review letters* 108, no. 15 (2012): 155501. https://doi.org/10.1103/PhysRevLett.108.155501.

[6] Zhu, Feng-feng, Wei-jiong Chen, Yong Xu, Chun-lei Gao, Dan-dan Guan, Can-hua Liu, Dong Qian, Shou-Cheng Zhang, and Jin-feng Jia. "Epitaxial growth of two-dimensional stanene." *Nature materials* 14, no. 10 (2015): 1020-1025. https://doi.org/10.1038/nmat4384.

[7] Johari, Priya, and Vivek B. Shenoy. "Tuning the electronic properties of semiconducting transition metal dichalcogenides by applying mechanical strains." *ACS nano* 6, no. 6 (2012): 5449-5456. https://doi.org/10.1021/nn301320r.

[8] Yan, Binghai, and Claudia Felser. "Topological materials: Weyl semimetals." *Annual Review of Condensed Matter Physics* 8 (2017): 337-354. https://doi.org/10.1146/annurev-conmatphys-031016-025458.

[9] Lv, Baiqing, Tian Qian, and Hong Ding. "Angle-resolved photoemission spectroscopy and its application to topological materials." *Nature Reviews Physics* 1, no. 10 (2019): 609-626. https://doi.org/10.1038/s42254-019-0088-5.

[10] Yang, Gang, ed. *Density Functional Calculations: Recent Progresses of Theory and Application*. BoD–Books on Demand, 2018. https://doi.org/10.5772/intechopen.68548.

[11] Tanaka, Toshiyuki. "A theory of mean field approximation." *Advances in Neural Information Processing Systems* (1999): 351-360. ISBN: 9780262561457.

[12] Seminario, Jorge M., ed. *Recent developments and applications of modern density functional theory*. Elsevier, 1996. ISBN: 9780444824042.

[13] Kolobov, Alexander V., and Junji Tominaga. *Two-dimensional transition-metal dichalcogenides*. Vol. 239. Springer, 2016. ISBN: 978-3-319-31450-1.

[14] Chen, Jun, and Jamie H. Warner. "Atomic structure of defects in transitional metal dichalcogenides using transmission electron microscopy." *2D Semiconductor Materials and Devices* (2020): 167-197. https://doi.org/10.1016/B978-0-12-816187-6.00006-6.

[15] Naito, Michio, and Shoji Tanaka. "Electrical transport properties in 2 H-NbS_2,-$NbSe_2$,-TaS_2 and-$TaSe_2$." *Journal of the Physical Society of Japan* 51, no. 1 (1982): 219-227. https://doi.org/10.1143/JPSJ.51.219.

[16] Naylor, Carl H., William M. Parkin, Jinglei Ping, Zhaoli Gao, Yu Ren Zhou, Youngkuk Kim, Frank Streller et al. "Monolayer single-crystal 1T′-$MoTe_2$ grown by chemical vapor deposition exhibits weak antilocalization effect." *Nano letters* 16, no. 7 (2016): 4297-4304. https://doi.org/10.1021/acs.nanolett.6b01342.

[17] Jana, Manoj K., Anjali Singh, Dattatray J. Late, Catherine R. Rajamathi, Kanishka Biswas, Claudia Felser, Umesh V. Waghmare, and C. N. R. Rao. "A combined experimental and theoretical study of the structural, electronic and vibrational properties of bulk and few-layer T_d-WTe_2." *Journal of Physics: Condensed Matter* 27, no. 28 (2015): 285401. https://doi.org/10.1088/0953-8984/27/28/285401.

[18] Huang, H. H., Xiaofeng Fan, David J. Singh, and W. T. Zheng. "Recent progress of TMD nanomaterials: phase transitions and applications." *Nanoscale* 12, no. 3 (2020): 1247-1268. https://doi.org/10.1039/C9NR08313H.

[19] Ekvall, Inger, Ju-Jin Kim, and Hakan Olin. "Atomic and electronic structures of the two different layers in $4H_b$-TaS_2 at 4.2 K." *Physical Review B* 55, no. 11 (1997): 6758. https://doi.org/10.1103/PhysRevB.55.6758.

[20] Zhong, Mianzeng, Yan Li, Qinglin Xia, Xiuqing Meng, Fengmin Wu, and Jingbo Li. "Ferromagnetism in VS_2 nanostructures: Nanoflowers versus ultrathin nanosheets." *Materials Letters* 124 (2014): 282-285. https://doi.org/10.1016/j.matlet.2014.03.110.

[21] Onuki, Yoshichika, Rumiko Inada, and Sei-ichi Tanuma. "T_2-dependence of electrical resistivity in TiS_2 and $ZrSe_2$." *Journal of the Physical Society of Japan* 51, no. 4 (1982): 1223-1227. https://doi.org/10.1143/JPSJ.51.1223.

[22] Rimmington, H. P. B., and A. A. Balchin. "The growth by iodine vapour transport techniques and the crystal structures of layer compounds in the series TiS_xSe_{2-x}, TiS_xTe_{2-x}, $TiSe_xTe_{2-x}$." *Journal of Crystal Growth* 21, no. 2 (1974): 171-181. https://doi.org/10.1016/0022-0248(74)90001-3.

[23] Wang, Fengmei, Tofik A. Shifa, Peng Yu, Peng He, Yang Liu, Feng Wang, Zhenxing Wang et al. "New frontiers on van der Waals layered metal phosphorous trichalcogenides." *Advanced Functional Materials* 28, no. 37 (2018): 1802151. https://doi.org/10.1002/adfm.201802151.

[24] Susner, Michael A., Marius Chyasnavichyus, Michael A. McGuire, Panchapakesan Ganesh, and Petro Maksymovych. "Metal thio-and selenophosphates as multifunctional van der Waals layered materials." *Advanced Materials* 29, no. 38 (2017): 1602852. https://doi.org/10.1002/adma.201602852.

[25] Samal, Rutuparna, Gopal Sanyal, Brahmananda Chakraborty, and Chandra Sekhar Rout. "Two-dimensional transition metal phosphorous trichalcogenides (MPX_3): a review on emerging trends, current state and future perspectives." *Journal of Materials Chemistry A* 9, no. 5 (2021): 2560-2591. https://doi.org/10.1039/D0TA09752G.

[26] Margadonna, Serena, Yasuhiro Takabayashi, Martin T. McDonald, Karolina Kasperkiewicz, Yoshikazu Mizuguchi, Yoshihiko Takano, Andrew N. Fitch, Emmanuelle Suard, and Kosmas Prassides. "Crystal structure of the new $FeSe_{1-x}$ superconductor." *Chemical communications* 43 (2008): 5607-5609. https://doi.org/10.1039/B813076K.

[27] Ge, Jian-Feng, Zhi-Long Liu, Canhua Liu, Chun-Lei Gao, Dong Qian, Qi-Kun Xue, Ying Liu, and Jin-Feng Jia. "Superconductivity above 100 K in single-layer FeSe films on doped $SrTiO_3$." *Nature materials* 14, no. 3 (2015): 285-289. https://doi.org/10.1038/nmat4153.

[28] Zhang, Haijun, Chao-Xing Liu, Xiao-Liang Qi, Xi Dai, Zhong Fang, and Shou-Cheng Zhang. "Topological insulators in Bi_2Se_3, Bi_2Te_3 and Sb_2Te_3 with a single Dirac cone on the surface." *Nature physics* 5, no. 6 (2009): 438-442. https://doi.org/10.1038/nphys1270.

[29] Poplavko, Yuriy M. *Electronic materials: Principles and applied science*. Elsevier, 2018. ISBN 978-0-12-815780-0.

[30] Kasap, Safa O. *Principles of electronic materials and devices*. Vol. 2. New York: McGraw-Hill, 2006. ISBN13: 9780078028182.

[31] Abrikosov, A. A. "Quantum magnetoresistance." *Physical Review B* 58, no. 5 (1998): 2788. https://doi.org/10.1103/PhysRevB.58.2788.

[32] Ramirez, A. P. "Colossal magnetoresistance." *Journal of Physics: Condensed Matter* 9, no. 39 (1997): 8171. https://doi.org/10.1088/0953-8984/9/39/005.

[33] Grasso, Vincenzo, ed. "*Electronic structure and electronic transitions in layered materials*." (2012). ISBN: 978-94-009-4542-5.

[34] Yang, Heejun, Sung Wng Kim, Manish Chhowalla, and Young Hee Lee. "Structural and quantum-state phase transitions in van der Waals layered

materials." *Nature Physics* 13, no. 10 (2017): 931-937. https://doi.org/10.1038/nphys4188.

[35] Wilson, Jl A., and A. D. Yoffe. "The transition metal dichalcogenides discussion and interpretation of the observed optical, electrical and structural properties." *Advances in Physics* 18, no. 73 (1969): 193-335. https://doi.org/10.1080/00018736900101307.

[36] Rasmussen, Filip A., and Kristian S. Thygesen. "Computational 2D materials database: electronic structure of transition-metal dichalcogenides and oxides." *The Journal of Physical Chemistry C* 119, no. 23 (2015): 13169-13183. https://doi.org/10.1021/acs.jpcc.5b02950.

[37] Motizuki, Kazuko, ed. *Structural phase transitions in layered transition metal compounds*. Vol. 8. Springer Science & Business Media, 1986. ISBN: 978-90-277-2171-6.

[38] Sun, Xingdan, Wanying Li, Xiao Wang, Qi Sui, Tongyao Zhang, Zhi Wang, Long Liu et al. "Room temperature ferromagnetism in ultra-thin van der Waals crystals of 1T-$CrTe_2$." *Nano Research* 13, no. 12 (2020): 3358-3363. https://doi.org/10.1007/s12274-020-3021-4.

[39] Ali, Mazhar N., Jun Xiong, Steven Flynn, Jing Tao, Quinn D. Gibson, Leslie M. Schoop, Tian Liang et al. "Large, non-saturating magnetoresistance in WTe_2." *Nature* 514, no. 7521 (2014): 205-208. https://doi.org/10.1038/nature13763.

[40] Keum, Dong Hoon, Suyeon Cho, Jung Ho Kim, Duk-Hyun Choe, Ha-Jun Sung, Min Kan, Haeyong Kang et al. "Bandgap opening in few-layered monoclinic $MoTe_2$." *Nature Physics* 11, no. 6 (2015): 482-486. https://doi.org/10.1038/nphys3314.

[41] Cho, Suyeon, Se Hwang Kang, Ho Sung Yu, Hyo Won Kim, Wonhee Ko, Sung Woo Hwang, Woo Hyun Han et al. "Te vacancy-driven superconductivity in orthorhombic molybdenum ditelluride." *2D Materials* 4, no. 2 (2017): 021030. https://doi.org/10.1088/2053-1583/aa735e.

[42] Ataca, Can, Hasan Sahin, and Salim Ciraci. "Stable, single-layer MX_2 transition-metal oxides and dichalcogenides in a honeycomb-like structure." *The Journal of Physical Chemistry C* 116, no. 16 (2012): 8983-8999. https://doi.org/10.1021/jp212558p.

[43] Fang, C. M., R. A. De Groot, and Cornelis Haas. "Bulk and surface electronic structure of 1T− TiS_2 and 1T−$TiSe_2$." *Physical Review B* 56, no. 8 (1997): 4455. https://doi.org/10.1103/PhysRevB.56.4455.

[44] Ghafari, A., L. Petaccia, and C. Janowitz. "Splitting of the Ti-3d bands of $TiSe_2$ in the charge-density wave phase." *Applied Surface Science* 396 (2017): 1649-1656. https://doi.org/10.1016/j.apsusc.2016.12.016.

[45] Suga, Shigemasa, Christian Tusche, Yu-ichiro Matsushita, Martin Ellguth, Akinori Irizawa, and Jürgen Kirschner. "Momentum microscopy of the layered semiconductor TiS_2 and Ni intercalated $Ni_{1/3}TiS_2$." *New Journal of Physics* 17, no. 8 (2015): 083010. https://doi.org/10.1088/1367-2630/17/8/083010.

[46] Strocov, Vladimir N., Ming Shi, Masaki Kobayashi, Claude Monney, Xiaoqiang Wang, Juraj Krempasky, Thorsten Schmitt, Luc Patthey, Helmuth Berger, and Peter Blaha. "Three-dimensional electron realm in VSe_2 by soft-x-ray

photoelectron spectroscopy: Origin of charge-density waves." *Physical review letters* 109, no. 8 (2012): 086401. https://doi.org/10.1103/physrevlett.109.086401.

[47] Nakata, Yuki, Katsuaki Sugawara, Ryota Shimizu, Yoshinori Okada, Patrick Han, Taro Hitosugi, Keiji Ueno, Takafumi Sato, and Takashi Takahashi. "Monolayer 1T-$NbSe_2$ as a Mott insulator." *NPG Asia Materials* 8, no. 11 (2016): e321-e321. https://doi.org/10.1038/am.2016.157.

[48] Ekvall, Inger, Ju-Jin Kim, and Hakan Olin. "Atomic and electronic structures of the two different layers in 4Hb-TaS_2 at 4.2 K." *Physical Review B* 55, no. 11 (1997): 6758. https://doi.org/10.1103/PhysRevB.55.6758.

[49] Pandey, Shishir K., Ruma Das, and Priya Mahadevan. "Layer-dependent electronic structure changes in transition metal dichalcogenides: the microscopic origin." *ACS omega* 5, no. 25 (2020): 15169-15176. https://doi.org/10.1021/acsomega.0c01138.

[50] Yeh, Po-Chun, Wencan Jin, Nader Zaki, Datong Zhang, Jonathan T. Liou, Jerzy T. Sadowski, Abdullah Al-Mahboob et al. "Layer-dependent electronic structure of an atomically heavy two-dimensional dichalcogenide." *Physical Review B* 91, no. 4 (2015): 041407. https://doi.org/10.1103/PhysRevB.91.041407.

[51] Chen, Peng, Y-H. Chan, M-H. Wong, X-Y. Fang, Mei-Y. Chou, S-K. Mo, Zahid Hussain, A-V. Fedorov, and T-C. Chiang. "Dimensional effects on the charge density waves in ultrathin films of $TiSe_2$." *Nano letters* 16, no. 10 (2016): 6331-6336. https://doi.org/10.1021/acs.nanolett.6b02710.

[52] Ghorbani-Asl, Mahdi, Agnieszka Kuc, Pere Miro, and Thomas Heine. "A single-material logical junction based on 2D crystal PdS_2." *Advanced Materials* 28, no. 5 (2016): 853-856. https://doi.org/10.1002/adma.201504274.

[53] Jing, Yu, Yandong Ma, Yafei Li, and Thomas Heine. "GeP_3: A small indirect band gap 2D crystal with high carrier mobility and strong interlayer quantum confinement." *Nano letters* 17, no. 3 (2017): 1833-1838. https://doi.org/10.1021/acs.nanolett.6b05143.

[54] Thorne, Robert E. "Charge-density-wave conductors." *Physics Today* 49, no. 5 (1996): 42-47. https://doi.org/10.1063/1.881498.

[55] Yoshida, Masaro, Ryuji Suzuki, Yijin Zhang, Masaki Nakano, and Yoshihiro Iwasa. "Memristive phase switching in two-dimensional 1T-TaS_2 crystals." *Science advances* 1, no. 9 (2015): e1500606. https://doi.org/10.1126/sciadv.1500606.

[56] Skolimowski, Jan, and Yaroslav Gerasimenko. "Mottness collapse without metallization in the domain wall of the triangular-lattice Mott insulator 1T− TaS_2." *Physical review letters* 122, no. 3 (2019): 036802. https://doi.org/10.1103/PhysRevLett.122.036802.

[57] Eichberger, Maximilian, Hanjo Schäfer, Marina Krumova, Markus Beyer, Jure Demsar, Helmuth Berger, Gustavo Moriena, Germán Sciaini, and RJ Dwayne Miller. "Snapshots of cooperative atomic motions in the optical suppression of charge density waves." *Nature* 468, no. 7325 (2010): 799-802. https://doi.org/10.1038/nature09539.

[58] Duong, Dinh Loc, Marko Burghard, and J. Christian Schön. "Ab initio computation of the transition temperature of the charge density wave transition in $TiSe_2$." *Physical Review B* 92, no. 24 (2015): 245131. https://doi.org/10.1103/PhysRevB.92.245131.

[59] Ishioka, J., Y. H. Liu, K. Shimatake, T. Kurosawa, K. Ichimura, Y. Toda, M. Oda, and S. Tanda. "Chiral charge-density waves." *Physical review letters* 105, no. 17 (2010): 176401. https://doi.org/10.1103/PhysRevLett.105.176401.

[60] Zhu, Xuetao, Yanwei Cao, Jiandi Zhang, E. W. Plummer, and Jiandong Guo. "Classification of charge density waves based on their nature." *Proceedings of the National Academy of Sciences* 112, no. 8 (2015): 2367-2371. https://doi.org/10.1073/pnas.1424791112.

[61] Peierls, Rudolf Ernst. *Quantum theory of solids.* Clarendon Press, 1996. ISBN: 978-0-471-62412-7.

[62] Peierls, Rudolf. *More surprises in theoretical physics*. Princeton University Press, 2020. ISBN 9780691025223.

[63] Chen, Peng, Y-H. Chan, X-Y. Fang, Yi Zhang, Mei-Yin Chou, S-K. Mo, Zahid Hussain, A-V. Fedorov, and T-C. Chiang. "Charge density wave transition in single-layer titanium diselenide." *Nature communications* 6, no. 1 (2015): 1-5. https://doi.org/10.1038/ncomms9943.

[64] Ritschel, T., J. Trinckauf, K. Koepernik, B. Büchner, M. V. Zimmermann, H. Berger, Y. I. Joe, P. Abbamonte, and J. Geck. "Orbital textures and charge density waves in transition metal dichalcogenides." *Nature physics* 11, no. 4 (2015): 328-331. https://doi.org/10.1038/nphys3267.

[65] Johannes, M. D., and I. I. Mazin. "Fermi surface nesting and the origin of charge density waves in metals." *Physical Review B* 77, no. 16 (2008): 165135. https://doi.org/10.1103/PhysRevB.77.165135.

[66] Cercellier, H., Claude Monney, F. Clerc, Corsin Battaglia, Laurent Despont, M. G. Garnier, H. Beck et al. "Evidence for an Excitonic Insulator Phase in 1T−$TiSe_2$." *Physical review letters* 99, no. 14 (2007): 146403. https://doi.org/10.1103/PhysRevLett.99.146403.

[67] Porer, Michael, Ursula Leierseder, J-M. Ménard, Hatem Dachraoui, L. Mouchliadis, I. E. Perakis, Ulrich Heinzmann, J. Demsar, K. Rossnagel, and Rupert Huber. "Non-thermal separation of electronic and structural orders in a persisting charge density wave." *Nature materials* 13, no. 9 (2014): 857-861. https://doi.org/10.1038/nmat4042.

[68] Ge, Yizhi, and Amy Y. Liu. "First-principles investigation of the charge-density-wave instability in 1T-$TaSe_2$." *Physical Review B* 82, no. 15 (2010): 155133. http://dx.doi.org/10.1103/PhysRevB.82.155133.

[69] Sturge, M. D. "The Jahn-Teller effect in solids." In *Solid state physics*, vol. 20, pp. 91-211. Academic Press, 1968. https://doi.org/10.1016/S0081-1947(08)60218-0.

[70] Kohn, W. "Image of the Fermi Surface in the Vibration Spectrum of a Metal." *Physical Review Letters* 2, no. 9 (1959): 393. https://doi.org/10.1103/PhysRevLett.2.393.

[71] Rietschel, H. "The giant Kohn phonon anomaly in a Peierls semiconductor." *Solid State Communications* 13, no. 11 (1973): 1859-1863. https://doi.org/10.1016/0038-1098(73)90745-X.

[72] Monceau, Pierre. "Electronic crystals: an experimental overview." *Advances in Physics* 61, no. 4 (2012): 325-581. https://doi.org/10.1080/00018732.2012.719674.

[73] Weber, F., S. Rosenkranz, J-P. Castellan, R. Osborn, R. Hott, R. Heid, K-P. Bohnen, T. Egami, A. H. Said, and D. Reznik. "Extended Phonon Collapse and the Origin of the Charge-Density Wave in 2H− $NbSe_2$." *Physical review letters* 107, no. 10 (2011): 107403. https://doi.org/10.1103/PhysRevLett.107.107403.

[74] Holt, Martin, Paul Zschack, Hawoong Hong, M. Y. Chou, and T-C. Chiang. "X-ray studies of phonon softening in $TiSe_2$." *Physical review letters* 86, no. 17 (2001): 3799. https://doi.org/10.1103/PhysRevLett.86.3799.

[75] Albertini, Oliver R., Rui Zhao, Rebecca L. McCann, Simin Feng, Mauricio Terrones, James K. Freericks, Joshua A. Robinson, and Amy Y. Liu. "Zone-center phonons of bulk, few-layer, and monolayer 1T− TaS_2: Detection of commensurate charge density wave phase through Raman scattering." *Physical Review B* 93, no. 21 (2016): 214109. https://doi.org/10.1103/PhysRevB.93.214109.

[76] He, Rui, Junichi Okamoto, Zhipeng Ye, Gaihua Ye, Heidi Anderson, Xia Dai, Xianxin Wu et al. "Distinct surface and bulk charge density waves in ultrathin 1T− TaS_2." *Physical Review B* 94, no. 20 (2016): 201108. https://arxiv.org/ct?url=https %3A%2F%2Fdx.doi.org%2F10.1103%2FPhysRevB.94.201108&v=e084604e.

[77] Duong, Dinh Loc, Gihun Ryu, Alexander Hoyer, Chengtian Lin, Marko Burghard, and Klaus Kern. "Raman characterization of the charge density wave phase of 1T-$TiSe_2$: from bulk to atomically thin layers." *ACS nano* 11, no. 1 (2017): 1034-1040. https://doi.org/10.1021/acsnano.6b07737.

[78] Yoshida, Masaro, Yijin Zhang, Jianting Ye, Ryuji Suzuki, Yasuhiko Imai, Shigeru Kimura, Akihiko Fujiwara, and Yoshihiro Iwasa. "Controlling charge-density-wave states in nano-thick crystals of 1T-TaS_2." *Scientific reports* 4, no. 1 (2014): 1-5. https://doi.org/10.1038/srep07302.

[79] Xi, Xiaoxiang, Liang Zhao, Zefang Wang, Helmuth Berger, László Forró, Jie Shan, and Kin Fai Mak. "Strongly enhanced charge-density-wave order in monolayer $NbSe_2$." *Nature nanotechnology* 10, no. 9 (2015): 765-769. https://doi.org/10.1038/nnano.2015.143.

[80] Bardeen, John, Leon N. Cooper, and John Robert Schrieffer. "Theory of superconductivity." *Physical review* 108, no. 5 (1957): 1175. https://doi.org/10.1103/PhysRev.108.1175.

[81] Mazin, Igor I. "Extraordinarily conventional." *Nature* 525, no. 7567 (2015): 40-41. https://doi.org/10.1038/nature15203.

[82] Chen, Xiao-Jia, Viktor V. Struzhkin, Yong Yu, Alexander F. Goncharov, Cheng-Tian Lin, Ho-kwang Mao, and Russell J. Hemley. "Enhancement of superconductivity by pressure-driven competition in electronic order." *Nature* 466, no. 7309 (2010): 950-953. https://doi.org/10.1038/nature09293.

[83] Wang, Cao, Linjun Li, Shun Chi, Zengwei Zhu, Zhi Ren, Yuke Li, Yuetao Wang et al. "Thorium-doping–induced superconductivity up to 56 K in $Gd_{1-x}Th_xFeAsO$." *EPL (Europhysics Letters)* 83, no. 6 (2008): 67006. https://doi.org/10.1209/0295-5075/83/67006.

[84] Morosan, Emilia, Henny W. Zandbergen, B. S. Dennis, J. W. G. Bos, Y. Onose, Tomasz Klimczuk, A. P. Ramirez, N. P. Ong, and Robert J. Cava. "Superconductivity in Cu_xTiSe_2." *Nature Physics* 2, no. 8 (2006): 544-550. https://doi.org/10.1038/nphys360.

[85] Kusmartseva, Anna F., B. Sipos, H. Berger, Laszlo Forro, and Eduard Tutiš. "Pressure Induced Superconductivity in Pristine 1T−$TiSe_2$." *Physical review letters* 103, no. 23 (2009): 236401. https://doi.org/10.1103/PhysRevLett.103.236401.

[86] Snow, C. S., J. F. Karpus, S. L. Cooper, T. E. Kidd, and T-C. Chiang. "Quantum Melting of the Charge-Density-Wave State in 1T− $TiSe_2$." *Physical review letters* 91, no. 13 (2003): 136402. https://doi.org/10.1103/PhysRevLett.91.136402.

[87] Leroux, Maxime, Ion Errea, Mathieu Le Tacon, Sofia-Michaela Souliou, Gaston Garbarino, Laurent Cario, Alexey Bosak, Francesco Mauri, Matteo Calandra, and Pierre Rodière. "Strong anharmonicity induces quantum melting of charge density wave in 2H−$NbSe_2$ under pressure." *Physical Review B* 92, no. 14 (2015): 140303. https://doi.org/10.1103/PhysRevB.92.140303.

[88] Calandra, Matteo, and Francesco Mauri. "Charge-density wave and superconducting dome in $TiSe_2$ from electron-phonon interaction." *Physical review letters* 106, no. 19 (2011): 196406. http://dx.doi.org/10.1103/PhysRevLett.106.196406.

[89] Yu, Yijun, Fangyuan Yang, Xiu Fang Lu, Ya Jun Yan, Yong-Heum Cho, Liguo Ma, Xiaohai Niu et al. "Gate-tunable phase transitions in thin flakes of 1T-TaS_2." *Nature nanotechnology* 10, no. 3 (2015): 270-276. https://doi.org/10.1038/nnano.2014.323.

[90] Li, L. J., E. C. T. O'Farrell, K. P. Loh, Goki Eda, B. Özyilmaz, and AH Castro Neto. "Controlling many-body states by the electric-field effect in a two-dimensional material." *Nature* 529, no. 7585 (2016): 185-189. https://doi.org/10.1038/nature16175.

[91] Frindt, R. F. "Superconductivity in ultrathin $NbSe_2$ layers." *Physical Review Letters* 28, no. 5 (1972): 299. https://doi.org/10.1103/PhysRevLett.28.299.

[92] Reshak, Ali Hussain, and S. Auluck. "Electronic and optical properties of the 1 T phases of TiS_2, $TiSe_2$, and $TiTe_2$." *Physical Review B* 68, no. 24 (2003): 245113. https://doi.org/10.1103/PhysRevB.68.245113.

[93] Bayliss, S. C., and W. Y. Liang. "Reflectivity, joint density of states and band structure of group IVb transition-metal dichalcogenides." *Journal of Physics C: Solid State Physics* 18, no. 17 (1985): 3327. https://doi.org/10.1088/0022-3719/18/17/010.

[94] Greenaway, David L., and Rudolf Nitsche. "Preparation and optical properties of group IV–VI_2 chalcogenides having the CdI_2 structure." *Journal of Physics and Chemistry of Solids* 26, no. 9 (1965): 1445-1458. https://doi.org/10.1016/0022-3697(65)90043-0.

[95] Murray, R. B., and A. D. Yoffe. "The band structures of some transition metal dichalcogenides: band structures of the titanium dichalcogenides." *Journal of Physics C: Solid State Physics* 5, no. 21 (1972): 3038. https://doi.org/10.1088/0022-3719/5/21/009.

[96] Dey, Aditya. "A First-Principles Study of TiX_2 (X= S, Se, and Te) Compounds Optical Properties under the Effect of Externally Applied Electric Field and Strain." *Physics of the Solid State* 62, no. 10 (2020): 1905-1915. https://doi.org/10.1134/S1063783420100042.

[97] Ghobadi, Amir, Turkan Gamze Ulusoy Ghobadi, Ali Kemal Okyay, and Ekmel Ozbay. "Emerging photoluminescence from defective vanadium diselenide nanosheets." *Photonics Research* 6, no. 4 (2018): 244-253. https://doi.org/10.1364/PRJ.6.000244.

[98] Reshak, Ali Hussain, and Sushil Auluck. "Theoretical investigation of the electronic and optical properties of ZrX_2 (X= S, Se and Te)." *Physica B: Condensed Matter* 353, no. 3-4 (2004): 230-237. https://doi.org/10.1016/j.physb.2004.10.001.

[99] Bayliss, S. C., and W. Y. Liang. "Symmetry dependence of optical transitions in group 4B transition metal dichalcogenides." *Journal of Physics C: Solid State Physics* 15, no. 6 (1982): 1283. https://doi.org/10.1088/0022-3719/15/6/021.

[100] Beal, A. R., J. C. Knights, and W. Y. Liang. "Transmission spectra of some transition metal dichalcogenides. II. Group VIA: trigonal prismatic coordination." *Journal of Physics C: Solid State Physics* 5, no. 24 (1972): 3540. https://doi.org/10.1088/0022-3719/5/24/016.

[101] Zhou, Kai-Ge, Min Zhao, Meng-Jie Chang, Qiang Wang, Xin-Zhi Wu, Yinglin Song, and Hao-Li Zhang. "Size-dependent nonlinear optical properties of atomically thin transition metal dichalcogenide nanosheets." *Small* 11, no. 6 (2015): 694-701. https://doi.org/10.1002/smll.201400541.

[102] Mermin, N. David, and Herbert Wagner. "Absence of ferromagnetism or antiferromagnetism in one-or two-dimensional isotropic Heisenberg models." *Physical Review Letters* 17, no. 22 (1966): 1133. https://doi.org/10.1103/PhysRevLett.17.1133.

[103] Huang, Bevin, Genevieve Clark, Efrén Navarro-Moratalla, Dahlia R. Klein, Ran Cheng, Kyle L. Seyler, Ding Zhong et al. "Layer-dependent ferromagnetism in a van der Waals crystal down to the monolayer limit." *Nature* 546, no. 7657 (2017): 270-273. https://doi.org/10.1038/nature22391.

[104] Ormaza, M., L. Fernández, M. Ilyn, A. Magana, Bin Xu, M. J. Verstraete, M. Gastaldo et al. "High temperature ferromagnetism in a $GdAg_2$ monolayer." *Nano letters* 16, no. 7 (2016): 4230-4235. https://doi.org/10.1021/acs.nanolett.6b01197.

[105] Wildes, Andrew R., Virginie Simonet, Eric Ressouche, Garry James Mcintyre, Maxim Avdeev, Emmanuelle Suard, Simon AJ Kimber et al. "Magnetic structure of the quasi-two-dimensional antiferromagnet $NiPS_3$." *Physical Review B* 92, no. 22 (2015): 224408. https://doi.org/10.1103/PhysRevB.92.224408.

[106] Ma, Yandong, Ying Dai, Meng Guo, Chengwang Niu, Yingtao Zhu, and Baibiao Huang. "Evidence of the existence of magnetism in pristine VX_2 monolayers (X= S, Se) and their strain-induced tunable magnetic properties." *ACS nano* 6, no. 2 (2012): 1695-1701. https://doi.org/10.1021/nn204667z.

[107] Gao, Daqiang, Qixin Xue, Xingze Mao, Weixiao Wang, Qiang Xu, and Desheng Xue. "Ferromagnetism in ultrathin VS_2 nanosheets." *Journal of Materials Chemistry C* 1, no. 37 (2013): 5909-5916. https://doi.org/10.1039/C3TC31233J.

[108] Qi, Jing, Yinghu Yang, Li Zhang, Junhong Chi, Daqiang Gao, and Desheng Xue. "Room-temperature ferromagnetism in Er-doped ZnO thin films." *Scripta Materialia* 60, no. 5 (2009): 289-292. https://doi.org/10.1016/j.scriptamat.2008.10.015.

[109] Bonilla, Manuel, Sadhu Kolekar, Yujing Ma, Horacio Coy Diaz, Vijaysankar Kalappattil, Raja Das, Tatiana Eggers, Humberto R. Gutierrez, Manh-Huong Phan, and Matthias Batzill. "Strong room-temperature ferromagnetism in VSe_2 monolayers on van der Waals substrates." *Nature nanotechnology* 13, no. 4 (2018): 289-293. https://doi.org/10.1038/s41565-018-0063-9.

[110] Zhang, Qingyun, Li-Yong Gan, Yingchun Cheng, and Udo Schwingenschlögl. "Spin polarization driven by a charge-density wave in monolayer 1T−TaS_2." *Physical Review B* 90, no. 8 (2014): 081103. https://doi.org/10.1103/PhysRevB.90.081103.

[111] Guo, Hongyan, Ning Lu, Lu Wang, Xiaojun Wu, and Xiao Cheng Zeng. "Tuning electronic and magnetic properties of early transition-metal dichalcogenides via tensile strain." *The Journal of Physical Chemistry C* 118, no. 13 (2014): 7242-7249. https://doi.org/10.1021/jp501734s.

[112] Lv, H. Y., W. J. Lu, D. F. Shao, Y. Liu, and Y. P. Sun. "Strain-controlled switch between ferromagnetism and antiferromagnetism in 1T−CrX_2 (X= Se, Te) monolayers." *Physical Review B* 92, no. 21 (2015): 214419. https://doi.org/10.1103/PhysRevB.92.214419.

[113] Rajca, Andrzej, Sumit Mukherjee, Maren Pink, and Suchada Rajca. "Exchange coupling mediated through-bonds and through-space in conformationally constrained polyradical scaffolds: Calix [4] arene nitroxide tetraradicals and diradical." *Journal of the American Chemical Society* 128, no. 41 (2006): 13497-13507. https://doi.org/10.1021/ja063567+.

[114] Jin, Hao, Ying Dai, BaiBiao Huang, and M-H. Whangbo. "Ferromagnetism of undoped GaN mediated by through-bond spin polarization between nitrogen dangling bonds." *Applied Physics Letters* 94, no. 16 (2009): 162505. https://doi.org/10.1063/1.3123169.

[115] Yun, Won Seok, and J. D. Lee. "Strain-induced magnetism in single-layer MoS_2: origin and manipulation." *The Journal of Physical Chemistry C* 119, no. 5 (2015): 2822-2827. https://doi.org/10.1021/jp510308a.

[116] Staley, Neal E., Jian Wu, Peter Eklund, Ying Liu, Linjun Li, and Zhuan Xu. "Electric field effect on superconductivity in atomically thin flakes of $NbSe_2$." *Physical Review B* 80, no. 18 (2009): 184505. https://doi.org/10.1103/PhysRevB.80.184505.

[117] Lebegue, S., and O. Eriksson. "Electronic structure of two-dimensional crystals from ab initio theory." *Physical Review B* 79, no. 11 (2009): 115409. http://dx.doi.org/10.1103/PhysRevB.79.115409.

[118] Zhou, Yungang, Zhiguo Wang, Ping Yang, Xiaotao Zu, Li Yang, Xin Sun, and Fei Gao. "Tensile strain switched ferromagnetism in layered NbS_2 and $NbSe_2$." *Acs Nano* 6, no. 11 (2012): 9727-9736. https://doi.org/10.1021/nn303198w.

[119] Wang, Yiren, Sean Li, and Jiabao Yi. "Electronic and magnetic properties of Co doped MoS_2 monolayer." *Scientific reports* 6, no. 1 (2016): 1-9. https://doi.org/10.1038/srep24153.

[120] Liu, Biao, Li-Juan Wu, Yu-Qing Zhao, Lin-Zhi Wang, and Meng-Qiu Cai. "A first-principles study of magnetic variation via doping vacancy in monolayer VS_2."

Journal of Magnetism and Magnetic Materials 420 (2016): 218-224. https://doi.org/10.1016/j.jmmm.2016.07.038.

[121] Boukhvalov, D. W., M. I. Katsnelson, and A. I. Lichtenstein. "Hydrogen on graphene: Electronic structure, total energy, structural distortions and magnetism from first-principles calculations." *Physical Review B* 77, no. 3 (2008): 035427. https://doi.org/10.1103/PhysRevB.77.035427.

[122] González-Herrero, Héctor, José M. Gómez-Rodríguez, Pierre Mallet, Mohamed Moaied, Juan José Palacios, Carlos Salgado, Miguel M. Ugeda, Jean-Yves Veuillen, Félix Yndurain, and Iván Brihuega. "Atomic-scale control of graphene magnetism by using hydrogen atoms." *Science* 352, no. 6284 (2016): 437-441. https://doi.org/10.1126/science.aad8038.

[123] Pan, Hui. "Electronic and magnetic properties of vanadium dichalcogenides monolayers tuned by hydrogenation." *The Journal of Physical Chemistry C* 118, no. 24 (2014): 13248-13253. https://doi.org/10.1021/jp503030b.

[124] Schlenker, C., C. Landee, R. Buder, and F. Lévy. "Magnetic properties of vanadium-doped titanium diselenide." *Journal of Magnetism and Magnetic Materials* 15 (1980): 91-92. https://doi.org/10.1016/0304-8853(80)90964-6.

[125] Freitas, Daniele C., Matias Nunez, Pierre Strobel, André Sulpice, Ruben Weht, Armando A. Aligia, and Manuel Nunez-Regueiro. "Antiferromagnetism and ferromagnetism in layered 1T-$CrSe_2$ with V and Ti replacements." *Physical Review B* 87, no. 1 (2013): 014420. http://dx.doi.org/10.1103/PhysRevB.87.014420.

[126] Colombet, P., and L. Trichet. "Ferro-antiferromagnetic phase transition in a diluted triangular chromium III lattice: The $Na_xCr_xTi_{1-x}S_2$ compounds." *Solid State Communications* 45, no. 4 (1983): 317-322. https://doi.org/10.1016/0038-1098(83)90887-6.

[127] Ohta, Satoru, Yusuke Narui, and Yoshitomo Sakayori. "Effect of Te-substitution on magnetic properties of $Cr_2Se_{3-y}Te_y$ ($0\leqslant y\leqslant 0.15$)." *Journal of magnetism and magnetic materials* 170, no. 1-2 (1997): 168-178. https://doi.org/10.1016/0304-8853(92)90408-G.

[128] Zhu, Zhiyong, Yingchun Cheng, and Udo Schwingenschlögl. "Band inversion mechanism in topological insulators: A guideline for materials design." *Physical Review B* 85, no. 23 (2012): 235401. https://doi.org/10.1103/PhysRevB.85.235401.

[129] Hasan, M. Zahid, Su-Yang Xu, Ilya Belopolski, and Shin-Ming Huang. "Discovery of Weyl fermion semimetals and topological Fermi arc states." *Annual Review of Condensed Matter Physics* 8 (2017): 289-309. https://doi.org/10.1146/annurev-conmatphys-031016-025225.

[130] Balents, Leon. "Weyl electrons kiss." *Physics* 4 (2011): 36. http://link.aps.org/doi/10.1103/Physics.4.36.

[131] Burkov, A. A. "Topological semimetals." *Nature materials* 15, no. 11 (2016): 1145-1148. https://doi.org/10.1038/nmat4788.

[132] Herring, Conyers. "Accidental degeneracy in the energy bands of crystals." *Physical Review* 52, no. 4 (1937): 365. https://doi.org/10.1103/PhysRev.52.365.

[133] Wan, Xiangang, Ari M. Turner, Ashvin Vishwanath, and Sergey Y. Savrasov. "Topological semimetal and Fermi-arc surface states in the electronic structure of

pyrochlore iridates." *Physical Review B* 83, no. 20 (2011): 205101. https://doi.org/10.1103/PhysRevB.83.205101.

[134] Soluyanov, Alexey A., Dominik Gresch, Zhijun Wang, QuanSheng Wu, Matthias Troyer, Xi Dai, and B. Andrei Bernevig. "Type-ii weyl semimetals." *Nature* 527, no. 7579 (2015): 495-498. https://doi.org/10.1038/nature15768.

[135] Deng, Ke, Guoliang Wan, Peng Deng, Kenan Zhang, Shijie Ding, Eryin Wang, Mingzhe Yan et al. "Experimental observation of topological Fermi arcs in type-II Weyl semimetal $MoTe_2$." *Nature Physics* 12, no. 12 (2016): 1105-1110. https://doi.org/10.1038/nphys3871.

[136] Di Sante, Domenico, Pranab Kumar Das, Chiara Bigi, Zeynep Ergönenc, N. Gürtler, Jonas A. Krieger, Thorsten Schmitt et al. "Three-dimensional electronic structure of the type-II Weyl semimetal WTe_2." *Physical review letters* 119, no. 2 (2017): 026403. https://doi.org/10.1103/PhysRevLett.119.026403.

[137] Wang, Chenlu, Yan Zhang, Jianwei Huang, Simin Nie, Guodong Liu, Aiji Liang, Yuxiao Zhang et al. "Observation of Fermi arc and its connection with bulk states in the candidate type-II Weyl semimetal WTe_2." *Physical Review B* 94, no. 24 (2016): 241119. https://doi.org/10.1103/PhysRevB.94.241119.

[138] Belopolski, Ilya, Daniel S. Sanchez, Yukiaki Ishida, Xingchen Pan, Peng Yu, Su-Yang Xu, Guoqing Chang et al. "Discovery of a new type of topological Weyl fermion semimetal state in $Mo_xW_{1-x}Te_2$." *Nature communications* 7, no. 1 (2016): 1-9. https://doi.org/10.1038/ncomms13643.

[139] Shekhar, Chandra, Ajaya K. Nayak, Yan Sun, Marcus Schmidt, Michael Nicklas, Inge Leermakers, Uli Zeitler et al. "Extremely large magnetoresistance and ultrahigh mobility in the topological Weyl semimetal candidate NbP." *Nature Physics* 11, no. 8 (2015): 645-649. https://doi.org/10.1038/nphys3372.

[140] Huang, Xiaochun, Lingxiao Zhao, Yujia Long, Peipei Wang, Dong Chen, Zhanhai Yang, Hui Liang et al. "Observation of the chiral-anomaly-induced negative magnetoresistance in 3D Weyl semimetal TaAs." *Physical Review X* 5, no. 3 (2015): 031023. https://doi.org/10.1103/PhysRevX.5.031023.

[141] Yang, F. Y., Kai Liu, C. L. Chien, and P. C. Searson. "Large magnetoresistance and finite-size effects in electrodeposited single-crystal Bi thin films." *Physical review letters* 82, no. 16 (1999): 3328. https://doi.org/10.1103/PhysRevLett.82.3328.

[142] Lin, Zhong, Bruno R. Carvalho, Ethan Kahn, Ruitao Lv, Rahul Rao, Humberto Terrones, Marcos A. Pimenta, and Mauricio Terrones. "Defect engineering of two-dimensional transition metal dichalcogenides." *2D Materials* 3, no. 2 (2016): 022002. https://doi.org/10.1088/2053-1583/3/2/022002.

[143] Wang, Shanshan, Alex Robertson, and Jamie H. Warner. "Atomic structure of defects and dopants in 2D layered transition metal dichalcogenides." *Chemical Society Reviews* 47, no. 17 (2018): 6764-6794. https://doi.org/10.1039/C8CS00236C.

[144] Wang, Ziying, Runlai Li, Chenliang Su, and Kian Ping Loh. "Intercalated phases of transition metal dichalcogenides." *SmartMat* 1, no. 1 (2020): e1013. https://doi.org/10.1002/smm2.1013.

[145] Hu, Wei, and Jinlong Yang. "Two-dimensional van der Waals heterojunctions for functional materials and devices." *Journal of Materials Chemistry C* 5, no. 47 (2017): 12289-12297. https://doi.org/10.1039/C7TC04697A.

[146] Mali, D., K. Kumar, P. K. Jangid, P. K. Joshi, G. Arora, and B. L. Ahuja. "Influence of Te doping in titanium dichalcogenides: LCAO calculations and Compton spectroscopy." In *Journal of Physics: Conference Series*, vol. 1849, no. 1, p. 012031. IOP Publishing, 2021. https://doi.org/10.1088/1742-6596/1849/1/012031.

[147] Zhao, Xu, Xiaonan Zhang, Tianxing Wang, Shuyi Wei, and Lin Yang. "Electronic structures and optical properties of ZrS_2 monolayer by n-and p-type doping." *Journal of Alloys and Compounds* 748 (2018): 798-803. https://doi.org/10.1016/j.jallcom.2018.03.240.

[148] Wang, Yong, Zdeněk Sofer, Jan Luxa, and Martin Pumera. "Exfoliated Vanadium Dichalcogenides (VS_2, VSe_2, VTe_2) By Lithium Intercalation Exhibit Dramatically Different Properties from Their Bulk Counterparts." In *ECS Meeting Abstracts*, no. 10, p. 861. IOP Publishing, 2018. http://dx.doi.org/10.1002/admi.201600433.

[149] Sanchez-Perez, Clara, Caroline E. Knapp, Ross H. Colman, Carlos Sotelo-Vazquez, Sanjayan Sathasivam, Raija Oilunkaniemi, Risto S. Laitinen, and Claire J. Carmalt. "Iron-Intercalated Zirconium Diselenide Thin Films from the Low-Pressure Chemical Vapor Deposition of [Fe (η5-C_5H_4Se) 2Zr(η5-C_5H_5)$_2$]$_2$." *ACS omega* 5, no. 26 (2020): 15799-15804. https://doi.org/10.1021/acsomega.0c00413.

[150] Popov, Z. I., N. S. Mikhaleva, M. A. Visotin, A. A. Kuzubov, S. Entani, H. Naramoto, S. Sakai, P. B. Sorokin, and P. V. Avramov. "The electronic structure and spin states of 2D graphene/VX_2 (X= S, Se) heterostructures." *Physical Chemistry Chemical Physics* 18, no. 48 (2016): 33047-33052. https://doi.org/10.1039/C6CP06732H.

[151] Vu, Van Tu, Thi Thanh Huong Vu, Thanh Luan Phan, Won Tae Kang, Young Rae Kim, Minh Dao Tran, Huong Thi Thanh Nguyen, Young Hee Lee, and Woo Jong Yu. "One-Step Synthesis of $NbSe_2$/Nb-Doped-WSe_2 Metal/Doped-Semiconductor van der Waals Heterostructures for Doping Controlled Ohmic Contact." *ACS nano* (2021). https://doi.org/10.1021/acsnano.1c02038.

Chapter 3

Theoretical and DFT Related Studies of Metallic TMDs

Heera T. Nair[1] **and Brahmananda Chakraborty**[2,3,*]
[1]School of Pure and Applied Physics, Mahatma Gandhi University, Kerala, India
[2]High Pressure and Synchrotron Radiation Physics Division,
Bhabha Atomic Research Centre, Trombay, Mumbai, India
[3]Homi Bhabha National Institute, Mumbai, India

Abstract

Transition metal dichalcogenides (TMDCs), exhibit excellent properties like direct bandgap, atomic-scale thickness, mechanical, electrical, and electronic properties, making them potent candidates in spintronics, energy harvesting, electronics, optoelectronics, DNA sequencing, medicines, etc. Theoretical studies like Density Functional Theory helps in exploring these benefits and choosing the right one among the group for the demand concerned. Most of these are combined with literature data and experimental results to refine the information as well as the simulation technology and methods used. Due to this continuous evolution, DFT has various methodologies suiting the system in consideration and can even improve the calculations with the efficiency of fast and sophisticated supercomputers used. DFT enables us to get even the most minute and fundamental information, that may be a trademark for a particular material. In this chapter, we discuss how DFT is used in investigating the properties of 2D Transition Metal Dichalcogenides (TMDs) and exploring its applications as sensors, supercapacitors, and water-splitting catalysts.

Keywords: density functional theory, TMDs, sensors, supercapacitors

* Corresponding Author's Email: brahma@barc.gov.in.

In: 2D Metallic Transition Metal Dichalcogenides
Editors: Chandra Sekhar Rout and Brahmananda Chakraborty
ISBN: 978-1-68507-965-9

3.1. Introduction

The immense number of TMDs and their numerous qualities have demanded theoreticians to study these systems and infer the right contender for various applications. Quantum Theory has opened a new era in science where Quantum Mechanics can be applied to real-world objects, systems, or processes. The application of theoretical simulations to such a vista of materials and their related applications makes its value priceless and validity limitless. The major part of theoretical simulations relies on the Schrodinger equation, which demands the use of very sophisticated numerical methods. This ab initio nature of simulation work helps in solving the Schrodinger equation without any data from the experiments, thus giving information regarding the physical properties that are often inaccessible to the experimental researchers. This has eased the job of analyzing TMDs and expanding its domain.

There are various simulation packages like the VASP [1] or Quantum Espresso [2] which uses a plane-wave basis set. Also, when pseudo-potentials fail, we generally use codes such as WIEN2K, or ELK [3] where the all-electron potential is considered instead of the valence electrons only. These versatile packages and procedures have helped to define various phenomena as well as been able to find out new materials, or combinations in the form of hybrids. Even though theoretical simulations and predictions may involve errors, they have been a guide for many to initiate the quest and find out the underlying causes of a problem.

In this chapter, we will provide a brief overview of the root ideas and laws governing Density Functional Theory and how it evolved to be one of the best methods in computational studies. In detail, we will discuss how Density Functional Theory is useful in exploring the wide variety of TMDs, their structural properties, including lattice parameters and forces, electronic properties investigated from DOS data, as well as the optical characteristics. In the final sections, we focus on the applications of TMDs, sensing, and energy storage.

3.2. Density Functional Theory: An Overview

Density Functional Theory (DFT) [4] lends accuracy in investigating the electronic structure (specifically) of many-body systems particularly, atomic, molecular, solid-state, and even nuclear physics. Many computational physicists use it merely as a recipe, not binding much on their logical basis. The prime part of quantum simulations is to calculate the electronic structure of a system by solving the many-body Schrodinger equation:

$$H\psi = E\psi \tag{3.1}$$

For a solid with positively charged nuclei and negatively charged electrons, we can write the non-relativistic, electromagnetically interacting many-particle Hamiltonian as

$$\hat{H} = T_{nuclei} + T_{elect} + V_{n\text{-}e} + V_{e\text{-}e} + V_{n\text{-}n} \tag{3.2}$$

Using the above Hamiltonian, we can get the wave function **ψ** of the many particles system. In the above equation, the first two terms are the kinetic energy operators for the nuclei and electrons and the last three terms describe the Coulomb interaction. To solve the above equation, certain approximations can be utilized at various levels.

3.2.1. Born Oppenheimer Approximation

Though the electromagnetic forces acting on the nuclei and the electrons are similar, since the mass of the nuclei is high compared to the mass of the electron, we can assume that the nuclei are frozen at a fixed position and the electrons are in instantaneous equilibrium with nuclei moving comparatively faster than them. Thus, we can say electrons are moving in a potential provided by the nuclei, an external particle. So, the last term $\mathbf{V_{n\text{-}n}}$ becomes a constant, and $\mathbf{T_{nuclei}}$ can be neglected.

Thus, the new Hamiltonian becomes

$$\hat{H} = T_{elect} + V_{ext} + V_{e\text{-}e} \tag{3.3}$$

where $\mathbf{T_{elect}}$ is the kinetic energy of the electron gas, $\mathbf{V_{ext}}$ is the external potential i.e., the potential energy of the nuclei, and $\mathbf{V_{e\text{-}e}}$ is the potential energy due to electrons interacting with each other. Thus, now the problem has been reduced to solving the electronic wave function part only. Density functional Theory (DFT) is based on the two theorems proposed by Hohenberg and Kohn, which were then improved by Kohn and Sham equations. DFT considers electron density, unlike Hartree-Fock (HF) method which uses many-body wavefunctions. This is the reason why DFT is considered the more accurate and feasible method.

3.2.2. The Hohenberg and Kohn Theorems

Theorem 1: There is a one-to-one correspondence between the ground-state density $\rho(r)$ of a many-electron system (atom, molecule, solid) and the external potential V_{ext}. An immediate consequence is that the ground-state expectation value of any observable O is a unique function of the exact ground-state electron density.

$$<\psi|O|\psi> = O[\rho] \tag{3.4}$$

Theorem 2: For O being the Hamiltonian H, the ground state energy functional,

$$\hat{H}(\rho) = E_{Vext}(\rho) \tag{3.5}$$

$$E_{Vext}(\rho) = <\psi|T+V|\psi> + <\psi|V_{ext}|\psi> \tag{3.6}$$

$$= F_{HK}(\rho) + \rho(r)\, Vext(\rho)\, dr \tag{3.7}$$

$\mathbf{F_{HK}(\rho)}$ is the Hohenberg – Kohn density functional which is universal for a many-electron system.

$\mathbf{E_{Vext}(\rho)}$ tends to be the minimum value for ground-state density corresponding to $\mathbf{Vext(\rho)}$. Thus, there is a one-to-one correspondence between external potential and electron density.

3.2.3. Kohn - Sham Equations

In the above Hohenberg Kohn equation, the actual form of FHK is unknown making it practically inapplicable. To make DFT applicable, Kohn and Sham put forward equations to construct suitable energy functions. They considered a fictitious system of non-interacting particles that has the same ground-state characteristics as a real system of interacting particles.

The Hamiltonian is defined as,

$$H_{KS} = To + V_H + V_{xc} + V_{ext} \tag{3.8}$$

Here, **To** is the kinetic energy operator, $\mathbf{V_H}$ is the Hartree term, $\mathbf{V_{xc}}$ is the exchange-correlation and $\mathbf{V_{ext}}$ is the external potential.

Similarly, the electron density of the non-interacting system of N particles is defined as:

$$\rho(r)= \sum_{i=1}^{N} \Psi i * (r)\, \Psi i(r) \tag{3.9}$$

Kinetic energy functional is

$$To(\rho) = \hbar^2/\, 2m_e \sum_{i=1}^{N} <\Psi i\, |\nabla 2|\Psi i> \tag{3.10}$$

$|\Psi i>$ is the one-electron orbital, thus the electron density is obtained by summing up all the occupied orbitals. Similarly, the energy functional is written as

$$E_{Vext}(\rho) = To(\rho) + E_H(\rho) + E_{xc}(\rho) + \int\rho(r)\, Vext\, (r)\, dr \tag{3.11}$$

where,

$$To(\rho) + E_H(\rho) + E_{xc}(\rho) = F_{HK}(\rho) \tag{3.12}$$

Thus, the F_{HK} (ρ) got a new form comprising the kinetic energy, Hartree potential, and the exchange-correlation term. Now, the ground state energy can be calculated from

$$[-\hbar^2/\, 2m_e\;\; \nabla + V_{eff}(r)]\; \Psi_i(r) = \varepsilon_i \Psi_i(r) \tag{3.13}$$

where,

$$V_{eff}(r) = V_H + Vxc + Vext \tag{3.14}$$

$$V_H = e^2\, /4\Pi\varepsilon o \int\rho(r^I)\, /|r - r^I| \quad \text{and} \quad Vxc = \delta Exc(\rho)/\delta n$$

ε_i and Ψ_i are the single-particle energy and single-particle wave function. Stationary density $\rho_o(r)$ is a guess from which Hks is calculated, which in turn is calculated from V_H and Vxc. Then comes the eigenvalue problem from which Ψ_i and ε_i are obtained. Using new density, the same procedure is repeated until convergence is reached. Thus, the only unknown term is the **Exc(ρ)** which is to be approximated. The two most common approximations are Local Density Approximation (LDA) [5] which is the oldest and the simplest, and the Generalised Gradient Approximation (GGA) [6]. In LDA, the exchange-correlation functional is defined as

$$E_{xc}(\rho) = \int \rho(r)\, E_{xc}\,(\rho(r))dr \tag{3.15}$$

Whereas in GGA, even the gradient of density is considered. There are two most popular forms of GGA called PBE and BYLP [7].

In some cases, especially, for example, the bandgap of the system, due to lack of self-interaction correction, methods like Hubbard correction, hybrid functionals, and GW are considered to describe the system properly.

With the advancing technology, supercomputers with very large memory and extremely high speed open the chance to conduct quantum simulations more easily depending on the available versatile codes the material demands. Density Functional Theory is used to explore and explain the electronic, optical, magnetic properties of Transition Metal Dichalcogenides. Also, simulations can provide a better foundation for the experimental results and also a deep insight into different applications such as sensing, energy storage, conversion, etc. The following section discusses the methods of DFT used and various other parameters that could give vast information about the TMD properties.

3.3. Structural and Electronic Properties of Various TMDs

Layered TMDs have emerged as unique materials with exceptional properties that are quite rare and difficult to acquire in other 2D materials such as graphene. In this section, we will discuss the structure and electronic properties of the TMDs explained under First-Principles Calculations. Also, we will peep into the DFT studies conducted in bandgap engineering that plays a major role in tunable nano-electronic devices.

TMDs constitute a wide spectrum of materials ranging from insulators to semiconductors to semi-metals to metals. They have a sandwich structure (X-M-X) where the metal atoms are bound between two layers of chalcogen atoms. The layers of these TMDs are attached by van der Waals interaction while the atoms within the layers are bonded covalently. Talking about their electronic nature, unlike graphene which has a zero-band gap, the TMDs like MoX_2 and WX_2, where X = S, Se, and Te, are semiconducting. By varying the number of layers, they can be tuned on demand; as the number of layers increases, the bandgap also increases. Also, the properties of TMDs can be varied by chemical treatment, nano-patterning, application of external electric field, or strain engineering [8].

3.3.1. Structural Analysis

DFT Calculations commence with optimizing the overall lattice geometry by minimizing the interatomic forces, potential energy, and internal stress tensor. DFT simulations by the VASP package use a plane wave basis set to expand the Kohn-Sham orbitals and cutoff energy is chosen above which the plane waves are truncated. Also, the energy of the super-cell is integrated over the k-points chosen, where the Monkhorst- Pack Scheme [9] is used for generating the k-points that expand the periodic function in reciprocal space. The structural relaxation can provide information on cell parameters, total energy, and forces on atoms. The calculations can be performed using the density functional method as implemented in the Quantum Espresso code, Generalized Gradient Approximation using Perdew-Wang 91 functional (GGA-PW91) [6], etc.

Javed Rehman et al. [8] have explored the possibilities of strain-engineered WSe_2 as anode material for Li-ion batteries. Just like other TMDs such as MoS_2, WS_2, $NbSe_2$, $MoSe_2$, etc., the WSe_2 monolayer also has a hexagonal structure, crystallizing in two phases- trigonal (1T) and hexagonal (2H), where the 2H phase is more stable. The optimized 2H structure has lattice constants, a = b = 3.3 Å. The top and side views are shown in Figure 3.1 (a and b). These optimized structures are the raw input for all the following steps in DFT Calculations. The parameters for the calculations should be changed accordingly to the systems and type of problems under consideration.

Ahmad et al. [10] has expounded the most important member of the dichalcogenide family, MoS_2. With the help of GGA based DFT calculations, he has defined both the structure and electronic properties of monolayer MoS_2 and bulk MoS_2. The only dissimilarity he found was in the electronic bandgap of the two.

MoS_2 has a hexagonal structure with S-Mo-S layers. Monolayer MoS_2 has a single layer of S-Mo-S whereas bulk MoS_2 has two such layers where the Mo atoms in one layer are straight above the sulfur atoms in the other layer and vice versa. They have compared the structural parameters of bulk and monolayer MoS_2 using GGA, with the experimental and theoretical findings where they could emphasize the inbuilt feature of the standard GGA Functional of overestimating the experimental values. The structural parameters for both, monolayer and bulk MoS_2 were almost the same [10].

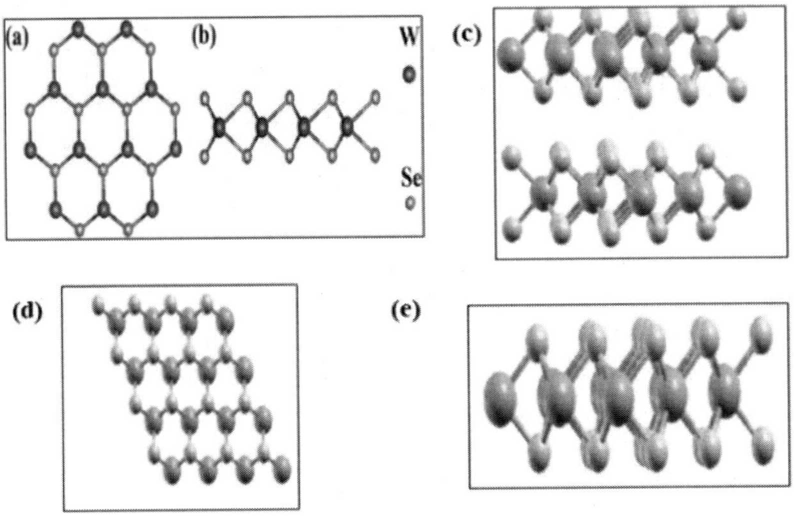

Figure 3.1. (a) Top view and side view (b) of the optimized strain engineered single-layer WSe_2. Yellowish is the Se atoms while reddish represents W atoms. (c) Side view of bulk MoS_2; (d) Top view of bulk and monolayer MoS_2; and (e) Side view of monolayer MoS_2. The Mo-atoms are denoted by gray and S-atoms by pale yellow balls. Reproduced with Permission from (a) and (b): Javed Rehman et al., "Theoretical Investigation of Strain-Engineered WSe_2 Monolayers as Anode Material for Li-Ion Batteries." Journal of Alloys and Compounds 804 (2019): 370–75 [8]. (c), (d) and (e): S. Ahmad et al., "A Comparative Study of Electronic Properties of Bulk MoS_2 and Its Monolayer Using DFT Technique: Application of Mechanical Strain on MoS_2 Monolayer." *Graphene* 03, no. 04 (2014): 52–59 [10].

3.3.2. Electronic Structure

The total Density of states in DFT calculations is the projection of the wave function of an atomic orbital. Interpreting the TDOS plots (illustrated as DOS data vs (E-Ef)), explains electronic properties including bandgap, bonding and hybridization, metallic and magnetic behavior, and doping information as well. We must provide a denser k-point grid for getting a consistent DOS result on par with band structure calculations. The results are available within the boundary of energy levels Emin and Emax, divided into grid points given in the input file. The contribution from each orbital of an atom can be elucidated using the Partial or the Projected Density of States (PDOS).

Figure 3.2 shows the TDOS and PDOS of tungsten diselenide compound (WSe_2) obtained using the mBJ method. Comparing the above three plots, we can infer that below the Fermi level, the maximum contribution to the valence band is from 4p orbitals of Se, and an average contribution from d- orbitals of W, whereas the conduction band above the Fermi level can be divided into two regions: The first in the energy range 1- 3.5 eV where the major contribution is from 5d orbitals of W and a small contribution from Se-4p states. The second is in 3.5-5 eV, due to Se-4p and W-5d states. There is no contribution from W-6s and Se-4s states in both the valence and conduction bands.

Also, DFT calculations suggested that WSe_2 has an indirect bandgap of 0.9 eV using the LDA method, 1 eV with GGA, and 1.18 eV by mBJ, where the last band gap value is the closest to the experimental result of 1.2 eV [11].

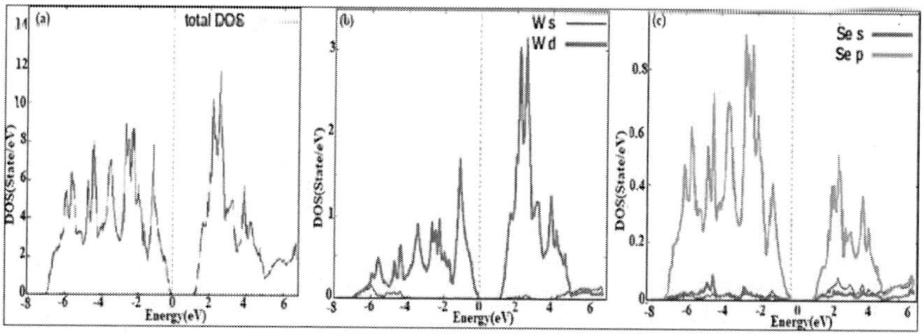

Figure 3.2. (a) Total DOS and Partial DOS (b) W and (c) Se, using mBJ for WSe_2 compound. Reproduced with Permission from Hajar Abbadi et al., "Ab Initio Calculations of Structural and Electronic Properties of WSE_2 Compound." *Materials Today: Proceedings* 31 (2020) [11].

3.3.3. Band Structure

The band structure data is the description of the energy levels of the electrons. The major difference in the band structure and DOS calculations lie in the k-points because in the band structure, choosing the right k-points is important as it depends on the structure and symmetry of the system under consideration.

The k- points should be adapted according to lines in the Brillouin Zone, along which we want the band structure. This is mainly supported by the klines directive. We can get the band gap value and the band character (metallic, semi-metallic, etc.) from the band structure.

Considering MoS_2, the indirect bandgap of bulk MoS_2 was 0.89 eV while in the monolayer-MoS_2 the direct bandgap was 1.57 eV. The band structure comparisons for monolayer and bulk MoS_2 are shown in Figure 3.3 (a). *Ahmad* et al.'s work [10] also suggests the possibilities of strain effects changing the bandgap pattern in all these systems. A uniform tensile strain from 0 to 10% was applied on monolayer MoS_2 to study the change in the bandgap. The bandgap versus strain is shown in Figure 3.3 (b). The bandgap is monotonic as the strain increases.

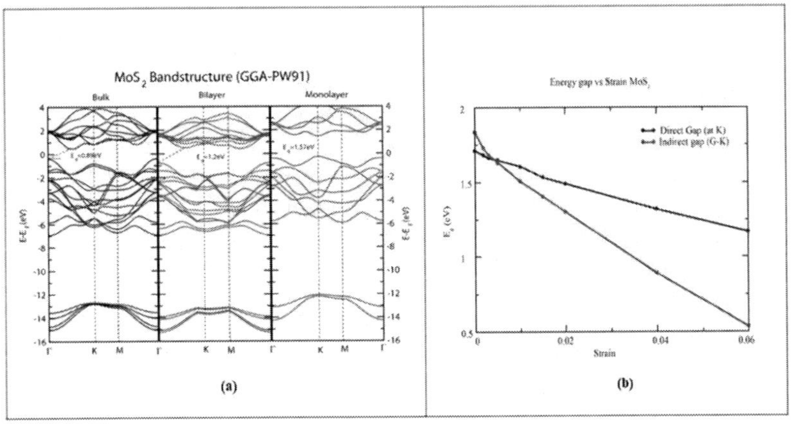

Figure 3.3. (a) Comparison of the calculated band structure of bulk, bilayer, and the monolayer of MoS_2 at high-symmetric points in the irreducible Brillouin zone. The position of valence band maxima, conduction band minima, and the bandgap (Eg) are also defined. Unlike other cases, for monolayer MoS_2 the direct bandgap occurs at K-point. (b) The variation of bandgap energy with the strain of monolayer MoS_2. (Reproduced with permission from Sohail Ahmad et al., "A Comparative Study of Electronic Properties of Bulk MoS_2 and Its Monolayer Using DFT Technique: Application of Mechanical Strain on MoS_2 Monolayer." *Graphene* 03, no. 04 (2014): 52–59 [10]).

3.3.4. A Journey through DFT Studies in Other TMDs

Pere Mir et al. [12] have explored some layered TMDs, known to be efficient in certain properties that even cannot be acquired in 2D materials like graphene. They investigated the structure and electronic properties of layered noble-transition-metal dichalcogenides MX_2 (M = Pt and Pd, X = S, Se, and Te) using DFT.

MS_2 is an indirect bandgap semiconductor whereas MSe_2 and MTe_2 have smaller bandgaps and can be semi-metallic or metallic. When strain is applied, it can even become a direct bandgap semiconductor. In this work, apart from the electronic structure of the mono and bilayers, conductance, as well as strain effects, have also been inspected using the First Principles Calculations.

With the dispersion corrected Perdew-Burke-Ernzerhof exchange-correlation functional (PBE-D) and the triple-z basis set plus a polarization function (TZP) on all atoms, they got the band gaps as 0.51, 1.11, and 1.75 electron volts for NiS_2, PdS_2, and PtS_2 respectively. The GLLB-SC model potential helped them in an improved band gap calculation better than PBE making it to 0.79, 1.67, and 2.72 electron volts with no change in the band structure shapes.

The band structures were calculated using the hybridHSE06 functional. The PdS_2 monolayer has an indirect bandgap of 1.60 eV and is semiconducting. PdS_2 has unique structural properties, unlike other 2D TMDs. PdS_2 has long electron and hole mobilities which makes them very useful. By DFT simulations, they could put forward the fact that the most stable configuration of PdS_2 is neither 1T nor 2H but tetracoordinate Pd atoms with covalent S-S bonds. Apart from the structural and electronic properties, they have also calculated the interlayer binding energy (Eb) of PdS_2 bulk which was about 855meV per unit cell [12-13] (Figure 3.4).

Some works were also done with the transition of a direct bandgap to an indirect one using h-BN heterostructure layers as the substrate by the DFT VASP package with GGA-PBE along with Vander Waal corrections. MoS_2 has a direct bandgap of 1.67 eV and the h-BN hetero-structure monolayer has an indirect bandgap of 4.64 eV. The MoS_2/1-h-BN system was found to have an indirect bandgap of 1.11 eV. The strain-induced on the MoS_2 monolayer was due to the lattice mismatch between MoS_2 and h-BN. To know the change in electronic and charge density properties with thickness, the same theoretical analysis using DFT with n = 2, 3, 4, 5 layers of h-BN as the substrate was also explored. It was found that the bandgaps of the MoS_2/n-h-BN systems decrease monotonically with increasing in the layer number of h-BN [14].

The studies broadened to other Dichalcogenides as well, like V_2O_5 and MoO_3. Since, the bulk structure involves Vander Waal forces between the layers, DFT+ VanderWaal calculations should be included for better results. Also, for much better visualization of the bandgap, DFT+U calculations can be done. In a study by *Bhandari* et al. [15], the V_2O_5 monolayer band gap was predicted as 4 eV with the help of GW calculations. Similarly taking the case of MoO_3, the study of monolayer and bulk MoO_3, has reported a small change in the indirect bandgap of monolayer (1.73 eV) compared to the bulk (1.71 eV), using the PBE + U/D2 DFT functional [16].

Figure. 3.5(a) depicts the DOS and PDOS for bulk V_2O_5. From the DOS plot, it is clear that the major contribution to the top of the valence band is essentially from p orbitals of the O atom, and the bottom of the conduction band is from a split V(d) band, forming a small band-gap for the V_2O_5 layered oxide. Similarly, for bulk MoO_3, we can infer that the bottom of the conduction band is formed mainly by the Mo d states, whereas the top of the valence band is mainly made up of O(p) levels. Here compared to V_2O_5, the split of the Mo(d) bands is larger than in V_2O_5, due to crystal field effects, indicating the stronger octahedral field.

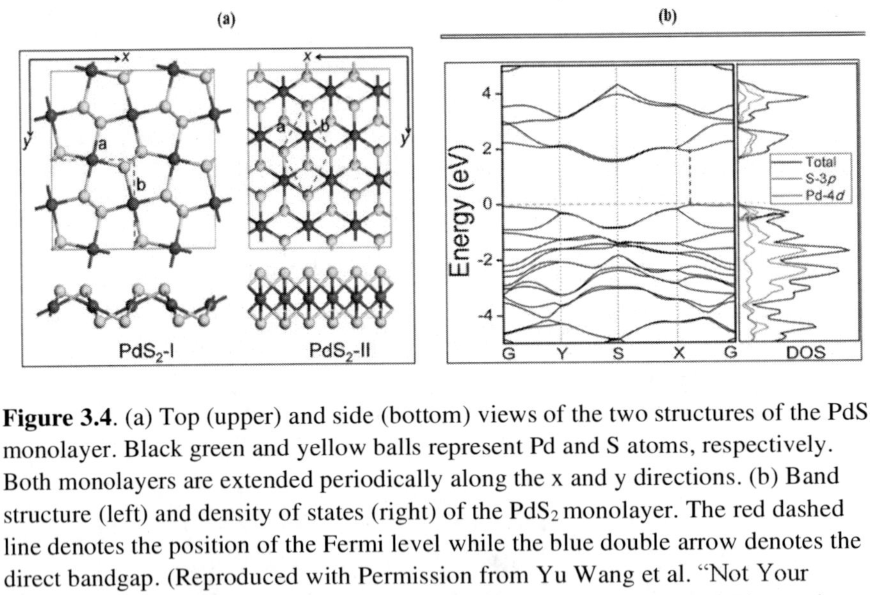

Figure 3.4. (a) Top (upper) and side (bottom) views of the two structures of the PdS_2 monolayer. Black green and yellow balls represent Pd and S atoms, respectively. Both monolayers are extended periodically along the x and y directions. (b) Band structure (left) and density of states (right) of the PdS_2 monolayer. The red dashed line denotes the position of the Fermi level while the blue double arrow denotes the direct bandgap. (Reproduced with Permission from Yu Wang et al. "Not Your Familiar Two-Dimensional Transition Metal Disulfide: Structural and Electronic Properties of the PdS_2 monolayer." *Journal of Materials Chemistry* C 3, no. 37 (2015): 9603–8 [13]).

Also, using DFT, we can look into the formation of energy of the bulk material as well as the dielectric constant apart from structural and electronic properties. The change in these properties on moving from a monolayer to an ultrathin slab with 3 layers and the bulk of V_2O_5 and MoO_3, can also be checked. *The most important point to be noted is that in DFT studies, exchange-correlation is significant especially while dealing with the transition metal oxides since there will be a strong correlation of the d-orbitals.* Since GGA, and LDA, have less accuracy we should improve the simulation techniques to an extent. The Hubbard U correction can be used for the correlated electrons and the hybrid functional for making it more practically applicable [16].

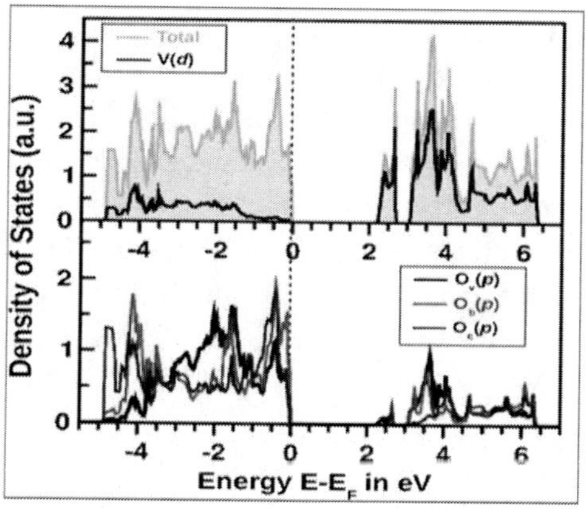

Figure 3.5. (a) TDOS and PDOS for V(d), and O(p) of bulk V_2O_5 (PBE + U/D3/SO). (b) Mo(d), and O(p) of bulk MoO_3 (PBE + U/D3/SO). The Fermi level is set to the top of the valence band. DOS is in arbitrary units (a.u.). (Reproduced with permission from Tilak Das et al., "Structural and Electronic Properties of Bulk and Ultrathin Layers of V_2O_5 and MoO_3." *Computational Materials Science* 163 (2019): 230–40 [16]).

So far, we have discussed the DFT-based investigations in properties of different TMDs. In the next section, we will explore the applications of TMDs studied using DFT.

3.4. Chemical Applications of TMDs

As mentioned in the previous sections, TMDs are highly useful materials with exceptional electronic, optical, and mechanical properties which makes them potent candidates in electronics, bio-sensing, gas sensing, and energy storage applications. Due to the large surface-volume ratio, sizable bandgaps, and reactive surface for the contact between the host and the gas molecules as well as surface diffusion, they are used for gas sensing and capturing. DFT simulations can explain the weak interaction between the sensing materials and the adsorbates reflecting the local interactions, and also describe variations of electronic structure as well as quantum transport in 2D material-based sensors, both before and after the interaction with gas molecules. This information like different possible adsorption sites and configurations, the adsorption energy involved, electronic and transport variation after adsorbing the gas molecules can support the experimental measurements, and also provide a theoretical understanding of gas sensing. It can also provide us with feasible approaches to improve adsorption or desorption in nanosensors.

Gas sensors rely on the change in the conductivity or electronic structure on the adsorption of gas molecules. When the gas species are adsorbed on the surface of sensing materials, a few electrons may be transferred to/from the gas molecules. The reactivity of adsorbates and adsorbents results in weak or strong interactions which in turn modifies the electronic properties of sensing materials and the new transport channels. The resistance of the sensors thus changes [17].

In a study conducted by *Maryam Nayeri* et al. [18], they exposed the monolayer TMDs like MoS_2, $MoSe_2$, WS_2, and WSe_2 to NO, NO_2, NH_3, and O_2 molecules and detected the variations in transfer and dielectric functions using SIESTA [19]. Initially, the supercell structures of all the monolayer TMDs were relaxed and the band gaps were calculated before exposing them to the gas molecules. The transmission spectra are plotted, which also give the bandgap energies, 1.84, 1.58, 1.89, and 1.59 eV for MoS_2, $MoSe_2$, WS_2, and WSe_2, respectively.

The next process is exposing the pristine STMDs (semi-conducting TMDs) to the gas molecules and studying the gas- STMDs interaction. The gas molecules are brought into the vicinity of the STMD substrates and allowed to relax, followed by plotting of the transmission spectrum. The change in the spectrum denotes the sensing merit of a system. *The rate at which the electrons having an energy E is transmitted from one electrode to another is the Transmission Function. Conductance is proportional to the average of the transmission function over a range of energy in the order of a few KBT around the Fermi energy; conductivity is the specific conductance.* Gas molecules can either accept or donate electrons. Figure 3.6 (a)–(d) illustrates the transmission spectrum comparison of the monolayer STMDs before and after exposure to the gas molecules.

These transmission plots can also help us identify the semiconducting nature of the STMDs on the adsorption of various gas molecules. For example, considering MoS_2, the NO molecule shifts the transmission spectrum to the negative energies, indicating charge transfer from NO to MoS_2, making the system behave as an n-type semiconductor. On the contrary, both NO_2 and O_2 shift the spectrum to positive energies making it a p-type semiconductor. Also, they found that NH_3 does not produce any remarkable change in any STMD spectra. In the case of $MoSe_2$, the same behavior is seen, but the shift is much more remarkable in the case of NO_2 than O_2. For all the STMDs, adsorption energies are calculated by

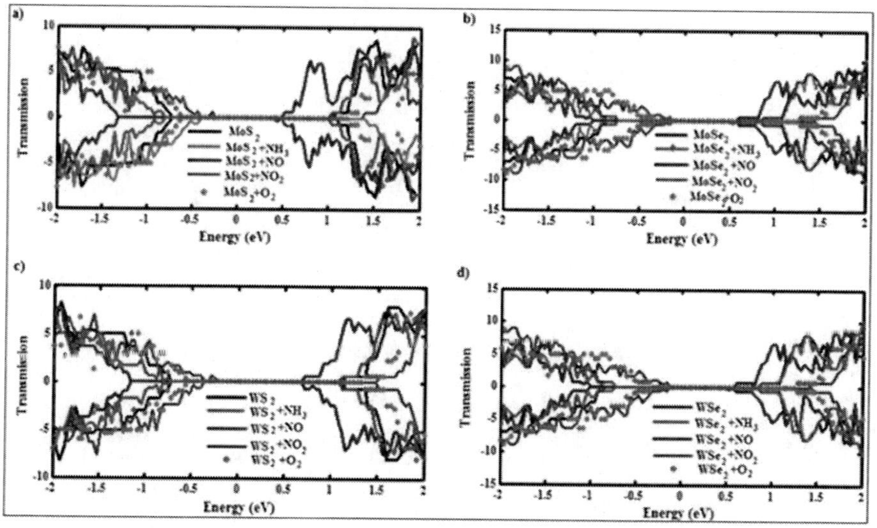

Figure 3.6. Transmission spectrum as a function of energy for (a) MoS_2, (b) $MoSe_2$, (c) WS_2, and (d) WSe_2 in the absence and presence of the gas molecules. (Reproduced with copyright permission from Maryam Nayeri et al., "The Transport and Optical Sensing Properties of MoS_2, $MoSe_2$, WS_2, and WSe_2 semiconducting Transition Metal Dichalcogenides." *Semiconductor Science and Technology* 33, no. 2 (2018): 025002 [18]).

$$E_{ads} = E_{gas\ on\ STMD} - E_{gas} - E_{STMD} \tag{3.16}$$

Here, $E_{gas\ on\ STMD}$ is the total energy of the gas molecule adsorbed STMD system. E_{STMD} is the energy of the monolayer pristine STMD system and E_{gas} is the energy of the gas molecule. This indicates the range of binding between the gas molecule and the substrate. Larger the adsorption value, the larger the binding, and consequently higher the sensitivity. Thus, they concluded that monolayer STMDs are more excellent in detecting NO, NO_2, and O_2

molecules than NH_3. Apart from adsorption energy, adsorption distance (d) can also be calculated, which is the shortest distance between the gas molecule and the STMD atom. Larger the distance, the lesser the binding which may indicate physisorption. These are conductivity-based gas sensors. We can also infer the sensing applications from investigating the absorption characteristics of these materials. This involves calculating the dielectric function in the exposure of gas molecules. They are the optical gas sensors where the imaginary part of the dielectric functions can give information about the absorption spectrum. Figure 3.7 shows the imaginary part of the dielectric functions.

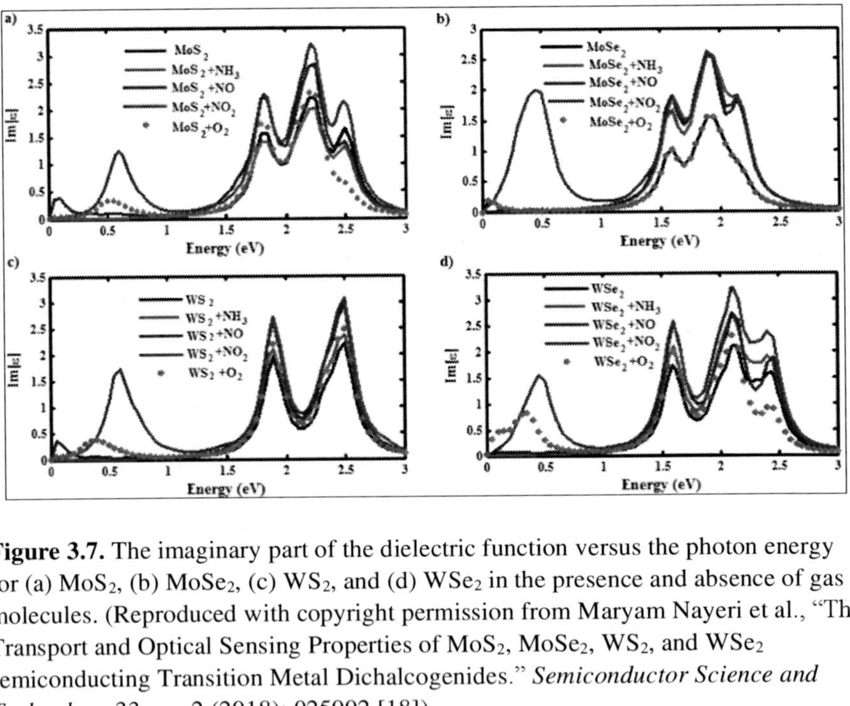

Figure 3.7. The imaginary part of the dielectric function versus the photon energy for (a) MoS_2, (b) $MoSe_2$, (c) WS_2, and (d) WSe_2 in the presence and absence of gas molecules. (Reproduced with copyright permission from Maryam Nayeri et al., "The Transport and Optical Sensing Properties of MoS_2, $MoSe_2$, WS_2, and WSe_2 semiconducting Transition Metal Dichalcogenides." *Semiconductor Science and Technology* 33, no. 2 (2018): 025002 [18]).

In the case of MoS_2 and WS_2, NO_2 and O_2 produce an extra peak at higher energies whereas NO produces a peak at lower energies. But in the case of $MoSe_2$ and WSe_2, NO enhances the intensity of the third peak. NH_3, on the other hand, does not give any significant changes, though $MoSe_2$ is an exception. Thus, the extra peaks and their energy can determine the type of gas molecules adsorbed [18].

In the study by *Dachang Chen* et al. [20] Au-doped and Pt-doped MoS_2 monolayer is exposed to SO_2, SOF_2, SO_2F_2 gas molecules. After the structural relaxation, the DOS calculations follow to understand the changes occurring in the energy diagrams of these systems that in turn can imply the composition of gas molecules adsorbed and its extent. Figure 3.8 represents the TDOS of SO_2 adsorbed Au -doped and Pt- doped MoS_2 systems. SO_2 initially has six peaks and a zero magnetic moment. But after adsorption, there is a re-arrangement of these peaks. The highest peak near −4 eV splits into two peaks located from − 6 to − 7 eV in the Pt-doped system. Also, the peak to the right of the Fermi level widens up. The three peaks from 0 to − 2 eV merge into a wider peak. Thus, from the DOS plots, it is clear that Pt-MoS_2 indeed changes the electronic orbitals of SO_2. The apparent overlaps of certain peaks indicate strong electron-orbital interaction between SO_2 and Pt. In the case of Au-MoS_2, apart from the intensity of peaks and width, magnetic nature is also introduced. Considering the magnetic moment, Au decreases from 0.56 to 0.16 μB after adsorption whereas SO_2 gains a magnetic moment of 0.57 μB. Thus, along with charge transfer, magnetism transfer also occurs here. Partial Density of States (PDOS) in DFT calculations implies the contribution from an atom constituting the system or an orbital of the atom to the DOS data. From the PDOS data we can infer that 3s orbitals contribute more than the 3p in the energy range −11 and −2 eV, whereas, for the other overlap region, the case is the opposite [14].

3.5. Biosensing Applications of TMDs

Similar to gas sensing applications, TMDs are also used for sensing biological molecules. Since the role of DNA is very significant, its detection is also vital in various fields like the diagnosis of a disease and treatments, forensic science, genomics, etc. The high surface-to-volume ratio always enables the 2D materials to be a good candidate among sensors. Density Functional Theory has been one of the best computational methods for detecting the adsorption of these molecules on TMD substrates based on the variations caused in the electronic structure and optical properties.

In a study by *Hakkim Vovusha and Biplab Sanyal* [21], monolayer MoS_2 and WS_2 systems were investigated on the adsorption of DNA/RNA nucleobases. They also compared the efficiency of adsorption in these systems to graphene. The nucleobases were placed on different adsorption sites of

MoS_2 and WS_2 systems and were relaxed using the vdW-DF method. The optimized structures of the WS_2 system are given in Figure 3.9 (a).

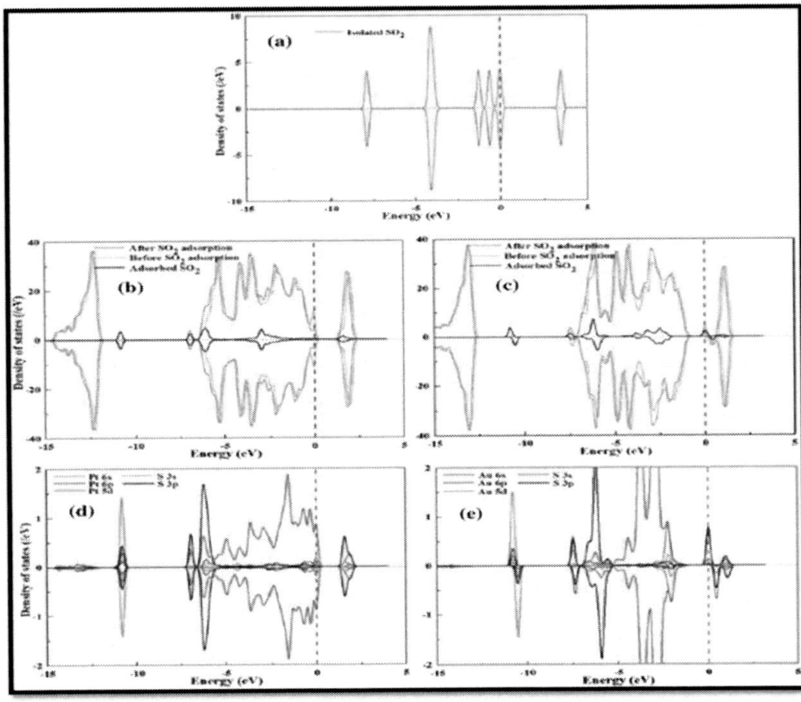

Figure 3.8. (a): TDOS of SO_2, (b), (d): TDOS and PDOS of SO_2 adsorbed on Pt-decorated MoS_2 monolayer, (c), and (e) TDOS and PDOS of SO_2 adsorbed on Au-decorated MoS_2 monolayer. (Reproduced with copyright permission from Dachang Chen et al., "Noble Metal (Pt or Au)-Doped Monolayer mos₂ as a Promising Adsorbent and Gas-Sensing Material to SO_2, SOF_2 and SO_2F_2: A DFT Study." *Applied Physics* A 124, no. 2 (2018) [20]).

They found that the TMD-Nucleobase complexes do not change in their geometrical parameters like bond distances and bond angles when compared to pristine TMDs, which indicates the physisorption of nucleobases on the TMD substrate. This is caused due to the van der Waals interactions between the adsorbate and adsorbent. To understand these interactions further, the DOS was calculated. Figure 3.9 (b) shows the TDOS plots for WS_2.

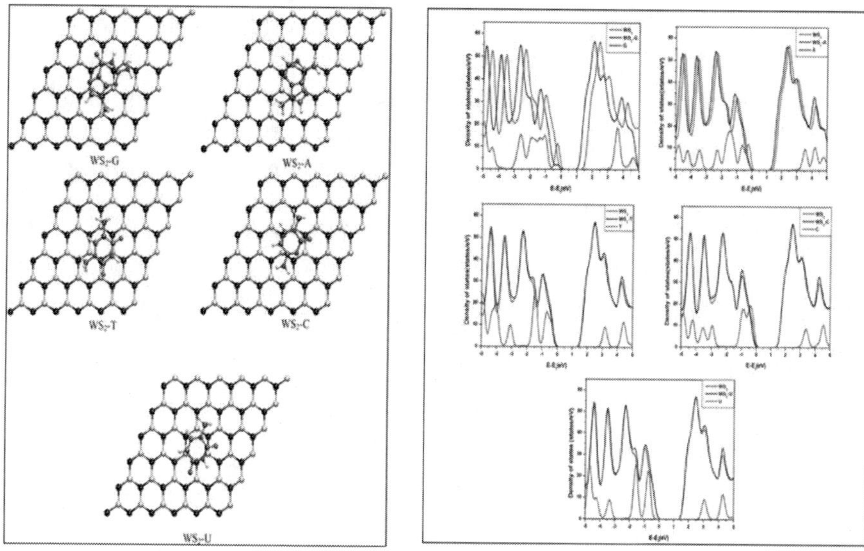

(a) (b)

Figure 3.9. (a) Optimized structures of nucleobases adsorbed on monolayer WS_2 (purple: Mo, yellow: S, red: O, blue: N, gray: C, and white: H). (b) Total densities of states of pristine WS_2, WS_2–nucleobase complexes, and S nucleobases in the gas phase. (Reproduced with copyright permission from HakkimVovusha et al., "Adsorption of Nucleobases on 2D Transition-Metal Dichalcogenides and Graphene Sheet: A First-Principles Density Functional Theory Study." *RSC Advances* 5, no. 83 (2015): 67427–34 [21]).

The TDOS plots suggested that bandgaps of pristine WS_2 and MoS_2 (1.85eV and 1.68eV respectively), changed on nucleobases adsorption. Among them, the bandgaps were reduced to a greater extent on adsorption of G (Guanine) and a lesser extent by A (Adenine), whereas T (Thymine) and U (Uracil) made no changes. Similarly, considering the optical properties, judging the dielectric functions of WS_2 and MoS_2, it was found that the addition of nucleobases does not have any remarkable changes in the peaks compared to the pristine system. To know the charge transfer in these systems, a Bader analysis can be used, which defines the quantity of charge transferred between the two materials [21].

3.6. Supercapacitor and Water Splitting Applications of TMDs

The increase in energy demands has always triggered the search for reliable alternate sources among which the 2D materials, especially TMDs have attracted attention. Due to quantum confinement and short path lengths for electron transport, fuel cells and supercapacitors built from TMDs and their hybrids are a major field of scientific studies now. DFT has always played a significant role in explaining the electronic properties, quantum capacitance, diffusion energy barrier of electrolytic ions, voltage induced, and so on. In this section, first, we will discuss supercapacitors from TMDs explored through Density Functional Theory.

For any DFT simulations, cell relaxation is the initial step that will optimize the lattice geometry and reduce the interatomic forces. All the cell parameters can be acquired from this process. This step is followed by the most important calculation, the DOS. Quantum Capacitance (CQ) is calculated from DOS using the equation,

$$CQ = e^2 \int D(E) F_T(E - e\varphi_G)\, dx \tag{3.17}$$

where D(E) is the DOS, e is the electronic charge, φ_G is the electrode potential, E is the energy, and $F_T(E)$ is the thermal broadening function.

$$F_T(E) = (4K_BT)^{-1} \sec h^2(E/2K_BT) \tag{3.18}$$

Here, K_B is the Boltzmann constant and T is the temperature. From the expression, CQ is directly proportional to the density of states near Fermi energy, which means denser states imply greater conductivity. Also, the states located far away from fermi energy would not contribute much to CQ.

The total capacitance is calculated by the formula,

$$\frac{1}{CT} = \frac{1}{CQ} + \frac{1}{CEDL} \tag{3.19}$$

CEDL is the electric double-layer capacitance. It depends on electrode-electrolyte interfacial interaction[22]

Q. Xu et al. have discussed the quantum capacitance in monolayer MoS_2 doped with Ti, Au, Ag, Cu, Al, B, N, and P doping pristine and single-vacancy. The high specific surface area in MoS_2 monolayer can increase the contact area with electrolytes thus having high double-layer capacitance (C_{EDL}), whereas a low quantum capacitance (C_Q), restricts the increase in C_T. This is because of the confinement of electron DOS near Fermi level in pristine MoS_2 monolayer. From the above equation for C_Q, we can say that by doping and introduction of defects, local states can be introduced near the Fermi level or there can be a shift in the Fermi level.

One substitute S atom with the metal atoms, Ti, Au, Ag, and Cu in pristine and single-vacancy vs MoS_2, and substitute S atom with nonmetal atoms- B, N, and P in pristine MoS_2, as shown in Figure 3.10.

It was found that Al and B doping enhances the electronic structure significantly, improving the quantum capacitance of MoS_2. When the doping concentration of Al and B atoms increased, the quantum capacitance also improved monotonously. This was explained with the help of DOS and other calculations from DFT analysis.

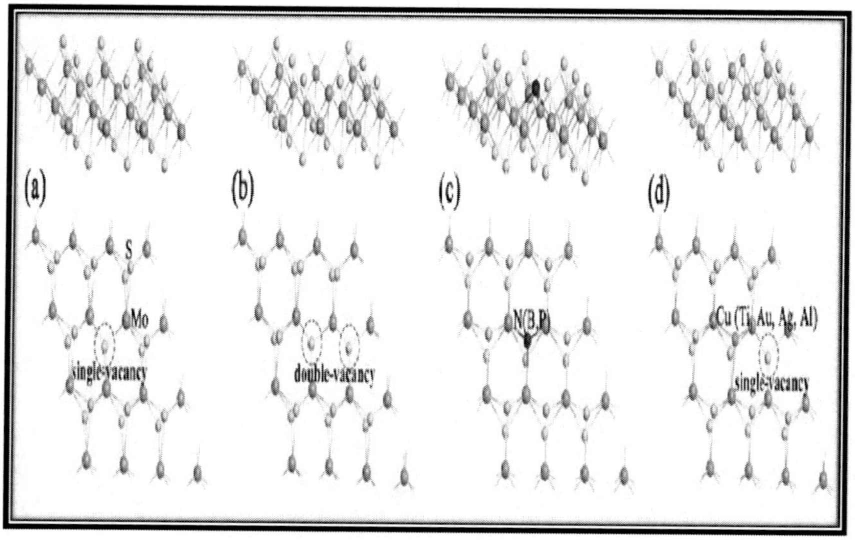

Figure 3.10. The proposed MoS_2 monolayer-based models including (a) single-vacancy VS MoS_2 monolayer, (b) double-vacancy V_2S MoS_2 monolayer, (c) B, N, P, Ti, Au, Ag, Cu, and Al substituting S atom of pristine MoS_2 monolayer, and (d) Ti, Au, Ag, Cu, and Al substituting S atom of single-vacancy VS MoS_2 monolayer. (Reproduced with permission from Q. Xua et al., "DFT Calculation for Stability and Quantum Capacitance of MOS2 Monolayer-Based Electrode Materials." *Materials Today Communications* 22 (2020): 100772 [23]).

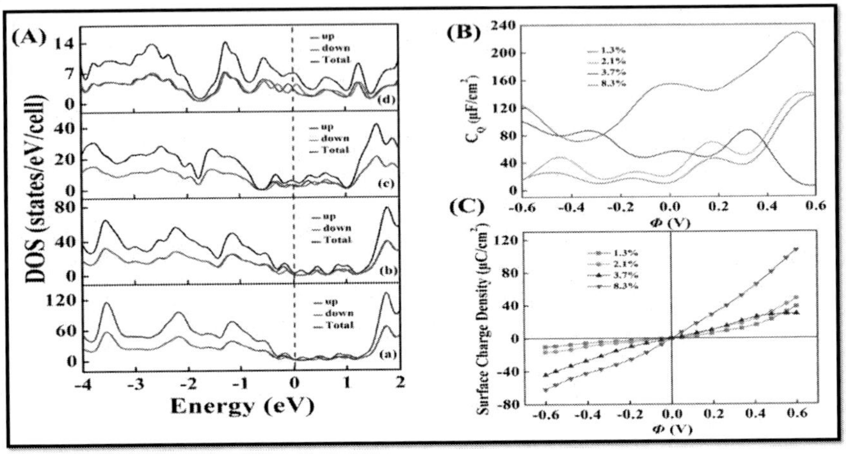

Figure 3.11. (A) Density of states (DOS) and local density of states (LDOS) of single vacancy VS MoS_2 monolayer with Al substituting S atom with different Al concentrations including (a) 1.3%, (b) 2.1%, (c) 3.7%, and (d) 8.3%, (B) calculated quantum capacitance CQ corresponding to Figure 7A, as a function of local electrode potential (Φ), and (C) surface charge vs potential drop and Φ between -0.6 to 0.6 eV (Reproduced with permission from Q. Xua et al., "DFT Calculation for Stability and Quantum Capacitance of MOS_2 Monolayer-Based Electrode Materials." *Materials Today Communications* 22 (2020): 100772 [23]).

The electronic structure of MoS_2 has a significant change on doping with increased concentration of Al. The DOS and LDOS for Al doping in different concentrations in single-vacancy v/s MoS_2 are given in Figure 3.11(a). Doping of Al at low concentration has no remarkable effect but for instance, at a concentration enhancement of 3.7%, the doping introduces higher and wider local states near the Fermi level which in turn causes high quantum capacitance. Figure 3. 11(b) depicts the variation of quantum capacitance with local electrode potential to different Al doping concentrations. When the doping concentration increases from 1.3% to 8.3%, the Quantum capacitance (CQ) value rises from 44.76 μF/cm^2 at 0.17 V to 227.85 μF/cm^2 at 0.53 V. Figure 3.11 (c) plots the relationship between surface charge density and potential varying from -0.6 to 0.6V at different doping concentration. With the positive bias along with Al doping concentration up to 8.3%, the charge accumulation increases [23].

Thus, with the help of DFT calculations, we can explain many such phenomena occuring in the system, at the same time, discover new capacitance improving methods, dopants, and hybrids. The calculations and the findings from this theoretical side can help to give a fundamental idea on choosing the systems for experimental works.

3.6.1. Water Splitting Applications

Finding out alternate fuel sources for fossil fuels has always been a challenging task. Hydrogen also called the 'Green fuel' is one of the potent candidates among the replacement options. Hydrogen is available in abundance as it can be produced from water by hydrolysis or by the reformation of hydrocarbons. The former involves water splitting and is an effective way to produce this fuel by photocatalysis or electrocatalysis.

The electrolysis of water consists of two half-cell reactions- Hydrogen Evolution Reaction (HER) and Oxygen Evolution Reaction (OER). In HER, there are two mechanisms- the Volmer–Tafel or Volmer–Heyrovsky mechanism. In the Volmer reaction, reduction of protons forms surfaces adsorbed hydrogen (+ + →− H(aq) e Hads). These surface-adsorbed hydrogen atoms combine to form a hydrogen molecule in the Volmer–Tafel mechanism whereas a surface adsorbed hydrogen reacts with a proton from water in the Volmer–Heyrovsky mechanism. Since both mechanisms involve adsorption of H atom on the surface (H_{ads}), H_{ads} can be attributed to checking upon the catalytic activity of the catalyst. The adsorption free energy of H (ΔG_H) is related to H_{ads} by the relation,

$$\Delta G_H = \Delta\ E_{ads} + \Delta E_{ZPE} + T\Delta S_H \tag{3.20}$$

where E_{ads} is the adsorption energy of hydrogen, ΔE_{ZPE} is the difference in zero-point energy between hydrogen in the adsorbed and gas-phase state, T is the temperature in K and ΔS_H is the change in entropy between hydrogen in the adsorbed and gas phase. Since $\mathbf{\Delta G_H}$ decides the rate of the reaction, an optimal catalyst should have $\Delta G_H \cong 0eV$.

Heterogeneous catalysts based on transition metals, especially Pt has been most widely studied. It is because $\Delta G_H \cong 0$ eV for Pt that makes it a near-perfect option. But due to high cost and scarcity, other replacements have been explored, including 2D materials like graphene and TMDs [24]. TMDs have been chosen because unlike graphene, they possess a finite bandgap making them suitable for photocatalytic splitting of water. In photocatalytic splitting, the two major requirements for water splitting are sunlight and a catalyst. On irradiation of sunlight, water can be reduced to oxygen and hydrogen on the condition that the valence band maximum (VBM) should be greater than the water oxidation potential (+1.23 eV) and the conduction band minimum (CBM) of the catalyst should be more negative than the hydrogen reduction potential (0 eV) [25].

In a work done by Showkat. H. Mir [26], they have found an improved catalytic activity by PtS_2 and WS_2 due to the influence of foreign elements like Ru, Rh, Pd, Ag, Ir, Au, and Hg in these systems. After the structural optimization of WS_2 and PtS_2, initially, H_2 molecules were placed at varying distances (3 – 4.2 Å) from the sandwiched transition metal layer. Figure 3.12 (a) gives the crystal structure of 2D PtS_2 and WS_2. PtS_2, in the ground state, exists in the 1T phase and WS_2 exists in the 2H phase. The most important criterion for a photocatalyst is its bandgap which is calculated using HSE06 functional and spin-orbit interaction was also considered. Figures 3.12 (c) and (f) show the band diagrams.

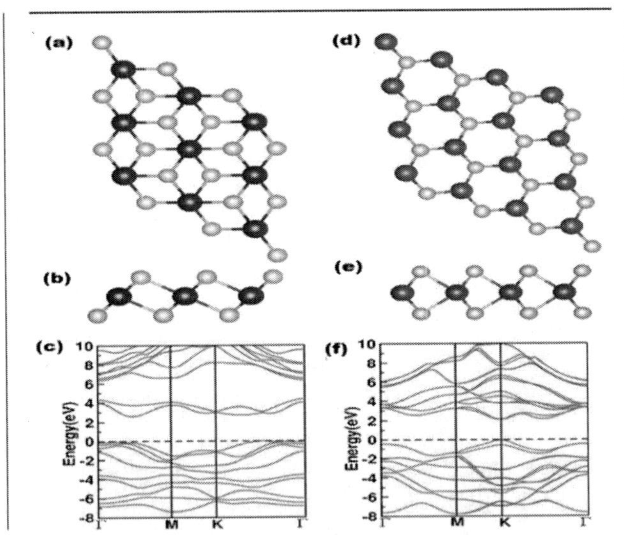

Figure 3.12. The crystal structure of PtS_2 in 1T phase (a) Top and (b) side view and that of WS_2 in 2H phase (d) Top and (e) side view. Band structure of (c) PtS_2 and (f) WS_2 calculated using HSE06 functional. The color of the elements is designated as follows: maroon for Platinum, Red for Tungsten, and green for Sulphide. (Reproduced with permission from Showkat H. Mira et al., "A Comparative Study of Hydrogen Evolution Reaction on Pseudo-Monolayer WS_2 and PtS_2: Insights Based on the Density Functional Theory." *Catalysis Science & Technology* 7, no. 3 (2017): 687–92 [26]).

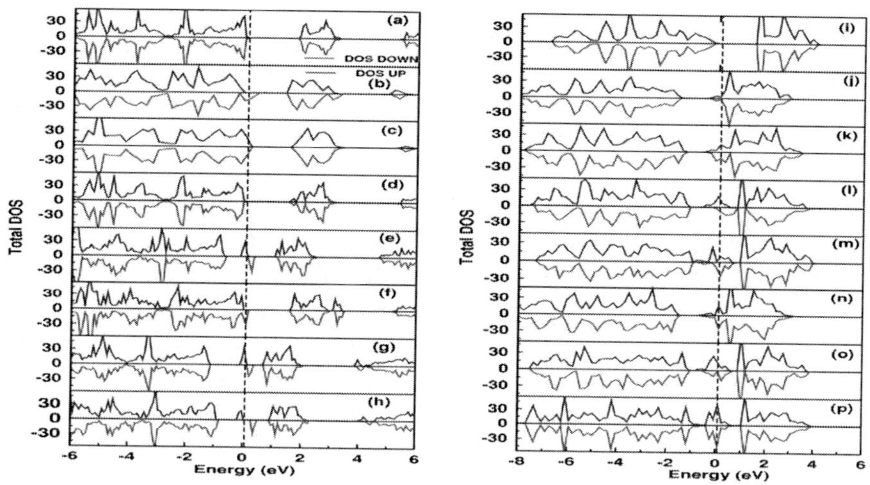

Figure 3.13. Total DOS of Pristine (a) PtS_2 (left panel) and (i) WS_2 (right panel). (b), (c), (d), (e), (f), (g), and (h) represents the total DOS of Ru, Rh, Pd, Ag, Ir, Au, and Hg doped functionalized PtS_2 respectively. (j), (k), (l), (m). (n), (o), and (p) represent total DOS for functionalized WS_2 with dopants in the same order as that of PtS_2. The Fermi energy is set to the zero-energy level. Showkat H. Mira et al., "A Comparative Study of Hydrogen Evolution Reaction on Pseudo-Monolayer WS_2 and PTS_2: Insights Based on the Density Functional Theory. "*Catalysis, Science & Technology* 7, no. 3 (2017): 687–92 [26].

From the above figure, we can say that PtS_2 is an indirect bandgap semiconductor having a bandgap of 2.60 eV, and the corresponding valence band maximum (VBM) is situated between Γ–K whereas the conduction band minimum (CBM) is located between Γ–M symmetry point. WS_2, on the other hand, is a monolayer with a direct bandgap of 2.14 eV as depicted in Figure 3.12 (f).

To find out the influence of dopants on the bandgap, the DOS has been determined. Figure 13 shows the DOS for pristine as well as for the functionalized systems. The DOS plots suggest that Au, Ag, and Hg induce electronic states near the Fermi level in PtS_2, whereas other dopants have no significant effects. Also, VBM and CBM shift towards the lower energy region in functionalized systems compared to the pristine PtS_2. In WS_2, all dopants shift the VBM towards a more negative energy value. These effects can tune the electronic properties of both WS_2 and PtS_2 systems and hence, improve the HER activity as well.

The adsorption energies of H_2 at the top and hollow site of pristine and functionalized pseudo monolayers were calculated using the following equation

$$Eads = E_T\,(system+ H_2) - E_T\,(system) - E_{H2} \tag{3.21}$$

where E_T (system+ H_2) is the total energy of PtS_2 or WS_2 system with the adsorbed molecule, E_T (system)is the energy without the adsorbed molecule, and E_{H2}represents the energy of hydrogen molecule in the gas phase. The calculations suggest that the Pd doped WS_2 system has the optimal binding energy of H_2 for HER activity among all the systems considered.[26]

Many such TMDs have been explored and also most of them have been tuned using dopants, strain, etc. to improve the HER efficiency. This influence can be studied and explained using DFT analysis.

Conclusion

In this chapter, we discussed the Density Functional Theory aspects in theoretical simulations to explore the properties and applications of TMDs. TMDs are growing to be one of the best candidates in relevant scientific research on a variety of society-relevant applications. We have summarized the basics of Density Functional Theory and the principles governing it that make it an excellent predictor and assessor. Various DFT studies on defining the structural, electronic, and optical properties have been discussed. We have also elaborated on the sensing applications, both chemical and biological, including how the Density of States, projection of atomic orbitals, and dielectric functions play a major role to find out the adsorption characteristics. DFT also finds applications in finding suitable replacements for energy storage using TMDs, with the help of super capacitance calculations and water splitting reactions. DFT involves computations on thousands of compounds, or even more or less than that. Such a change of scale demands sophisticated technologies and data management methodologies. A major challenge for high-level computations is that GGA errors are extremely dependent on the nature of chemistry. Also, DFT with GGA exchange-correlation functional may not accurately predict the excited states properties of d and f orbital. But advanced methods like GGA+U, GW approximations, hybrid functional calculations can solve these problems to an extent. The accuracy of prediction always depends on the functional chosen. DFT has always been found useful to produce innovations in several hundreds of atoms and their behavior, with highly powerful, versatile codes and manipulation of variables.

References

[1] Kresse, G., and J. Furthmüller. "Efficiency of AB-Initio Total Energy Calculations for Metals and Semiconductors Using a Plane-Wave Basis Set." *Computational Materials Science*, 6, no. 1 (1996): 15–50. https://doi.org/10.1016/0927-0256(96)00008-0.

[2] Giannozzi, Paolo, Stefano Baroni, Nicola Bonini, Matteo Calandra, Roberto Car, Carlo Cavazzoni, Davide Ceresoli et al. "QUANTUM ESPRESSO: a modular and open-source software project for quantum simulations of materials." *Journal of Physics: Condensed matter*, 21, no. 39 (2009): 395502.

[3] Schwarz, Karlheinz, and Peter Blaha. "Solid-state calculations using WIEN2k." *Computational Materials Science*, 28, no. 2 (2003): 259-273.

[4] Kohn, Walter. "Fundamentals of density functional theory." In *Density Functionals: Theory and Applications*, pp. 1-7. Springer, Berlin, Heidelberg, 1998.

[5] Kohn, Walter, and Lu Jeu Sham. "Self-consistent equations including exchange and correlation effects." *Physical Review*, 140, no. 4A (1965): A1133.

[6] Perdew, John P., John A. Chevary, Sy H. Vosko, Koblar A. Jackson, Mark R. Pederson, Dig J. Singh, and Carlos Fiolhais. "Atoms, molecules, solids, and surfaces: Applications of the generalized gradient approximation for exchange and correlation." *Physical Review B*, 46, no. 11 (1992): 6671.

[7] Perdew, John P., Kieron Burke, and Matthias Ernzerhof. "Generalized Gradient Approximation Made Simple." *Physical Review Letters*, 77, no. 18 (1996): 3865–68. https://doi.org/10.1103/physrevlett.77.3865.

[8] Rehman, Javed, Roshan Ali, Nisar Ahmad, Xiaodong Lv, and Chunlei Guo. "Theoretical investigation of strain-engineered WSe2 monolayers as anode material for Li-ion batteries." *Journal of Alloys and Compounds*, 804 (2019): 370-375.

[9] Monkhorst, Hendrik J., and James D. Pack. "Special points for Brillouin-zone integrations." *Physical Review B*, 13, no. 12 (1976): 5188.

[10] Ahmad, Sohail, and Sugata Mukherjee. "*A comparative study of electronic properties of bulk MoS2 and its monolayer using DFT technique: application of mechanical strain on MoS2 monolayer.*" (2014).

[11] Abbadi, Hajar, Siham Malki, and Larbi El Farh. "Ab initio calculations of structural and electronic properties of WSe2 compound." *Materials Today: Proceedings*, 31 (2020): S130-S133.

[12] Miró, Pere, Mahdi Ghorbani-Asl, and Thomas Heine. "Two dimensional materials beyond MoS2: noble-transition-metal dichalcogenides." *Angewandte Chemie International Edition*, 53, no. 11 (2014): 3015-3018.

[13] Wang, Yu, Yafei Li, and Zhongfang Chen. "Not your familiar two-dimensional transition metal disulfide: structural and electronic properties of the PdS 2 monolayer." *Journal of Materials Chemistry C*, 3, no. 37 (2015): 9603-9608.

[14] Huang, Zongyu, Chaoyu He, Xiang Qi, Hong Yang, Wenliang Liu, Xiaolin Wei, Xiangyang Peng, and Jianxin Zhong. "Band structure engineering of monolayer MoS2 on h-BN: first-principles calculations." *Journal of Physics D: Applied Physics*, 47, no. 7 (2014): 075301.

[15] Bhandari, Churna, Walter RL Lambrecht, and Mark Van Schilfgaarde. "Quasiparticle self-consistent gw calculations of the electronic band structure of bulk and monolayer V2O5." *Physical Review B*, 91, no. 12 (2015): 125116.

[16] Das, Tilak, Sergio Tosoni, and Gianfranco Pacchioni. "Structural and electronic properties of bulk and ultrathin layers of V2O5 and MoO3." *Computational Materials Science*, 163 (2019): 230-240.

[17] Tang, Xiao, Aijun Du, and Liangzhi Kou. "Gas sensing and capturing based on two-dimensional layered materials: Overview from a theoretical perspective." *Wiley Interdisciplinary Reviews: Computational Molecular Science*, 8, no. 4 (2018): e1361.

[18] Nayeri, Maryam, Mahdi Moradinasab, and Morteza Fathipour. "The transport and optical sensing properties of MoS2, MoSe2, WS2 and WSe2 semiconducting transition metal dichalcogenides." *Semiconductor Science and Technology*, 33, no. 2 (2018): 025002.

[19] Soler, José M., Emilio Artacho, Julian D. Gale, Alberto García, Javier Junquera, Pablo Ordejón, and Daniel Sánchez-Portal. "The SIESTA method for ab initio order-N materials simulation." *Journal of Physics: Condensed Matter*, 14, no. 11 (2002): 2745.

[20] Chen, Dachang, Xiaoxing Zhang, Ju Tang, Hao Cui, and Yi Li. "Noble metal (Pt or Au)-doped monolayer MoS2 as a promising adsorbent and gas-sensing material to SO2, SOF2 and SO2F2: a DFT study." *Applied Physics A*, 124, no. 2 (2018): 1-12.

[21] Vovusha, Hakkim, and Biplab Sanyal. "Adsorption of nucleobases on 2D transition metal dichalcogenides and graphene sheet: a first-principles density functional theory study." *RSC advances*, 5, no. 83 (2015): 67427-67434.

[22] Samal, Rutuparna, Soumen Mondal, Abhijeet Sadashiv Gangan, Brahmananda Chakraborty, and Chandra Sekhar Rout. "Comparative electrochemical energy storage performance of cobalt sulfide and cobalt oxide nanosheets: experimental and theoretical insights from density functional theory simulations." *Physical Chemistry Chemical Physics*, 22, no. 15 (2020): 7903-7911.

[23] Xu, Q., G. M. Yang, and W. T. Zheng. "DFT calculation for stability and quantum capacitance of MoS2 monolayer-based electrode materials." *Materials Today Communications*, 22 (2020): 100772.

[24] Ouma, Cecil Naphtaly Moro, Kingsley Onyebuchi Obodo, Moritz Braun, George Odhiambo Amolo, and Dmitri Bessarabov. "Insights on hydrogen evolution reaction in transition metal-doped monolayer TcS2 from density functional theory calculations." *Applied Surface Science*, 470 (2019): 107-113.

[25] Zulkifli, Nur'Adnin Akmar, Suhana Mohd Said, Mohamad Fariz Mohamad Taib, K. Arifin, S. M. Mahmood, Kai Lin Woon, Shobhit K. Patel, C. L. Tan, and Rozalina Zakaria. "Photocatalytic application of two-dimensional materials-based heterostructure based on molybdenum and tungsten disulfides and gallium nitride: a

density-functional theory study." *Materials Today Communications*, 25 (2020): 101646.

[26] Mir, Showkat H., Sudip Chakraborty, John Wärnå, Som Narayan, Prakash C. Jha, Prafulla K. Jha, and Rajeev Ahuja. "A comparative study of hydrogen evolution reaction on pseudo-monolayer WS 2 and PtS 2: insights based on the density functional theory." *Catalysis Science & Technology*, 7, no. 3 (2017): 687-692.

Chapter 4

Strategies for the Growth of Metallic TMDs and Their Characterizations

Sree Raj K. A. and Chandra Sekhar Rout*

Centre for Nano and Material Science, Jain University, Jain Global Campus, Jakkasandra, Ramanagaram, Bangalore, India

Abstract

In this chapter we are discussing various synthesis strategies of metallic TMDs and its characterization techniques. The TMDs synthesis route mainly include two type of approaches the top-down approach and the bottom-up approach. There are various methods confines under these two synthesis approaches. Compared to other TMDs metallic TMDs shows great promise in various applications due to its unique set of properties. We are also briefly discussing the growth and analysis of the polytypic structures of TMDs and the importance of the metallic phase of TMDs. The focus of the chapter is to provide an overview of the growth approaches of metallic TMDs along with the latter half discussing the various characterization techniques of them.

Keywords: metallic TMDs, CVD, molecular beam epitaxy, STEM

4.1. Introduction

In literature, several chemical and physical methods have been reported for the synthesis of metallic TMD materials. Some of the methods which have

* Corresponding Author's Email: r.chandrasekhar@jainuniversity.ac.in.

In: 2D Metallic Transition Metal Dichalcogenides
Editors: Chandra Sekhar Rout and Brahmananda Chakraborty
ISBN: 978-1-68507-965-9

been primarily employed to prepare TMDs include mechanical/chemical exfoliations, chemical vapour deposition (CVD), molecular beam epitaxy (MBE), hydrothermal and solution phase, etc. By considering the growth process, the synthesis methods can be classified into two categories i.e., top-down and bottom-up approaches. The reported synthesis methods for the metallic TMDs are summarised in **Figure 4.1**. Characterization techniques such as, atomic force microscopy (AFM), Raman spectroscopy, optical microscopy, Field emission scanning electron microscopy (FESEM), transmission electron microscopy (TEM) and scanning tunnelling microscopy (STM), etc. are commonly used to evaluate the quality of the prepared materials, which are beyond the scope this book and authors are suggested to refer to other relevant books for this purpose.

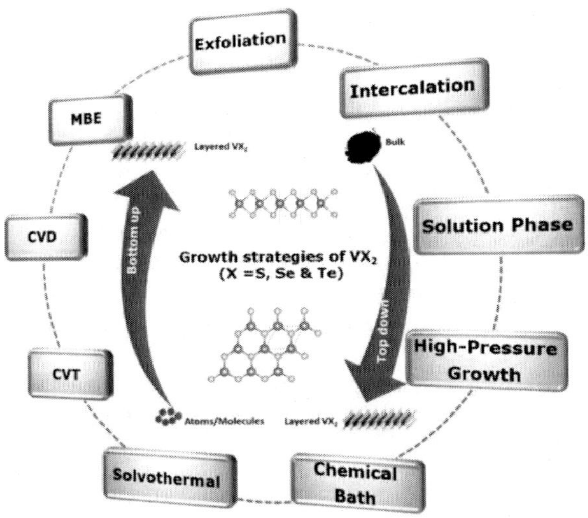

Figure 4.1. Schematic representation of the different synthesis top up and bottom down approaches used for the preparation of two dimensional layered VX_2 (X = S, Se, Te), (Reprinted with permission from [1]).

4.2. Top-Down Approaches

In this approach, bulk crystals of TMDs are used as the starting materials and mechanical force and chemical intercalation are employed to separate the layers of the 2D materials. The applied mechanical force or the chemical intercalation is sufficiently higher than the van der Waals force existing between the layers of the 2D metallic TMDs.

4.2.1. Mechanical Exfoliation and Liquid-Phase Exfoliation

These methods are the simplest of all the synthesis routes and can be used to prepare higher quality, with sizes up to a few microns. But the drawbacks of these methods are that it is at laboratory scale and low yield, this makes it impractical for large scale production. $NbSe_2$ monolayers are prepared by mechanical exfoliation of $NbSe_2$ bulk using the scotch tape-based method [1]. Xi et al. reported the preparation of metallic $NbSe_2$ monolayers by mechanical exfoliation method and studied their many-body collective-order phase diagrams by optical and electrical properties investigation [2]. Similarly, mono and few layer 2D materials can be obtained by exfoliating the bulk layered materials in a liquid medium. This approach offers the easiest method for scaling up the materials up to an industrial level. In this approach, ultra-sonication is a powerful and cost-effective approach to create mechanical waves through the liquid for the exfoliation.

The liquid phase-exfoliation is also known as chemical exfoliation when external ions or molecules (Li-ion, Li_2S, tetrabutylammonium bromide, etc.) are inserted into the van der Waals gaps of the MX_2 crystals to expand the interlayer-spacing and finally to achieve the mono or few layered materials. After the expansion of the interlayer spacing of MX_2, external force in the form of sonication, milling, agitation or shearing, etc. are introduced to laminate the swelled crystals. Zeng et al. reported the preparation of few layered metallic TaS_2, $NbSe_2$ and TiS_2 by the Li-ion intercalation followed by sonication approach [3]. Feng et al. performed the exfoliation of $VS_2.NH_3$ precursor in water by ultra-sonication in iced water followed by filtration to remove un-exfoliated product [4]. Formamide is widely used as the medium for the exfoliation of metallic VS_2 and VSe_2 [5-6]. Wong et al. reported the exfoliation of vanadium dichalcogenides by using n-butyl lithium dispersed in hexane [7]. Li_2CO_3 is used by Ghobadi et al. for the lithium intercalation of VSe_2 followed by sonication [8]. Peng et al. achieved the expansion of MX_2 up to 94% by using 1.6M n-BiLiin hexane at 100°C in a quartz-lined autoclave [9]. By this approach, monolayers of metallic TMDs such as TaS_2, $NbSe_2$ and $TiSe_2$ with unprecedented sizes up to 100 μm could be prepared [9]. Wu et al. reported the expansion of interlayers up to 155% by pre-treating the $LiTaS_2$ crystals in HCl and H_2SO_4 solutions [10]. The presence of H^+ etchant highly helped to greatly enlarge the inlayer spacing and large size metallic TaS_2 monolayers up to ten micrometer can be achieved (Figure 4.2) [11].

4.3. Bottom-Up Approach

Bottom-up approaches include chemical vapour deposition (CVD), physical vapour deposition (PVD), molecular beam epitaxy (MBE), chemical vapour transport (CVT), etc. These methods are powerful techniques for the growth of precisely controlled thin-films of layered metallic 2D TMDs.

4.3.1. Chemical Vapour Deposition (CVD)

Among all the bottom-up approach methods, CVD based techniques are known to be a powerful method since it can grow high quality and large-area films. Based on the reaction conditions and by tuning the parameters, single-to-few layered TMD films can be grown on different substrates. Various parameters include precursor, temperature, pressure, substrates and deposition rate and gas flow, etc. can significantly affect the thermodynamics and kinetics in the CVD growth of 2D materials.

In the CVD chamber, precursors serve as the reactants with possible reactions such as thermal decomposition, chemical transport and chemical synthesis reaction occur during the process of the conversion of the precursors to the desired products. For transition metal dichalcogenides growth, two sorts of metal precursors such as metal chloride, metal oxides and S, Se, Te powders are used as the starting materials and kept at different heating zones. The vapour sources of the precursors are carried out towards the substrates kept inside the chamber by the help of carrier gases such as Ar, N_2 or Ar+H_2. The presence of H_2 gas along with the inert gas in the reactor helps to terminate any broken or dangling bond to achieve high quality samples.

The temperature of the precursor positions and substrates in a CVD chamber plays a crucial role to achieve a high-quality product with uniform composition and crystallinity. Substrates on which the 2D materials are deposited sometimes act like catalyst and nucleation centre. The pressure of the chamber also a critical parameter to achieve controlled growth of 2D materials. By controlling pressure, the low concentration and high velocity of the mass feed of the precursor are available which makes the reaction more controllable.

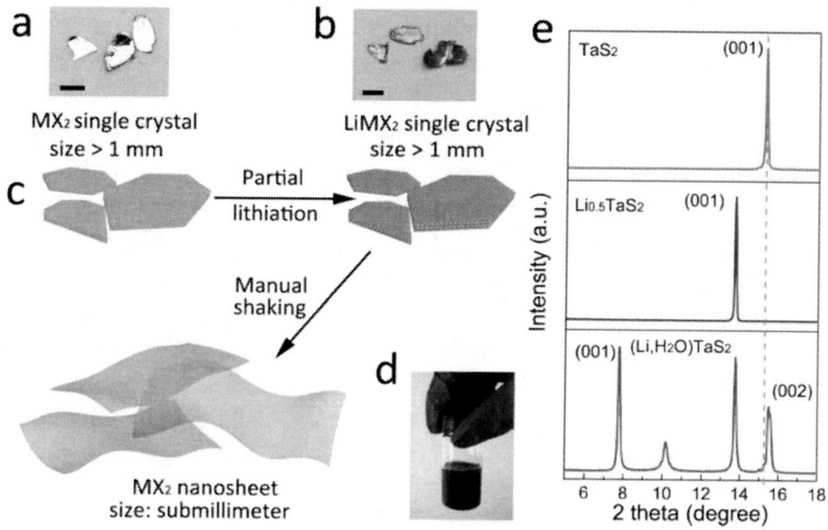

Figure 4.2. Process of the TSC exfoliation. (a and b) Photographs of the 1T-TaS_2 and Li$x$$TaS_2$ single crystals, respectively. (c) MX_2 single crystals are partially lithiated to form Li$x$$MX_2$ crystals that can be fast exfoliated into single-layer nanosheets by only manual shaking. (d) Photograph of the TaS_2 suspension. (e) XRD spectra of 1T-TaS_2, Li0.5TaS_2, and (Li, H_2O) TaS_2. The scale bars in (a) and (b) are 1 mm. (Adapted with permission from Ref. 10, American Chemical Society).

The CVD growth in a vacuum tube furnace is also termed as chemical vapour transport (CVT), where a condensed phase of solid gets volatilised in the pressure of carrier gas and the final product is deposited on the substrate. For TMDs growth, iodine and chloride are used as the transport agent [11, 12]. Transition metal chloride act as transport agents for it and final products of TMD films are formed on the substrates. Wang et al. reported the controlled growth of metallic 2D $TiSe_2$ with thickness below 4.8nm by a modified CVT method [13]. Ji et al. reported the size and thickness in the controlled growth of VS_2 by adjusting the temperature zone set for its precursor VCl_3 (275-300°C) [14]. Also, the H_2 flow rate triggered and determined the reaction rate of VCl_3 with S and no growth of VS_2 was observed without H_2. By maintaining the H_2 flow to 10% in N_2 for 10 min growth time uniform growth of VS_2 nanosheets was achieved on SiO_2/Si substrate by positioning VCl_3 precursor at 550°C with the substrates placed upside down and 'S' being placed in the upstream of the tube furnace at low temperature [15]. By controlling the temperature, source and substrate position, H_2: Ar ratio and growth time, desirable thickness and quality of VSe_2 films could be controlled by the CVD approach [16]. Growth of VTe_2 nanoplates was achieved by a modified CVD approach by taking a mixture of VCl_3 and Te powder (1:3) kept at a low temperature zone of (560-620°C) and the SiO_2/Si substrate was placed at 600°C [17]. By controlling the Ar-flow-rate different thickness of VTe_2 size and morphology could be achieved. Ge et al. reported the growth of vertical metallic NbS_2 nanosheets on SiO_2/Si substrates by using $NbCl_5$ and S-feedstock by a CVD method (Figure 4.3) [18]. Vertically aligned HfS_2 and ReS_2 nanosheets are achieved by utilizing the metal chloride precursor, rich dangling bonds and high surface roughness of the substrate mixture of transition metal oxides, alkali halides with chalcogens have been used for the CVD growth of metallic TMDs [19-20]. The alkali halides help in promoting the CVD growth rate and greatly decrease the melting point and facilitate the formation of intermediate products and increases the overall reaction rates. Wang et al. adopted their approach for the growth of metallic 2D $NbSe_2$ flakes on SiO_2/Si substrate by taking (NbO_x+ NaCl+ Se) as the precursor [19-20].

4.3.2. Molecular Beam Epitaxy (MBE)

Molecular Beam Epitaxy (MBE) is a scalable synthesis method for the direct growth of extremely smooth TMDs. MBE is modified form of vacuum evaporation in which collimated molecular or atomic beams impinge on a preheated substrate under ultra-high vacuum (typically pressure will be below 10^{-10} mbar) contains sources to form molecular beams [21-22]. The usage of high purity source and ultra-high-pressure conditions reduces the impurity of the films. The low growth temperature and slow growth rate (typically 1 Å/min) allow the growth of heterostructures with minimum degradation as well as excellent crystal perfection and precise control over thickness [23].

In a conventional solid source, MBE setup effusion cells are used for the metal evaporation but in the case of transition metals such as Mo and W, the high melting point demands the electron-beam evaporation. The sticking coefficient of S and Se on the growth surface under UHV will impact the growth structure in high growth temperature. These problems lead to the smaller domain sizes of TMDs. To avoid this and enhance surface diffusion, MBE growth of TMDs usually carried out on van der Waals materials such as epitaxial graphene, highly oriented pyrolytic graphite (HOPG) and mica. The growth of Te based TMDs is suited for MBE because of their preference in lower growth temperature than S or Se based TMDs [24-25].

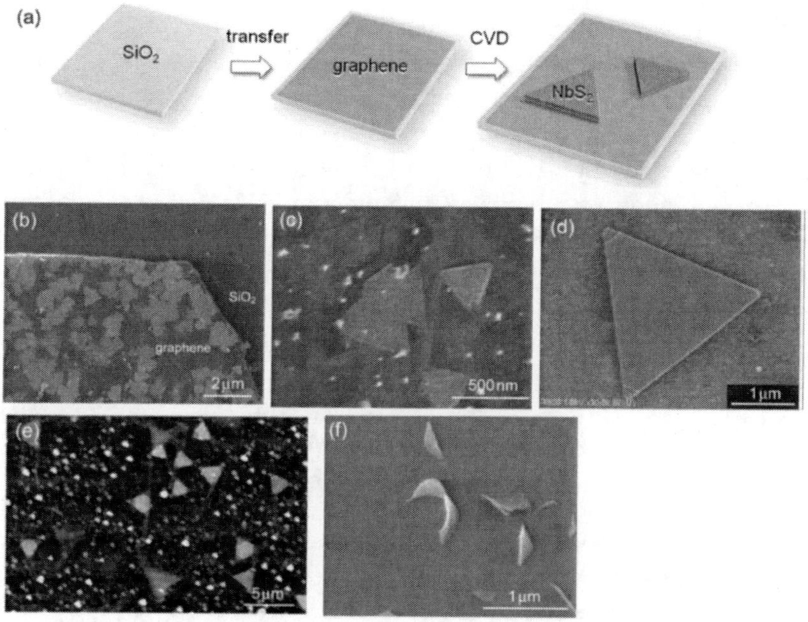

Figure 4.3. (a) Schematic of CVD growth of NbS_2 nanosheets on the transferred CVD-grown graphene. (b) SEM images of NbS_2 nanosheets grown on the transferred graphene. The density of NbS_2 nanosheets on SiO_2 is much lower than that on graphene. (c) NbS_2 nanosheets grown on graphene. (d) Large nanosheets grown with higher feedstock supply rate. (e) AFM image of the graphene surface with triangular NbS_2 nanosheets. (f) On the SiO_2 substrate without graphene, NbS_2 nanosheets grow perpendicular to the substrate surface, similar to those on Si. (Reprinted by permission from Ref. 19, Copyright (2016) Royal Society of Chemistry).

Ji et al. reported the MBE growth of monolayered 2H phase NbS_2 and a mixture of 2H and 1T phase TaS_2 on a 6H-SiC (0001) substrate terminated with monolayer or bilayer epitaxial graphene [26]. Yu et al. reported the fabrication of ultra-thin heterophase homojunction by the layer-by-layer MBE growth of $MoTe_2$ with a simultaneous stoichiometry of 2H and 1T' phases [27]. El Kazzi et al. synthesised MoS_2 layers by co-deposition using MBE [28]. Solid Mo was evaporated using electron gun of MBE and H_2S gas as the sulphur source was injected through a remote cell to the growth chamber [26]. Hinkle et al. reported the growth of WSe_2 layers using van de Waals epitaxy [29]. The rotational alignment to van der Waals substrate and absence of misfit dislocations in films were observed by this group. Ugeda et al. reported a high quality submonolayer growth of $NbSe_2$ films on epitaxial bilayer graphene on 6H-SiC (0001) substrate. With the combination scanning electron microscopy/

spectroscopy, the group tried to understand the intrinsic properties of $NbSe_2$ [30]. The thin films of $HfSe_2$ were reported by Hinkle et al. $HfSe_2$ thin films were grown on a 1x1cm^2 mechanically exfoliated HOPG and MoS_2 (Figure 4.4). During the growth, the pressure was maintained at ~ $1x10^{-9}$ mbar and the substrate temperature was kept at 450°C. The Se: Hf flux was maintained at a 5:1 ratio and the growth rate was 0.05 nm/min [31]. Hall et al. reported the MBE growth of monolayer transition metal disulphides such as MoS_2, TaS_2 and WS_2. For the growth process of transition metal disulphides Gr/Ir(111) substrate was exposed to a metal atom flux from an e-beam evaporator for a growth time tg, simultaneously a sulphur vapour flux out of a Knudsen cell filled with FeS_2 at a growth sample temperature in the range of $100\ K < T_g <$ 500K [32].

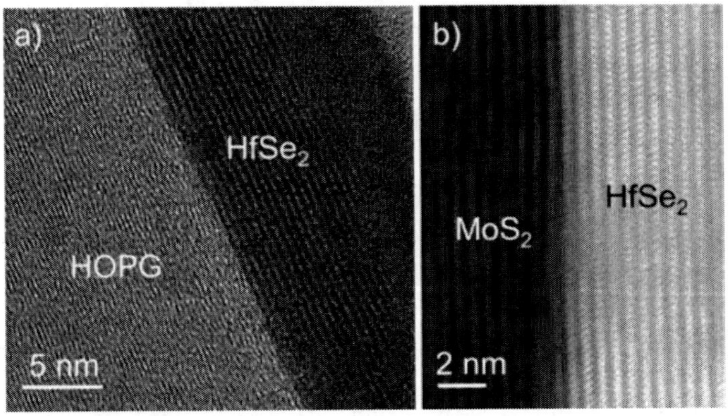

Figure 4.4. (a) TEM image of grown $HfSe_2$ on HOPG and (b) HADDF-STEM image of grown $HfSe_2$ on MoS_2 showing abrupt interfaces and layered crystalline films. (Adapted with permission from Ref. 28, American Chemical Society).

4.3.3. Hydrothermal/Solvothermal Method

Hydrothermal/solvothermal is a widely used low-cost method to prepare metallic transmission metal dichalcogenides. Hydrothermal can produce stable TMDs at elevated temperature with high vapour pressure and maximum yield. Typically hydrothermal synthesis was carried out in a temperature below 300°C. The solvent properties will change exponentially under supercritical conditions therefore for the supercritical nature of water or any other solvents (solvothermal method) provides a favourable reaction atmosphere for TMDs due to the enhancement in reaction rate [33-34].

Marri et al. prepared 3D cuboidal VSe_2 layers assisted with rGO in a 200°C temperature and 24h reaction time where sodium metavanadate and selenium dioxide used as the precursors [35] Rout et al. reported flower like metallic VS_2 sheets involving reaction of sodium orthovanadate as vanadium source and thioacetamide as a sulphur source. The group explained the formation of metallic VS_2 layers by the reduction of V^{5+} to V^{4+} with HS^- generated by the hydrolysis of thioacetamide [36]. Liu et al. developed a procedure to stabilize metallic 1T-MoS_2 by the intercalation of ammonium ions [37]. Where Ammonium heptamolybdate and thiourea acts as the Mo and S source respectively the reaction were carried out in 200°C at 24h [34]. Jiang et al. reported flower like 1T-$MoSe_2$ nanosheets synthesised by a one-pot solvothermal method by reacting $MoCl_5$ with SeO_2 in octylamine at 180°C [38]. Piao et al. synthesised stable metallic 1T phase of WS_2 nanosheets in a hydrothermal reaction with Ammonium tungsten oxide hydrate, thiourea, oxalic acid dihydrate and sodium cholate in 220°C for 48h [39].

4.4. Characterization Methods for Metallic TMDs

Characterization of synthesised metallic TMDs is an important step for the application of these materials in various fields. Here we are going to discuss different characterization techniques for the metallic TMDs.

To understand the fundamental properties and changes in the properties after the tuning of metallic TMDS for various applications microscopic techniques can shed light for these purposes. Microscopic techniques include scanning electron microscope (SEM), high resolution transmission electron microscope (HR-TEM), atomic force microscope (AFM), scanning tunnelling microscope (STM), optical microscopy, etc.

Electron microscopy uses electron beams to interact with the crystal lattices and explores the fundamental properties of TMDs. SEM provides detailed surface characteristics of the synthesised TMDs with great depth of field. One of the major characterization techniques to visualize the metallic TMDs is a transmission electron microscope (TEM). TEM techniques can analyse the defects and structural transformations in metallic TMDs. However, the introduction of electronic beam radiations sometimes leads to damage of the atomic structure of metallic TMDs. Garcia et al. studied this problem in MoS_2 structures and proposed the electronic beam of energy less than 80kev is favourable for the stable analysis of all TMDs [40]. Aberration corrected transmission electron microscopy (AC-TEM) and scanning transmission electron microscopy (STEM) imaging enables to visualize atomic structure of metallic TMDs. Due to the Z-sensitive contrast of annular dark field STEM (ADF-STEM), it is possible to see distinguishingly the metal and chalcogenide sites in the crystal structure of metallic TMDs [41]. AFM used to understand the lateral size, thickness and surface roughness of metallic TMD layers. Even though AFM gives a low throughput it is the only characterization technique gives the definitive identification of monolayer crystals [1].

Figure 4.5. FESEM images for VSe_2/RGO at (a) low and (b) high magnification. (Reprinted by permission from Ref. 31, Copyright (2016) Royal Society of Chemistry).

The crystalline phase and lattice structure properties of metallic TMDs can be understood by X-ray diffraction spectrometer. XRD spectrum can be used to differentiate polytypic structure transition metal dichalcogenides by the change in 2 theta peak positions of different phases. Acerce et al. reported the broadness of (002) peak and the formation of new (001) peak from XRD identifies the restacked (1T) phase of MoS_2 [42]. Huang et al. reported the shifting of (002) peak toward lower angle indicates the enhancement in interlayer spacing and the group attribute this result as one of the confirmation points for the formation of metallic 1T-MoS_2 [43]. Liu et al. prepared 1T-WS_2 and observed the peak shift of (002) towards a lower 2 theta angle in the XRD spectrum compared to its 2H-WS_2 counterpart [44].

X-ray photoelectron spectroscopy (XPS) is another widely used characterization technique for the analysis of metallic TMDs. XPS provides information about the chemical composition of the material, the stoichiometric ratio of the components and the chemical state of elements present in the material. Leng et al. observed a chemical shift in binding energy to 228.1 eV

(Mo $3d^{5/2}$) and 231.4 eV (Mo $3d^{3/2}$) for the conversion of 2H-MoS_2 to metallic MoS_2 and he further stated that with the increase in Li intercalation an enhancement in peak intensities observed for the metallic MoS_2 [45].

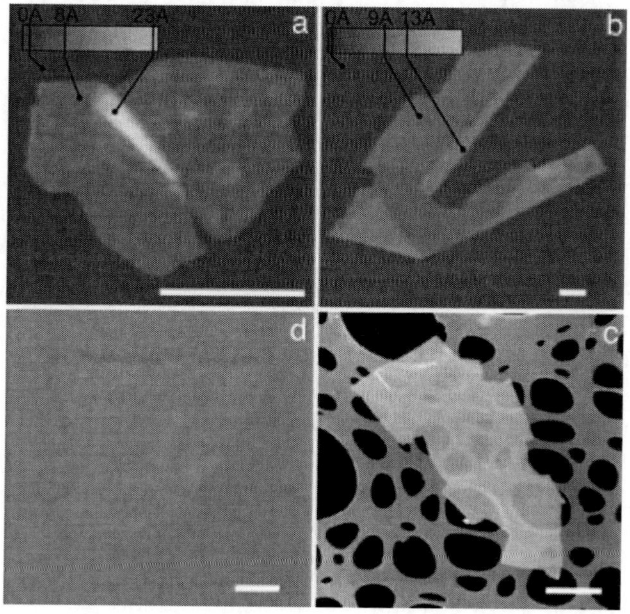

Figure 4.6. 2D crystal matter. Single-layer crystallites of $NbSe_2$ (a), graphite (b), $Bi_2Sr_2CaCu_2O_x$ (c), and MoS_2 (d) visualized by AFM (a and b), by scanning electron microscopy (c), and in an optical microscope (d). (All scale bars: 1μm.) The 2D crystallites are on top of an oxidized Si wafer (300 nm of thermal SiO_2) (a, b, and d) and on top of a holey carbon film (c). Note that 2D crystallites were often raised by an extra few angstroms above the supporting surface, probably because of a layer of absorbed water. In such cases, the pleated and folded regions seen on many AFM images and having the differential height matching the interlayer distance in the corresponding 3D crystals help to distinguish between double-layer crystals and true single sheets such as those shown here. (Reprinted by permission from Ref. 2. Copyright (2005) National Academy of Sciences, U.S.A.).

Sokolikova et al. reported the doublet of 4f W^{+4} peaks in 1T' WSe_2 (31.49 and 33.63 eV) is shifted from the positions of 2H-WSe_2 [46]. This shift of binding energies to the lower values might have caused due to the change in the metal coordination and a small enhancement in the W-Se bond lengths from 2.531 Å to 2.66 Å [44]. In a laser driven 2H – 1T' transformation of $MoTe_2$ observed by Tan et al. implies shifting of two peaks at 228.4 eV and 573.0 eV, corresponding to the Mo 3d and Te 3d (2H phase) to lower values of 227.7 eV and 572.5 eV [47]. Ma et al. reported binding energies of V-$2p_{3/2}$ at 512.84 eV and V-$2p_{1/2}$ at 520.45 eV in 1T-VTe_2 which are close to the values of 1T-VSe_2 and 1T-VS_2 [48]. Ji et al. observed that the XPS data of VS_2 exhibits a nonstoichiometric value of ~1.1:2 which indicates the presence of possible interstitial V atoms in the vdW gaps [14]. Zhao et al. found that the VSe_2 has two marked peaks at 517.0 ev and 524.0 eV respective to the V $2p_{3/2}$ and V $2p_{1/2}$, which indicate the +4 valence state and the peaks at 53.2 eV represents to Se [49].

The success of the characterization of graphene using Raman spectroscopy paved the way to study the structure of metallic TMDs also [50-51]. A monolayer MX_2 contains 3 atoms in its primitive unit cell with three of the nine translational acoustic vibration modes and the other six are optical modes [52]. For the metallic phase of TMDs, these optical modes at Γ point have $D^3{}_{3d}$ space group and semiconducting 2H phase has $D^1{}_{3h}$ space group [53]. Gupta et al. reported the distinction of Raman spectrum 1T- $MoSe_2$ from its 2H counterpart [54]. Geng et al. observed a strong Raman at 146 cm^{-1} of Mo–Mo stretching vibrations in metallic MoS_2 [55]. The Raman shifts at ~219, 283 and 326 cm^{-1} are associated with the phonon modes in metallic MoS_2. Meanwhile, the semiconducting 2H-MoS_2 exhibits typical Raman shifts of 378 and 404 cm^{-1} for E_1 2_g and A_{1g}, respectively, which are substantially different from those of metallic 1T-MoS_2 [53]. Jiang et al. discussed the formation of metallic $MoSe_2$ with the resonance peaks at 106, 149 and 221 cm^{-1} can be constitutes as the J1, J2 and J3 phonon modes of 1T-$MoSe_2$ [38].

Metallic TMDs can also be characterized using photoluminescence spectroscopy (PL). PL measurements provide band gap information with the change in quantum efficiency. The band gap transitions usually occur at the K point of the at the Brillouin zone which is the primary reason behind PL emission [56-57]. Eda et al. studied the PL properties of MoS_2 films and observed a direct band gap with the exciton relaxation at the K point points the PL emission with the energy of 1.9 eV [58]. Benoit et al. characterized the 1T and 2H phase of WS_2 using absorption spectra [59]. The 1T-WS_2 absorbs light almost in full spectral range and the group proposed the metallic behaviour of 1T-WS_2 with the observed monotonic decrease in absorbance [57]. Friedman et al. studied the phase transition of 1T-$MoSe_2$ from its 2H counterpart with the help of PL studies [60]. They observed two components neutral exciton and trion [58]. When the film immersed in tripropylamine (TPA) the intensity of both the PL components decreased and observed a general shift in the spectrum from neutral exciton to trion and they arrived a conclusion this behaviour is consistent with the 1T phase of $MoSe_2$ [58]. Zhou et al. observed enhanced PL activity in 2D amorphous VS_2 compared to the VS_2 crystals [61]. The broad photoluminescence peak observed at 491 nm corresponds to the 2.52 eV photon energy of 2D amorphous VS_2. The PL spectrum of $HfSe_2$ from the study done by Tsai et al. observed two peaks A at = 536 nm and B at =548 nm. With the help of DFT calculations, they observed two groups of direct bands located at E_{dir} = 2.21– 2.22 eV (558–561 nm) (A) and E_{dir} = 2.18– 2.20 eV (564–569 nm) (B) and these direct bands are the reason behind the two PL peaks [59].

Conclusion

The detailed literature analysis based on the synthesis of metallic TMDs provides there are numerous ways to grow the TMDs with metallic nature such as chemical exfoliation, chemical vapor transport, epitaxial growth, etc. Techniques like CVD and MBE can provide high quality metallic TMDs but these techniques need very sophisticated and costly instrumentation facilities. The bulk production of layered metallic TMDs is also difficulty these techniques facing. Other techniques such as hydrothermal/solvothermal suffers from the non-uniformity in the size of synthesized metallic TMDs. The thermodynamic stability of some of the metallic TMDs is another issue to be resolved. The proper characterizations to understand the properties of the metallic TMDs are also complicated and costly. One of the interesting aspects of the characterization of TMDs is the usage of Raman spectroscopy. Many of the researchers are using the Raman signals to analyses the metallic nature of the TMDs. The Raman signals of A_{1g} and E_{2g} bands of the metallic TMDs is a key aspect in the analysis of these materials. metallic phase of many of the TMDs yet to be characterize using techniques like STEM, and XPS which provides key information in the lattice fringe positions and the chemical state information of these TMDs.

References

[1] A. K. G. K. S. Novoselov, D. Jiang, F. Schedin, T. J. Booth, V. V. Khotkevich, S. V. Morozov, "Two-dimensional atomic crystals," *PNAS*, vol. 102, p. 10451, 2005, doi: 10.1063/1.4797258.

[2] X. Xi et al., "Ising pairing in superconducting NbSe2 atomic layers," *Nature Physics*, vol. 12, no. 2, pp. 139–143, 2016, doi: 10.1038/nphys3538.

[3] Z. Zeng et al., "An effective method for the fabrication of few-layer-thick inorganic nanosheets," *Angewandte Chemie - International Edition*, vol. 51, no. 36, pp. 9052–9056, 2012, doi: 10.1002/anie.201204208.

[4] J. Feng et al., "Metallic Few-Layered VS 2 Ultrathin Nanosheets : High Two-Dimensional Conductivity for In-Plane Supercapacitors," *Journal of American Chemical Society*, vol. 133, pp. 17832–17838, 2011.

[5] K. Xu et al., "Ultrathin nanosheets of vanadium diselenide: A metallic two-dimensional material with ferromagnetic charge-density-wave behavior," *Angewandte Chemie - International Edition*, vol. 52, no. 40, pp. 10477–10481, 2013, doi: 10.1002/anie.201304337.

[6] J. Feng et al., "Giant moisture responsiveness of VS 2 ultrathin nanosheets for novel touchless positioning interface," *Advanced Materials*, vol. 24, no. 15, pp. 1969–1974, 2012, doi: 10.1002/adma.201104681.

[7] Y. Wang, Z. Sofer, J. Luxa, and M. Pumera, "Lithium Exfoliated Vanadium Dichalcogenides (VS2, VSe2, VTe2) Exhibit Dramatically Different Properties from Their Bulk Counterparts," *Advanced Materials Interfaces*, vol. 3, no. 23, pp. 1–8, 2016, doi: 10.1002/admi.201600433.

[8] A. Ghobadi, T. G. U. Ghobadi, A. K. Okyay, and E. Ozbay, "Emerging photoluminescence from defective vanadium diselenide nanosheets," *Photonics Research*, vol. 6, no. 4, p. 244, 2018, doi: 10.1364/prj.6.000244.

[9] J. Peng et al., "Very Large-Sized Transition Metal Dichalcogenides Monolayers from Fast Exfoliation by Manual Shaking," *Journal of the American Chemical Society*, vol. 139, no. 26, pp. 9019–9025, 2017, doi: 10.1021/jacs.7b04332.

[10] J. Wu et al., "Acid-Assisted Exfoliation toward Metallic Sub-nanopore TaS2 Monolayer with High Volumetric Capacitance," *Journal of the American Chemical Society*, vol. 140, no. 1, pp. 493–498, 2018, doi: 10.1021/jacs.7b11915.

[11] J. Shi, Min Hong, Zhepeng Zhang, Qingqing Ji, Yanfeng Zhang, "Physical properties and potential applications of two-dimensional metallic transition metal dichalcogenides," *Coordination Chemistry Reviews*, vol. 376. Elsevier B.V., pp. 1–19, 2018. doi: 10.3390/cryst7100298.

[12] M. Hossain, Z. Zhao, W. Wen, X. Wang, J. Wu, and L. Xie, "Recent advances in two-dimensional materials with charge density waves: Synthesis, characterization and applications," *Crystals*, vol. 7, no. 10, pp. 1–19, 2017, doi: 10.3390/cryst7100 298.

[13] J. Wang et al., "Controlled Synthesis of Two-Dimensional 1T-TiSe2 with Charge Density Wave Transition by Chemical Vapor Transport," *Journal of the American Chemical Society*, vol. 138, no. 50, pp. 16216–16219, 2016, doi: 10.1021/jacs. 6b10414.

[14] Q. Ji et al., "Metallic vanadium disulfide nanosheets as a platform material for multifunctional electrode applications," *Nano Letters*, vol. 17, no. 8, pp. 4908–4916, 2017, doi: 10.1021/acs.nanolett.7b01914.

[15] J. Yuan et al., "Facile Synthesis of Single Crystal Vanadium Disulfide Nanosheets by Chemical Vapor Deposition for Efficient Hydrogen Evolution Reaction," *Advanced Materials*, vol. 27, no. 37, pp. 5605–5609, 2015, doi: 10.1002/adma. 201502075.

[16] Z. Zhang et al., "Van der Waals Epitaxial Growth of 2D Metallic Vanadium Diselenide Single Crystals and their Extra-High Electrical Conductivity," *Advanced Materials*, vol. 29, no. 37, pp. 1–9, 2017, doi: 10.1002/adma.201702359.

[17] J. Li et al., "Synthesis of Ultrathin Metallic MTe2 (M = V, Nb, Ta) Single-Crystalline Nanoplates," *Advanced Materials*, vol. 30, no. 36. p. 1801043, 2018. doi: 10.1002/adma.201801043.

[18] W. Ge, K. Kawahara, M. Tsuji, and H. Ago, "Large-scale synthesis of NbS2 nanosheets with controlled orientation on graphene by ambient pressure CVD," *Nanoscale*, vol. 5, no. 13, pp. 5773–5778, 2013, doi: 10.1039/c3nr00723e.

[19] H. Wang et al., "High-quality monolayer superconductor NbSe2 grown by chemical vapour deposition," *Nature Communications*, vol. 8, no. 1, pp. 1–8, 2017, doi: 10.1038/s41467-017-00427-5.

[20] H. Wang et al., "Large-Area Atomic Layers of the Charge-Density-Wave Conductor TiSe2," *Advanced Materials*, vol. 30, no. 8. 2018. doi: 10.1002/adma.201704382.

[21] B. A. Joyce, "Molecular beam epitaxy," in *Reports on Progress in Physics*, vol. 48, 1985, p. 1637. doi: 10.1016/C2010-0-68986-3.

[22] L. A. Walsh, R. Addou, R. M. Wallace, and C. L. Hinkle, "Molecular Beam Epitaxy of Transition Metal Dichalcogenides," in *Molecular Beam Epitaxy*, Elsevier Inc., 2018, pp. 515–531. doi: 10.1016/b978-0-12-812136-8.00024-4.

[23] S. Manzeli, D. Ovchinnikov, D. Pasquier, O. V. Yazyev, and A. Kis, "2D transition metal dichalcogenides," *Nature Reviews Materials*, vol. 2, no. 17033, pp. 1–15, 2017, doi: 10.1038/natrevmats.2017.33.

[24] T. H. Choudhury, X. Zhang, Z. Y. Al Balushi, M. Chubarov, and J. M. Redwing, *Epitaxial Growth of 2D Layered Transition Metal Dichalcogenides*. 2019. doi: 10.1146/annurev-matsci-090519-113456.

[25] S. Vishwanath et al., "Comprehensive structural and optical characterization of MBE grown MoSe2 on graphite, CaF2 and graphene," *2D Materials*, vol. 2, no. 2, 2015, doi: 10.1088/2053-1583/2/2/024007.

[26] H. Lin et al., "Growth of atomically thick transition metal sulfide filmson graphene/6H-SiC(0001) by molecular beam epitaxy," *Nano Research*, vol. 11, no. 9, pp. 4722–4727, 2018, doi: 10.1007/s12274-018-2054-4.

[27] Y. Yu et al., "Molecular beam epitaxy growth of atomically ultrathin MoTe2 lateral heterophase homojunctions on graphene substrates," *Carbon*, vol. 115, pp. 526–531, 2017, doi: 10.1016/j.carbon.2017.01.026.

[28] S. El Kazzi et al., "MoS2 synthesis by gas source MBE for transition metal dichalcogenides integration on large scale substrates," *Journal of Applied Physics*, vol. 123, no. 13, pp. 1–6, 2018, doi: 10.1063/1.5008933.

[29] D. B. Ruoyu Yue, Yifan Nie, Lee A Walsh, Rafik Addou, Chaoping Liang, Ning Lu, Adam T Barton, Hui Zhu, Zifan Che, "Nucleation and growth of WSe2: enabling large grain transition metal dichalcogenides," *2D materials*, vol. 4, no. 4, p. 045019, 2017, doi: 10.1016/j.neuropsychologia.2007.03.010.

[30] M. M. Ugeda et al., "Characterization of collective ground states in single-layer NbSe 2," *Nature Physics*, vol. 12, no. 1, pp. 92–97, 2016, doi: 10.1038/nphys3527.

[31] R. Yue et al., "HfSe2 thin films: 2D transition metal dichalcogenides grown by molecular beam epitaxy," *ACS Nano*, vol. 9, no. 1, pp. 474–480, 2015, doi: 10.1021/nn5056496.

[32] M. K. and T. M. Joshua Hall, Borna Pielić, Clifford Murray, Wouter Jolie, Tobias Wekking, Carsten Busse, "Molecular beam epitaxy of quasi-freestanding transition metal disulphide monolayers on van der Waals substrates: a growth study," *2D materials*, vol. 5, no. 2, 2018.

[33] D. O'Hare, *Hydrothermal Synthesis*. 2001. doi: 10.1016/b0-08-043152-6/00701-4.

[34] S. H. Feng and G. H. Li, *Hydrothermal and Solvothermal Syntheses*. 2017. doi: 10.1016/B978-0-444-63591-4.00004-5.

[35] S. R. Marri, S. Ratha, C. S. Rout, and J. N. Behera, "3D cuboidal vanadium diselenide embedded reduced graphene oxide hybrid structures with enhanced supercapacitor properties," *Chemical Communications*, vol. 53, no. 1, pp. 228–231, 2017, doi: 10.1039/c6cc08035a.

[36] C. S. Rout et al., "Metallic few-layer flowerlike VS2 nanosheets as field emitters," *European Journal of Inorganic Chemistry*, vol. 2014, no. 31, pp. 5331–5336, 2014, doi: 10.1002/ejic.201402448.

[37] L. Cai et al., "Vacancy-induced ferromagnetism of MoS2 nanosheets," *Journal of the American Chemical Society*, vol. 137, no. 7, pp. 2622–2627, 2015, doi: 10.1021/ja5120908.

[38] M. Jiang et al., "Synthesis of 1T-MoSe2 ultrathin nanosheets with an expanded interlayer spacing of 1.17 nm for efficient hydrogen evolution reaction," *Journal of Materials Chemistry A*, vol. 4, no. 39, pp. 14949–14953, 2016, doi: 10.1039/c6ta07020e.

[39] M. Piao et al., "Hydrothermal synthesis of stable metallic 1T phase WS2 nanosheets for thermoelectric application," *Nanotechnology*, vol. 29, no. 2, 2018, doi: 10.1088/1361-6528/aa9bfe.

[40] A. Garcia et al., "Analysis of electron beam damage of exfoliated MoS2 sheets and quantitative HAADF-STEM imaging," *Ultramicroscopy*, vol. 146, pp. 33–38, 2014, doi: 10.1016/j.ultramic.2014.05.004.

[41] S. Wang, A. Robertson, and J. H. Warner, "Atomic structure of defects and dopants in 2D layered transition metal dichalcogenides," *Chemical Society Reviews*, vol. 47, no. 17, pp. 6764–6794, 2018, doi: 10.1039/c8cs00236c.

[42] M. Acerce, D. Voiry, and M. Chhowalla, "Metallic 1T phase MoS2 nanosheets as supercapacitor electrode materials," *Nature Nanotechnology*, vol. 10, no. 4, pp. 313–318, 2015, doi: 10.1038/nnano.2015.40.

[43] L. Huang et al., "Methyl-functionalized MoS2 nanosheets with reduced lattice breathing for enhanced pseudocapacitive sodium storage," *Physical Chemistry Chemical Physics*, vol. 19, no. 21, pp. 13696–13702, 2017, doi: 10.1039/c7cp00990a.

[44] Q. Liu et al., "Stable Metallic 1T-WS2 Nanoribbons Intercalated with Ammonia Ions: The Correlation between Structure and Electrical/Optical Properties," *Advanced Materials*, vol. 27, no. 33, pp. 4837–4844, 2015, doi: 10.1002/adma.201502134.

[45] K. Leng et al., "Phase Restructuring in Transition Metal Dichalcogenides for Highly Stable Energy Storage," *ACS Nano*, vol. 10, no. 10, pp. 9208–9215, 2016, doi: 10.1021/acsnano.6b05746.

[46] M. S. Sokolikova, P. C. Sherrell, P. Palczynski, V. L. Bemmer, and C. Mattevi, "Direct solution-phase synthesis of 1T' WSe 2 nanosheets," *Nature Communications*, vol. 10, no. 1, 2019, doi: 10.1038/s41467-019-08594-3.

[47] Y. Tan et al., "Controllable 2H-to-1T′ phase transition in few-layer MoTe2," *Nanoscale*, vol. 10, no. 42, pp. 19964–19971, 2018, doi: 10.1039/c8nr06115g.

[48] X. Ma et al., "Charge Density Wave Phase Transitions in Large-Scale Few-Layer 1T-VTe 2 Grown by Molecular Beam Epitaxy," *ACS Applied Materials and Interfaces*, vol. 11, no. 11, pp. 10729–10735, 2019, doi: 10.1021/acsami.8b21442.

[49] W. Zhao et al., "Colloidal synthesis of VSe2 single-layer nanosheets as novel electrocatalysts for the hydrogen evolution reaction," *Chemical Communications*, vol. 52, no. 59, pp. 9228–9231, 2016, doi: 10.1039/c6cc03854a.

[50] R. Saito, M. Hofmann, G. Dresselhaus, A. Jorio, and M. S. Dresselhaus, "Raman spectroscopy of graphene and carbon nanotubes," *Advances in Physics*, vol. 60, no. 3, pp. 413–550, 2011, doi: 10.1080/00018732.2011.582251.

[51] A. C. Ferrari and D. M. Basko, "Raman spectroscopy as a versatile tool for studying the properties of graphene," *Nature Nanotechnology*, vol. 8, no. 4, pp. 235–246, 2013, doi: 10.1038/nnano.2013.46.

[52] M. A. Pimenta, E. Del Corro, B. R. Carvalho, C. Fantini, and L. M. Malard, "Comparative study of raman spectroscopy in graphene and MoS2-type transition metal dichalcogenides," *Accounts of Chemical Research*, vol. 48, no. 1, pp. 41–47, 2015, doi: 10.1021/ar500280m.

[53] S. Jiménez Sandoval, D. Yang, R. F. Frindt, and J. C. Irwin, "Raman study and lattice dynamics of single molecular layers of MoS2," *Physical Review B*, vol. 44, no. 8, pp. 3955–3962, 1991, doi: 10.1103/PhysRevB.44.3955.

[54] U. Gupta et al., "Characterization of few-layer 1T-MoSe2 and its superior performance in the visible-light induced hydrogen evolution reaction," *APL Materials*, vol. 2, no. 9, 2014, doi: 10.1063/1.4892976.

[55] X. Geng et al., "Pure and stable metallic phase molybdenum disulfide nanosheets for hydrogen evolution reaction," *Nature Communications*, vol. 7, pp. 1–7, 2016, doi: 10.1038/ncomms10672.

[56] H. Zeng et al., "Optical signature of symmetry variations and spin-valley coupling in atomically thin tungsten dichalcogenides," *Scientific Reports*, vol. 3, pp. 2–6, 2013, doi: 10.1038/srep01608.

[57] K. P. Dhakal et al., "Confocal absorption spectral imaging of MoS2: Optical transitions depending on the atomic thickness of intrinsic and chemically doped MoS2," *Nanoscale*, vol. 6, no. 21, pp. 13028–13035, 2014, doi: 10.1039/c4nr 03703k.

[58] G. Eda, H. Yamaguchi, D. Voiry, T. Fujita, M. Chen, and M. Chhowalla, "Photoluminescence from chemically exfoliated MoS 2," *Nano Letters*, vol. 11, no. 12, pp. 5111–5116, 2011, doi: 10.1021/nl201874w.

[59] B. Mahler, V. Hoepfner, K. Liao, and G. A. Ozin, "Colloidal synthesis of 1T-WS2 and 2H-WS2 nanosheets: Applications for photocatalytic hydrogen evolution,"

Journal of the American Chemical Society, vol. 136, no. 40, pp. 14121–14127, 2014, doi: 10.1021/ja506261t.

[60] A. L. Friedman, A. T. Hanbicki, F. K. Perkins, G. G. Jernigan, J. C. Culbertson, and P. M. Campbell, "Evidence for Chemical Vapor Induced 2H to 1T Phase Transition in MoX2 (X = Se, S) Transition Metal Dichalcogenide Films," *Scientific Reports*, vol. 7, no. 1, pp. 1–9, 2017, doi: 10.1038/s41598-017-04224-4.

[61] Y. Zhou, Q. Xu, T. Ge, X. Zheng, L. Zhang, and P. Yan, "Accurate Control of VS2 Nanosheets for Coexisting High Photoluminescence and Photothermal Conversion Efficiency," *Angewandte Chemie - International Edition*, vol. 59, no. 8, pp. 3322–3328, 2020, doi: 10.1002/anie.201912756.

Chapter 5

Chemical, Gas and Bio Sensors Based on Metallic 2D TMDs

Aaryashree[1], Chandra Sekhar Rout[2,*] and Dattatray J. Late[3]

[1]Shibaura Institute of Technology, Toyosu, Koto City, Tokyo, Japan
[2]Centre for Nano and Material Sciences, Jain University, Jain Global Campus, Ramanagaram, Bangalore, India
[3]Centre for Nanoscience and Nanotechnology Amity University Maharashtra, Mumbai-Pune Expressway, Bhatan, Post, Somathne, Panvel, Mumbai, Maharashtra, India

Abstract

Recently, two-dimensional metallic materials such as VS_2, VSe_2, NbS_2, HfS_2, ReS_2, and MXenes have evoked great interest among the research community for their diverse electronics applications catalysts, energy storage/conversion devices, and bio-medical. These materials have also been employed as a sensing layer in various sensors, such as gas and vapor sensors, humidity sensors, biosensors, and electrochemical sensors. Hence, insight is required on the recent developments in these sensors. In this context, the present chapter focuses on the recent growth of various sensors based on 2D metallic materials, such as chemical sensors, gas sensors, and biosensors.

Keywords: 2D materials, gas sensors, biosensor

---

* Corresponding Author's Email: r.chandrasekhar@jainuniversity.ac.in.

In: 2D Metallic Transition Metal Dichalcogenides
Editors: Chandra Sekhar Rout and Brahmananda Chakraborty
ISBN: 978-1-68507-965-9

5.1. Introduction

Recently significant gain in research interests in the area of two-dimensional (2D) nanostructured materials has been focused on utilizing the unusual properties intrinsic to 2D materials for use in numerous technological fields of applications, including energy storage devices, nanoelectronics, etc. chemical sensing are the few among them [1, 2]. A 2D material can be a one or few atomically thin layered nano-crystalline materials, which morphologically/structurally exist as nanosheets, nano-membranes, thin-film and nanoflakes [3]. The 2D nanostructure materials can be divided into either allotropic elements or covalently bonded compounds [4]. The atomically thin 2D nanostructured materials possess unique chemical and physical properties, which show the great potential for gas sensing applications. Notably, the single layer to few-layer atomic-scale thickness of a 2D material possesses the unusual surface-to-volume ratio, which has more active, reactive sites on the surfaces to attach the gases. The van der Waals gaps between the layers and layer-dependent physical, chemical, mechanical and optical properties are completely different from those of bulk materials, making atomically thin 2D nanomaterials ideal and more suitable candidates for the gas sensors [5, 6].

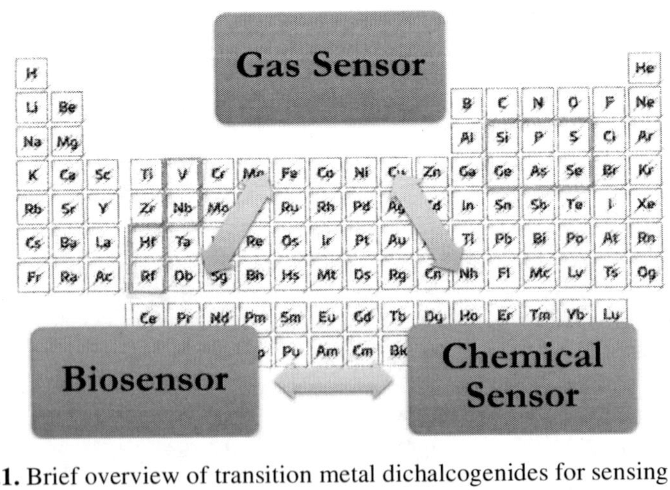

Figure 5.1. Brief overview of transition metal dichalcogenides for sensing applications.

Further, it has been identified that the capability of atomically thin 2D nanomaterials to detect target gas at room temperature offers the possibility of fabricating the wearable gas sensors on a flexible substrate [7]. Among all the 2D materials widely studied materials for sensing applications are graphene,

TMDs, phosphorene, MXenes, etc. These materials have been widely studied due the fact that these materials possess extra ordinary properties that are helpful for sensing applications. On the other hand, the rapidly developing research field of metallic TMD research immobile needs to resolve some of the issues such as selectivity, gas response, response time, recovery time, and longtime stability. The response time of 2D TMDs sensors is quite slow, and recovery is habitually incomplete. Over time, it has been learn that the surface of layered metallic 2D TMDs can also be some extent covered by the oxygen or water molecules at an ambient condition due to various other kinetics of adsorption and desorption, which leads the slow debase in the sensing performance and long-term cyclic stability of the sensor device [8]. The unique gas sensing performance of few chosen metallic 2D TMDs have been presented in this chapter. Finally, the challenges and future directions to develop the sensors based on this unique structure have been addressed. The current book chapter will provide great insight into the evolution of metallic TMD materials and their unravel sensing properties.

5.1.1. Two-Dimensional Transition Metal Dichalcogenides as the Sensing Material

Two-dimensional Transition Metal Dichalcogenides (2D TMDs) are an evolving class of materials that offer a high degree of interest to fundamental investigations of novel physical phenomena and for applications ranging from micro- and nano-photonics to nanoscales. TMDs are a class of materials with properties that make them extremely appealing. In recent years, there have been significant resurgences in 2D TMDs, driven by the first transistor demonstration and by the discovery of powerful photoluminescence in MoS_2 monolayers. TMDs have the unique combination of atomic-scale thicken, direct bandgap, good spin-orbit relation, and favorable electronic and mechanical characteristics that allow fundamental and high-end electronics, spintronic, optoelectronics, energy-capacity array, flexible electronics, DNA sequencing, and personalized medicine applications. The 2D nature in combination with the appealing electronic and optical properties, which can also be tailored by chemical functionalization, makes such materials extremely appealing in the field of sensing, ranging from the quantitative detection of gases to numerous biologically relevant molecules. This chapter describes the use of TMDs for nano-electronic devices with a particular emphasis on chemical sensing, gas sensing, and bio-sensing.

5.2. Gas Sensing Applications

TMD structures have drawn renewed interest because of their superior molecular sensing capabilities, unusual physical and chemical properties, including semiconducting properties and the high surface-to-volume ratio [9–14]. The variations between the typical surface structure of the layers and the thickness-dependent physical and chemical properties are distinct attributes that make 2D nanomaterials ideal for gas sensors [15–17]. The intrinsic 2D materials such as grapheme [18–20], and other carbon-based materials are chemically inert, and often require surface modifications (or functionalization) for gas sensing applications [21–25]. 2D TMDs, on the other hand, display flexible interactions and tunable bandgaps, which are much more beneficial in designing practical gas sensing devices. Layer-dependent properties with surface modification using chemical doping and ligand conjugation have been reported to enhance the sensing efficiency of the TMDs [26–28]. However, challenges such as selectivity, gas response, response time, recovery time, and stability in TMDs based gas sensors are still to be addressed. Similar to metal oxide sensors, the 2D TMD based sensors are yet to have high selectivity and suffer from sluggish response time [16, 28]. Many a time the recovery of 2D TMDs based gas sensors have incomplete recovery. The surface of layered 2D TMDs often gets oxidized in room conditions due to various adsorption and desorption kinetics, gradually degrading the sensing performance along with the long-term stability [29, 30]. Therefore, further research is being conducted in this area to improvise the sensing properties of 2D TDMs by surface modification, hybridization, etc.

5.2.1. The Sensing Mechanism

The primary gas sensing principle in 2D TMDs is mainly based on the charge transfer between gas molecules and the surface of sensing materials [28]. The change in resistance or conductance of the system occurs due to the charge transfer in 2D TMDs, which either act as a charge acceptor or as a charge donor. When exposed to an analyte gas, gas molecules are absorbed on the surface of 2D TMDs by electrostatic force. The analyte gas (reducing or oxidizing) determines the direction of the electronic charge transfer. The resistance of the sensing material is recovered to its initial value due to the desorption of gas molecules upon exposure to air. The resistance variation is determined by the charge affinity of reactive gas to release or withdraw electrons. For example, in an *n*-type 2D TMD, the resistance of the sensing material typically increases under the exposure of an oxidizing gas.

5.2.2. Gas Sensing Properties of 2D TMDs

The two-dimensional transition metal dichalcogenides have strong molecular intra-layer bonds and weak van der Waals inter-layer bonds [31]. The weak interlayer bonding facilitates the tailoring of TMDs down to not only multi-layers but also to single layers through simple processes mechanical or chemical intercalation [32–34]. Typically, the thickness of the layered TMDs and the length of the M-M bond is approximately 6-7 Å and 3.15-4.03 Å, respectively, depending on the atomic size of the metal and chalcogen [35]. The exceptional functionalities powered by the high specific surface area-based structure of 2D TMDs make them promising candidates for sensor applications. The coordination number of the transition metal and its d-electrons decides the electronic and magnetic characteristics of TMDs. In TMDs, the transition metal (M) atoms have half-filled electronic orbital, and the chalcogen (X) atoms naturally exist with an oxidation state of -2. Thus, they have various oxidation states such as +2, +4, and +6, which often manipulate electronic properties. As a result of these electronic variations, the TMDs exist as a) semiconductors like VS_2, VSe_2, b) metals like $NbTe_2$, MXenes, and c) superconductors like NbS_2, and $NbSe_2$ [36, 37].

The chalcogen atoms with lone-pair electrons act as the terminators at the surface of the layered TMDs. The absence of dangling bonds on the surface of TMDs makes it stable in various ambient conditions and also provides high mobility between contacts (metals and substrates). For instance, the mobility of vanadium diselenide monolayer (VSe_2) is about 307 cm^2/Vs and 96 cm^2/Vs in the "zigzag" and "armchair" directions at 300 K, which increases to 1611 cm^2/Vs and 5085 cm^2/Vs under 3% and 5% compressive uniaxial strain, respectively [38]. Yet another TMD with high mobility is Hafnium disulfide (HfS_2), having a mobility of 1800 cm^2/Vs [39]. Another critical characteristic of gas sensing is the tunable bandgap. TMDs show both direct and indirect band gaps depending upon layer or bulk.

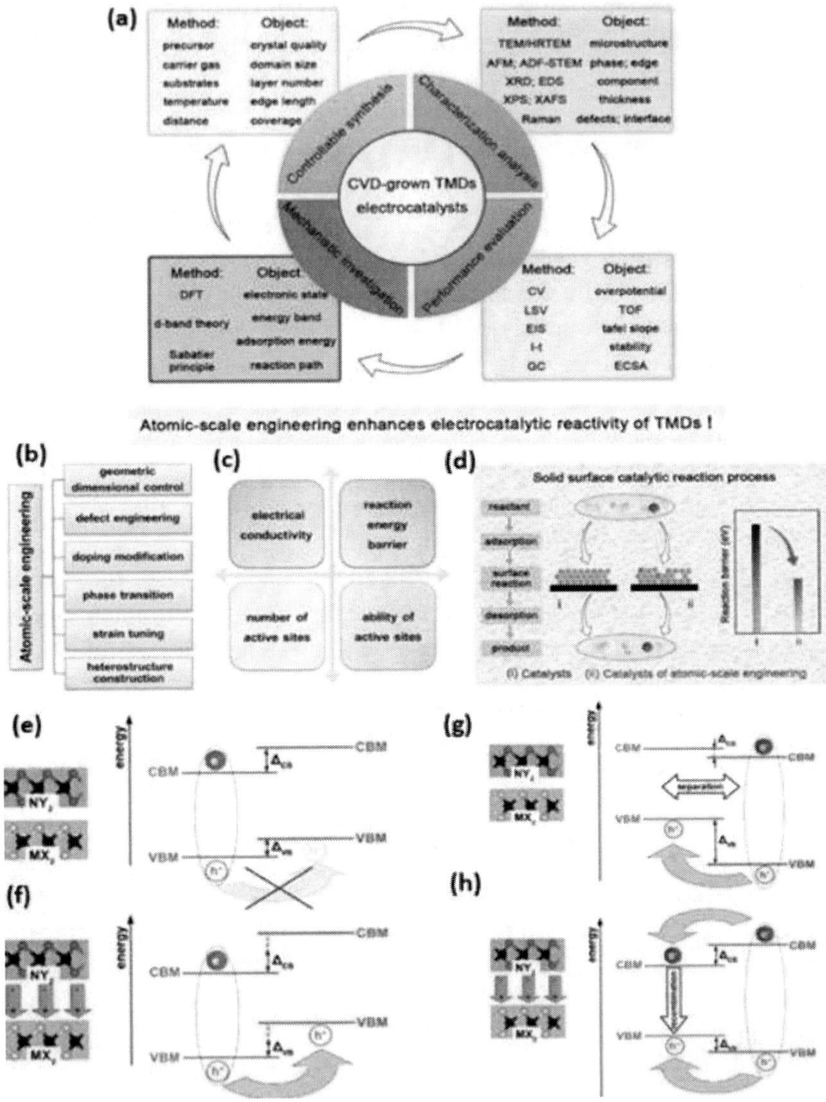

Figure 5.2. a) CVD grown 2D TMDs, b)Strategies of atomic-scale engineering, c) Factors influencing electrocatalysis, d) Catalytic reaction process, ((a)-(d) reproduced with permission Wang et al. © RSC Publishing), e) van der Waals heterostructure of two TMDs showing symmetric type II band alignment, f) Valance band offset change in type II band alignment by applying electrostatic design, g) van der Waals heterostructure with type II band alignment with distinct band offsets, h) change in valance band offset of (g) by applying electrostatic design. (Reproduced from Winkler et al. *2D Materials* 5.3 (2018): 035019, © 2018 IOP Publishing Ltd.).

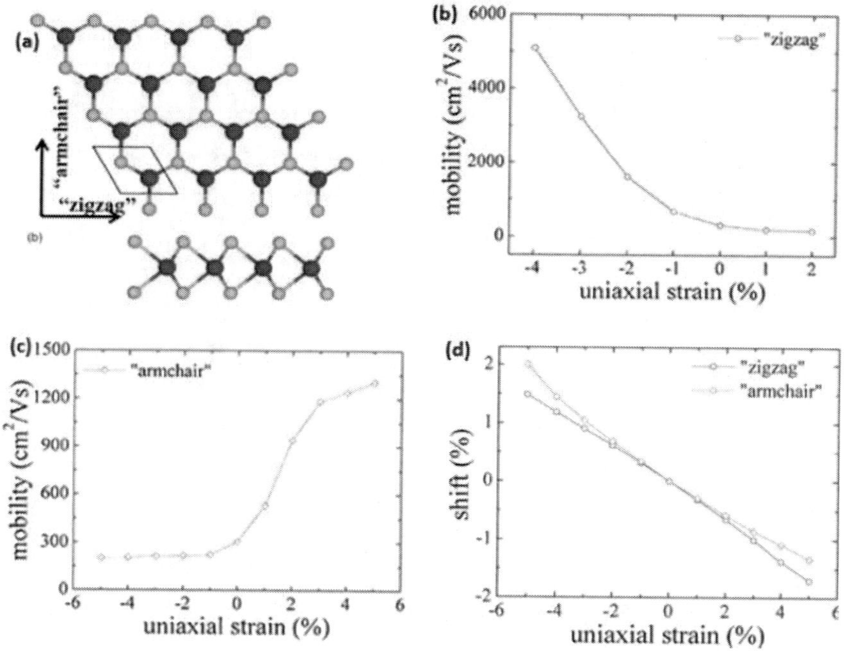

Figure 5.3. a) The lattice structure of monolayer 2H-$VSe_2$2H-VSe_2, top and side view, where red and green balls represent the V atom and Se atom, respectively, b) and c) The in-plane uniaxial strain is applied along the "zigzag" and "armchair" directions, respectively, d) The change as a function of uniaxial strain in the directions "zigzag" and "armchair" in the transverse uniaxial strain would contribute to the vertical change. (Rights and permissions by AIP Publishing through license number 4843380274421).

For example, vanadium disulfide is an intrinsic semiconductor with a direct bandgap of 0.3 eV [40], MXenes have been shown to have a bandgap ranging from 0.24 eV to 0.55 eV [41]. Most of the TMDs have low band gaps around 0.3-2 eV. The quantum-mechanical confinement occurring when moving from bulk to monolayer can be the reason for the transition of an indirect bandgap to a direct bandgap in TMDs, proving them with unusual electrical and optical properties. An additional advantage of TMDs to be considered for their applications in gas sensors is their mechanical flexibility and strength. In the TMD class, multiple reports are available on disulfides of molybdenum, tungsten, and tin (MoS_2, WS_2, and SnS_2) based TMDs for gas sensing. However, there have been recent increasing reports on the gas sensors based on MXenes and few reports on Rhenium disulfide (ReS_2) and niobium disulfide (NbS_2).

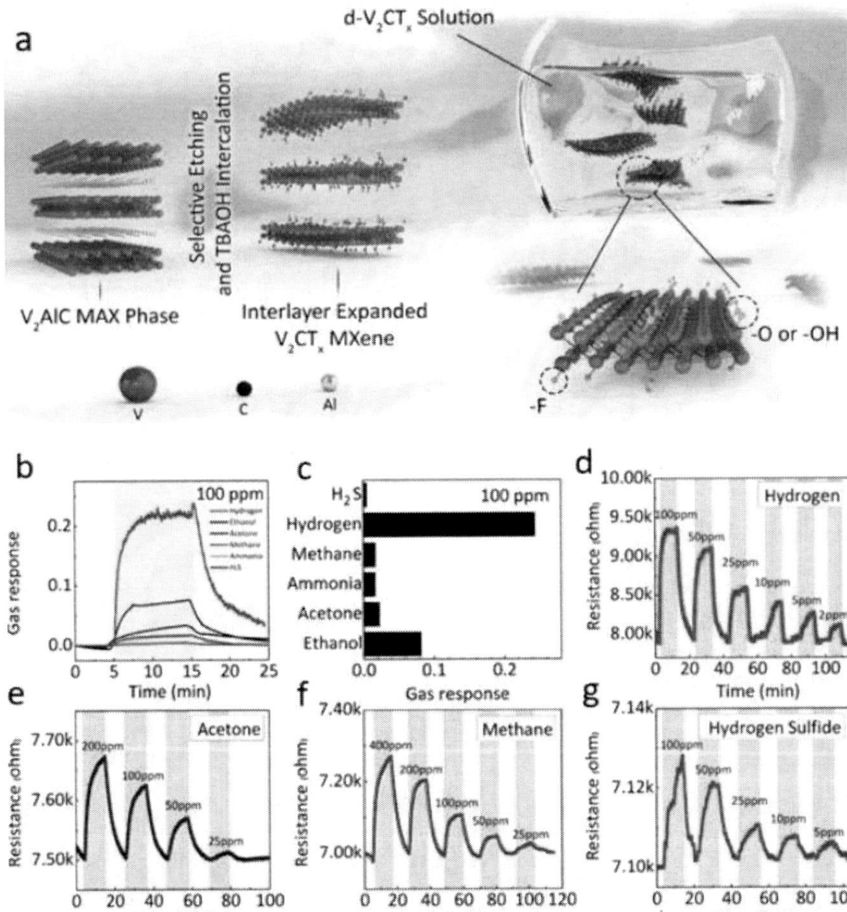

Figure 5.4. a) Schematic of the synthesis of V2CTx MXene. b) – g) Gas sensing properties of a V2CTx sensor at RT (23°C) towards various gases in real-time. (Reprinted (adapted) with permission from *ACS Sensors* 2019 4 (6), 1603-1611. Copyright 2019 American Chemical Society).

As discussed earlier, the outer layer of TMDs is generally functionalized. In the case of MXenes, these layers are either chemically terminated or functionalized with functional groups containing O, F, or OH [42]. This surface functionalization contributes substantially to the electronic and ion transport properties of MXenes directly associated with their surface conductivity and complex charge transfer processes. Therefore, a variable electronic activity can be seen in MXenes, depending upon their functionalization [42, 43]. Despite a short research span, MXenes have been shown to have high sensitivity and selectivity detection in sensing applications.

The accelerating environmental pollution is a global concern now, and its monitoring has become a focus for present-day researchers. Many new efforts and user-friendly techniques have been reported for the detection of hazardous gases that deteriorate the environmental conditions and are also lethal for human health [44, 45]. Gas sensors are among the key technologies for rapid, selective, sensitive, and efficient detection of gases [46–49]. The ultra-trace level (ppb) detection of certain gases is extremely critical for pollution analysis, detection of toxic gases, and in fields of biomedical diagnoses. For example, the detection of ammonia (NH_3) is highly recommended for the diagnosis of peptic ulcers caused by H. pylori and also to detect lung disorders. Likewise, the early symptoms of diabetes can be determined by acetone detection in the breath of the patient. MXenes are being exploited to detect various environmentally hazardous gases. A recent report suggested a theoretical study on the absorption of carbon monoxide (CO), carbon dioxide (CO_2), ammonia (NH_3), nitrous oxide (NO), and nitrogen dioxide (NO_2) gas molecules on a monolayer phosphorene, an MXene [50]. The report claimed that phosphorene sensing performance was better when compared to other 2D materials. Phosphorene showed high selectivity towards gases like NO and NO_2. However, the calculations, majorly done by the first-principle, restricted their findings. Later, a FET-based gas sensor was demonstrated, which had high performance and selectivity towards NO_2 detection at the p.p.b level in dry air [51]. The gas sensitivity of the sensor was found to depend on the thickness of the nanosheets, and a 4.8 nm thick device showed a sensitivity up to 190% at 20 p.p.b. of NO_2 at room temperature. The FET gas sensor also exhibited high selectivity towards NO_2 in the presence of other gases such as hydrogen gas (H_2), CO, and hydrogen sulfide (H_2S), which substantiates the theoretical results published on the selectivity of the gas sensing of phosphorene towards NO and NO_2 gases. In general, the higher sensitivity of the gas sensor towards certain gas molecules can be attributed to the changes in the magnitudes of the transport properties when a specific gas molecule is attached to it. Another report claimed a similar NO_2 gas sensing using FETs [52]. These sensors showed sensitivity towards NO_2 at a concentration as low as 5 ppb. In the NO_2 ambiance, the hole conductivity increases because the NO_2 molecule withdraws electrons from the phosphorene. The phosphorene FET showed a systematic increase in conductance with varying concentrations, which fitted well with the Langmuir isotherm for molecules adsorbed on a surface, confirming charge transfer as the dominant mechanism for NO_2 sensing.

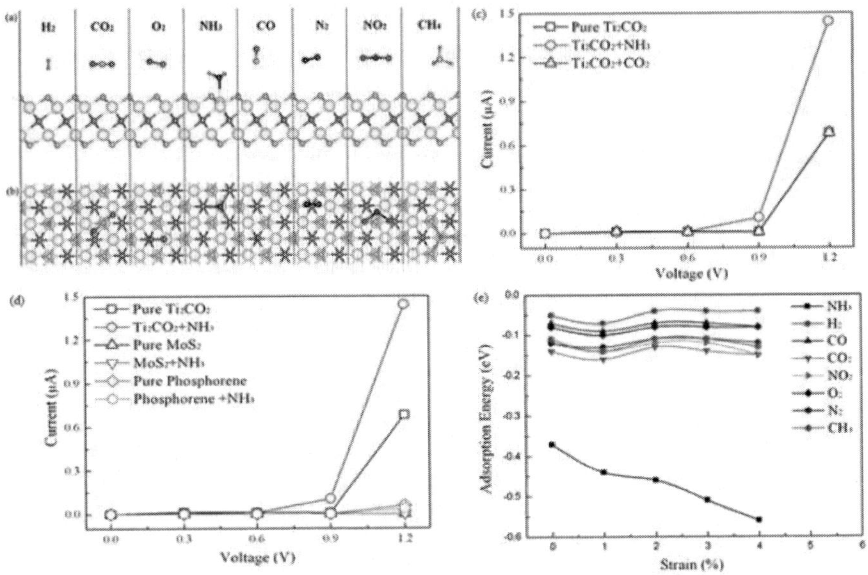

Figure 5.5. A schematic illustration of (a) side, and (b) top view of the adsorption of various molecules on monolayer Ti_2CO_2. c) I−V characteristics of adsorption of NH_3 or CO_2on Ti_2CO_2, d) Theoretically calculated I−V curve before and after the adsorption of NH_3 on Ti_2CO_2, MoS_2, and phosphorene, e) Adsorption energies of NH_3, H_2, CH_4, CO, CO_2, N_2, NO_2, or O_2 on Ti_2CO_2 as a function of applied biaxial strains (from 0 to 4%). (Reprinted (adapted) with permission from *ACS Appl. Mater. Interfaces* 2015, 7, 24, 13707–13713. Copyright 2015 American Chemical Society).

In a typical study, the phosphorene surface was functionalized Pt nanoparticles to improve its hydrogen sensing efficiency [52]. The functionalized MXene sensor exhibited a significant electrical response to H_2 gas with the drain-source current (IDS) decreasing from 2.46 to 1.17 μA at the drain-source voltage (VDS) of +10 mV. Further, the μeff and ION/IOFF ratio was also found to be increased by 15.6% and 115%, respectively, when exposing the sensor to low concentration, i.e., 4% of H_2 gas. Yet another study showed the improved H_2 gas sensing by similar Pt nanoparticle injections into phosphorene [53]. The Pt/phosphorene sensor exhibited high sensitivity and selective response to H_2 gas of 1% concentration with a channel resistance variation (R/Rb) of ≈500%. Although these types of research have achieved huge success in the H_2 gas sensor, the mechanism of phosphorene obtaining

H_2 gas sensing after Pt modification is yet to be explained. Apart from such surface modifications, constructing different heterostructures on phosphorene base is also a possible way in which differential change of charge transfer and ionic conduction targeting specific molecules can be achieved. Further, a report was published on the NH_3 sensing using B-, Si-, and Ni- doped phosphorene using DFT method [54]. They reported a theoretical increase in the adsorption of NH_3 when phosphorene was doped with a heteroatom. An inference was deduced that doping significantly influences the bandgap and the overall shape of density of states of the systems. The gas sensing properties and performances of the chemically exfoliated phosphorene at room temperature (RT) have also been studied [55]. The study claims that phosphorene flakes show good response both at RT and in a dry air environment, with a sensitivity as low as 20 ppb for NO_2 and 10 ppm for NH_3, with extrapolated theoretical detection limits of 7 ppb and 1 ppm, respectively. A tentative explanation of the gas sensing mechanism is also proposed, based on the binding energies of the adsorbed gas molecules and the morphology of the phosphorene exfoliated flakes. However, phosphorene flakes did not show any response to CO and CO_2, which confirmed the previously theoretically predicted gas sensing properties of exfoliated phosphorene flakes and their capability to detect target gases at RT.

MXenes have been considered gas sensing materials with high sensitivity, high substrate flexibility, and low signal noise. They have also been considered suitable for portability. In a report, the gas sensing performance of $Ti_3C_2OH_2$ MXene was studied over gases like acetone (CH_3COCH_3), ethanol (C_2H_5OH), propanol (C_2H_5CHO), sulfur dioxide (SO_2) along with NH_3 and NO_2 at room temperature. The $Ti_3C_2OH_2$ MXene sensor showed a positive resistance change on absorbing the analyte gases. Here, the sensing response varied depending upon the electron-donating or accepting properties of gases and the charge carrier (p or n-type) of the sensing channel [56, 57]. In a typical study, Ti_3C_2 MXene-based sensors were fabricated from the single-layer Ti_3C_2 MXene suspension. Both the theoretical and the experimental studies validated the high-performing NH_3 sensing capability of Ti_3C_2 MXene at room temperature. The sensor's selectivity was determined by collective resistive responses for methane (CH_4), H_2S, H_2O, ethanol, methanol, acetone, NH_3, and NO [58]. Among these gases, selectivity for NH_3 was dominant. The response to NH_3 was 6.13%, almost four times higher than the response to ethanol, which was only 1.5%. The selectivity for NH3 is due to the cumulative effects of adsorption energy, adsorbed geometry, and charge transfer. In another study, single-/few-layer vanadium carbide (V_2C) MXene flakes-based gas sensors were able to detect trace amounts of various gases, including H_2 and CH_4, with a theoretical limit of detection as low as 1 ppm for H_2 and 9 ppm for CH_4 at RT [59]. The excellent gas-sensing performance of the 2D V_2C MXene was is attributed to the surface oxygen functional groups present on the surface of V_2C MXene nanoflakes.

Other MXenes are made from Group IVA, germanene, silicene, and Group VA phosphorene, arsenene, and antimonene, which are the analogs of graphene; however, they lack atomic aligned as in the grapheme [60–62]. In a study, the first-principle calculation was implemented to determine the adsorption of CO, CO_2, N_2, NH_3, NO, and NO_2 molecules on pristine arsenene [63]. The arsenene and analyte gas interaction were in between the physisorbed and chemisorbed states. This theoretical study predicts that arsenene has excellent potential as a gas sensing material [64]. Antimonene has also been claimed to have gas-adsorption behaviors [65]. The authors discovered that NH_3, SO_2, NO, and NO_2 showed a stronger affinity toward antimonene with a significant amount of charge transfers and high absorption energies [66]. The monolayered antimonene seems a promising candidate for gas sensing material as its electronic properties are prone to the activities of absorbed molecules.

Further, it was also predicted to apply for acetone sensors due to the sensitivity and high selectivity observed [67]. Like antimonene sensing behavior, silicene has also shown good sensing capabilities with high binding energy and excessive charge transfer capacity [68]. A study showed the highly selective NO, NO_2, NH_3, and CO sensing of silicene using DFT and the non-equilibrium Green's function method [69]. The physisorption of these toxic gases has also been seen in germanene, and a highly stable and sensitive gas sensor has been theoretically established using the DFT method [70].

Recently, ReS_2 has started gaining interest as a sensing material due to its strong interaction with non-metals such as H, N, P, O, S, F, etc. [71]. In a study, a transistor built from scotch tape exfoliated ReS_2, enhanced gas sensing capabilities have been attributed to the sulfur vacancies and interlayer interactions [72]. Additionally, ReS_2 has also been reported as an O_2 sensor, and their application in humidity sensors has been predicted [73]. In a study, ReS_2 has shown a positive response to NH_3 and H_2O [72]. This is elucidated by the physisorption of the gas molecules that act as electron donors. The various outcomes in terms of sensitivity and responsivity achieved for ReS_2 by varying the functionalization molecules argument the significance of i) the probable binding sites for gas molecules; ii) the charge transfer efficiency for enhanced gas sensitivity, and iii) a good conductivity of the sensing film, which is favorable for a fast response and recovery.

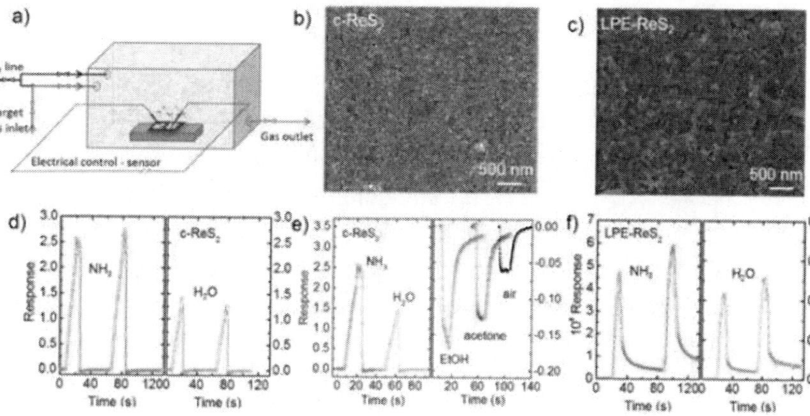

Figure 5.6. a) Sketch of the gas sensing chamber for experiments, b) and c) SEM images of the colloidal and LPE-ReS_2-based gas sensors, respectively, d)-f) Representative gas-induced time response of the device built with ann-ReS_2 and LPE-ReS_2 exposed to various gases. © 2019 Martín-García et al. Small 15.52 (2019): 1904670. (Published by WILEY-VCH Verlag GmbH & Co. KGaA, Weinheim).

Another important 2D TMD, niobium disulfide (NbS_2), has been speculated to have a great potential in gas sensing applications, as various studies have demonstrated that 2D NbS_2 holds numerous active edge sites which facilitate chemical detection [74, 75]. A recent study demonstrated the chemo resistive gas sensing mechanism of 2D NbS_2 nanosheets for a stable room temperature NO_2 sensing [76]. The sensing mechanisms were investigated by first-principle calculations based on density functional theory (DFT). First, the adsorption free energy (G_{ad}) of the NO_2 molecule on a clean surface at temperature T and partial pressure P was calculated as follows:

$$G_{ad}\,(T,P) \;=\; G\,(NbS_2 \;+\; NO_2\,) \;-\; G(NbS_2\,) - \mu_{NO_2}\,(T,P)$$

where G(NbS_2 + NO_2) and G(NbS_2) indicate the free energies of the NbS_2 surface with and without a NO_2 molecule adsorbed on clean NbS_2, respectively, and μ_{NO_2}(T, P) is the gas-phase chemical potential of the NO_2 molecule. The experimental conditions that were considered in μ_{NO_2}(T, P) as follows:

$$\mu_{NO_2}(T,P) = \mu_{NO_2}(T,p^0) + k_bT\ln\left(\frac{P}{p^0}\right)$$

where P^0 is 1 atm. Additionally,

$$\mu_{NO_2}(T,P^0) = \Delta H - T\Delta S\mu_{NO_2}(T,P^0) + \mu_{NO_2}(0\ K,P^0)$$

where ΔH and ΔS correspond to the enthalpy and entropy changes per molecule between T and 0 K at the standard pressure, respectively, which are obtained from thermodynamical tables [77]. And, $\mu_{NO_2}(0\ K, P^0)$ is equal to the total energy of the NO_2 molecule. The calculations from the above equations revealed that modification in edge configurations depending on synthetic conditions can change the sensing properties of 2D NbS_2. The various 2D TMDs can, therefore, present a promising solution for the fabrication of gas sensors with still considerable room for development by not only surface modifications but also improvisation of device designs.

5.3. Chemical Sensing Applications

2D TMDS have unprecedented material properties leading to their utilization in various sensor applications owing to their tunable physical and electrical properties. In the last few sections, we saw the gas sensing mechanism and properties of some the 2D TMDS. This section will cover some of the chemical sensing properties of these materials, such as vapor sensing, humidity sensing, and electrochemical sensing.

5.3.1. Vapor Sensing

Methanol is a widely used chemical agent in areas pertaining to food fermentation, biomedicines, color-manufacturing, perfumes, dyes, brewery, etc. Apart from this, methanol also finds extensive usage in automotive industries and is also used in wastewater treatment plants and for biodiesel preparation. It is known to have lethal effects on the environment and hence needs to be monitored [78–80]. In recent times, the escalations on the safety concerns in industries and laboratories have put immense demand for fast, efficient, and cost-effective room temperature detection of methanol. In a recent study, a methanol sensor based on HfS_2 flake has been demonstrated for fast response methanol sensing at a low temperature of 300 K [81]. The response and recovery times have a range from 12.12-21.14s and 23.72-39s, respectively in the methanol concentration range of 100–500 ppm. The possible mechanism of the sensing has been explained through charge transfer. When a reducing vapor (such as methanol) comes to the contact of the sensing surface, charge transfer takes place between the sensing material and targeted species. First, methanol gets adsorbed to the surface and forms methoxy (CH_3O) species by partial oxidation as follows,

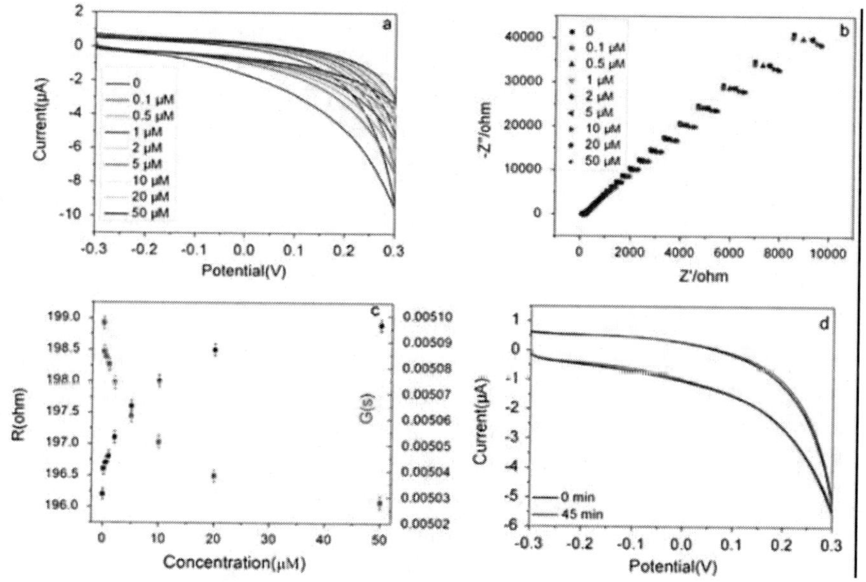

Figure 5.7. a) Cyclic voltammograms of the few-layered black phosphorus nanosheet sample in N_2-saturated phosphate buffer at varying concentration, b) Variation in electroimpedance of the BP nanosheet sample in buffer solution with frequency range from 1 Hz to 100 kHz, c) relationships between the electrochemical impedance, electrochemical conductance and different concentrations of H_2O_2, and d) cyclic voltammograms of BP nanosheet in N_2-saturated buffer solution with respect to time. (Rights and permissions by AIP Publishing through license number 4843361286839).

$$CH_3OH(gas) \Leftrightarrow CH_3OH(ads) \Leftrightarrow CH_3O(ads)$$

Adsorbed CH_3O species then forms CH_2O and CHO species and finally CO by C-H bond scission as in the below equation,

$$CH_3O(ads) \rightarrow CH_2O \rightarrow CHO \rightarrow CO(C-H\ bond\ scission)$$

On the other hand, CH_3O species also forms carbon species by C-O bond scission with complete dehydrogenation, as below.

$$CH_3O(ads) \rightarrow CHX \rightarrow C(C - O \text{ bond scission and dehydrogenation})$$

Generated CO and C species also get oxidized and form CO_2 that also enhances the sensitivity of the sensor with low operating temperature.

$$C + O^-_{(ads)} + h^+ \rightarrow CO_{(gas)}$$

$$CO_{(gas)} + O^-_{(ads)} + h^+ \rightarrow CO_{2\,(gas)}$$

Apart from HfS_2, some MXenes have also been reported as vapor sensors. A study recently explored phosphorene nanosheets as hydrogen peroxide (H_2O_2) vapor sensor with a significantly lower detection limit of 1×10^{-7} m compared to the general detection limit of 1×10^{-7} to 5×10^{-5} m by means of electrochemical methods [82]. When the phosphorene sensor exposed to trace amounts of H_2O_2, its impedance would evidently increase due to surface oxidation. As we know, phosphorene tends to degenerate under ambient conditions, and many types of research have been reported toward effective passivation of exfoliated phosphorenes [83–87]. But sometimes the deficiency of materials can also be its gift for specific applications. A study reported a methanol vapor sensor based on the sensitivity of phosphorene to the surrounding atmosphere [88]. In yet another report, by monitoring the changes in the electronic properties of phosphorene, researchers found a good response at a resonance frequency of approximately 1 kHz in the impedance phase spectra in the presence of methanol vapor [89]. The high specificity for methanol enabled it to be distinguished from toluene, acetone, chloroform, dichloromethane, ethanol, isopropyl alcohol, and water because of different dielectric constants of these molecules. Besides, such a selective sensor can detect a low concentration of 28 ppm, which is well below the approved exposure limit of 200 ppm with long-term stability. However, its intrinsic mechanism of how vapor molecules with different dielectric constants affect phosphorene's impedance among specific resonance frequency remaining obscure, and thus, may require further studying for broader applications.

5.3.2. Humidity Sensor

Humidity sensors are among the most widely used sensors not only in environmental monitoring applications but also in industries such as human health, food processing, pharmaceuticals, agriculture, etc. Many recent efforts have considerably enhanced the performance of humidity sensors in terms of response times by using 2D TMDs as sensing materials. Nonetheless, for feasible mass production of such 2D TMDs based sensors, a simple fabrication process is required while also minimizing sensor hysteresis, which is a standard limitation for any commercial humidity sensor.

We know that the sensing mechanism of 2D TMDs is based on electronic charge transfers. In a study on ReS_2 film with n-type behavior, the humidity sensing mechanism was explained by the n-doping of the ReS_2 sheets [90]. The water molecules absorbed on the surfaces of ReS_2 sheets will donate electrons to the n-type ReS_2, which increases the carrier concentration (electrons) in the channel between the electrodes, thereby increasing the conductance. Apart from the resistive output, humidity sensors can also have capacitive output. The relationship between the capacitive response of the sensors and the physical adsorption process under different moisture conditions can be well explained by the equation $C = (\varepsilon_\gamma - i_\gamma/\omega\varepsilon_0)C_0$ [91, 92]. In this equation, γ is the conductance level, ω is the excitation frequency, and is the imaginary constant. γ is determined by the adsorption level of water molecules in the sensitive layer at different physisorption processes. γ and ω are directly and inversely proportional to output capacitance C, respectively. One such mechanism has been demonstrated in a study on Ti_3C_2 MXene-Derived TiO_2 Nanowires [93].

MXenes, such as phosphorenes, have also been studied as material for humidity sensors. In a study, the ultrasensitive and selective response of phosphorene to humid air with a trace-level detection capability was observed [94]. The drain current of the phosphorene sensor increased by four orders of magnitude as the relative humidity (RH) varied from 10% to 85%.

The operation mechanism of the phosphorene sensors could be based on modulation in the ionic current leakage through the absorbed moist media. The stability tests revealed that the response of phosphorene sensors remained nearly unchanged after exposure to ambient conditions for three months. An electrochemically exfoliated phosphorene nanosheet thick film device was also studied for gas-sensing applications with exposure to various levels of humidity [95]. In an interesting report, a kitchen blender was used for the exfoliation and size-reduction of 2D phosphorene sheets to nanoparticles, which were then used as sensing material for the humidity sensors [96].

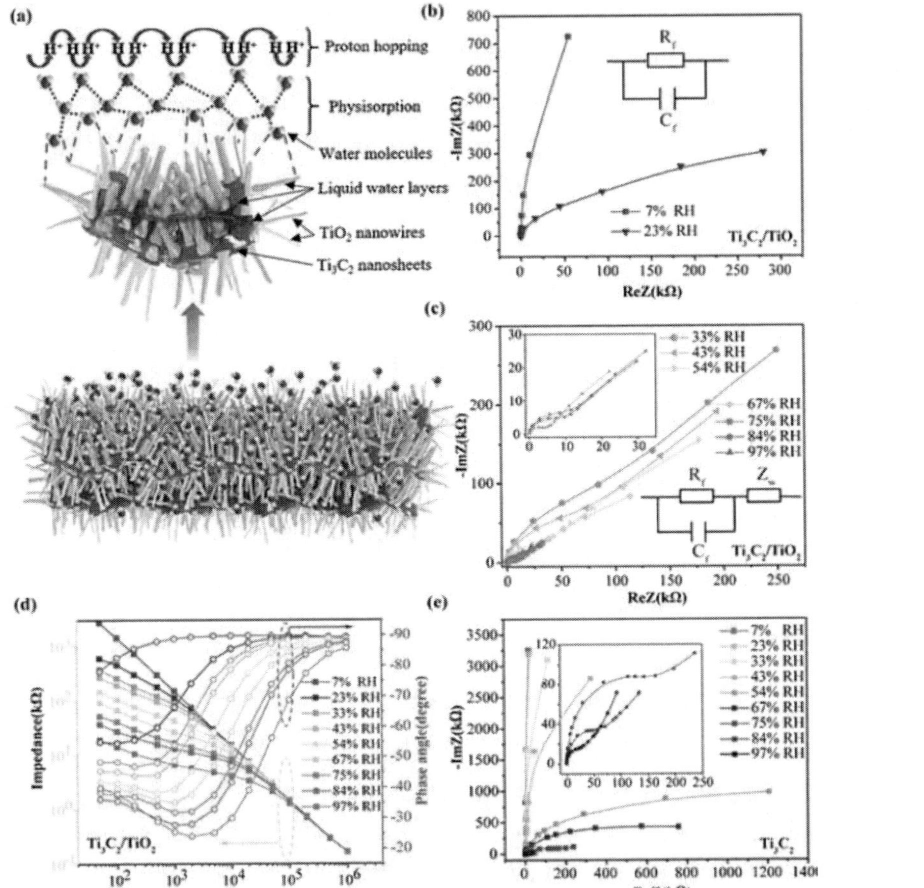

Figure 5.8. (a) Schematic of the water molecule adsorption process at the $Ti3C_2/TiO_2$ composite film. The impedance plots of Ti_3C_2/TiO_2 composite film at b) 7−23% RH and c) 33−97% RH. Inset of (b, c) correspond to the electric circuit generated by the impedance spectroscopy spectrum, d) Bode diagrams of the Ti_3C_2/TiO_2 composite in different relative humidity conditions, and, e) Complex impedance plots of the pure Ti_3C_2 film. (Reprinted (adapted) with permission from *ACS applied materials & interfaces* 11.41 (2019): 38116-38125. Copyright 2019 American Chemical Society).

It was inferred that the change in resistivity could be attributed to the changing contact resistance between the sheets during the change in humidity. Owing to the large specific area and abundant active sites of phosphorene, it is seen to capture water molecules having weak hydrogen bonds at ambient conditions [97]. The sensing mechanism in phosphorene is also based on ionic current modulation. In a study, phosphorene thin film made from 2D nanoflakes by vacuum filtration directly as the humidity sensor [98]. When this sensor exposed to different gases and vapors such as H_2, O_2, CO_2, benzenes, toluene, ethanol, and also water vapor, it reflected an enhancement of drain current only in a water ambiance. To investigate the mechanism of the humidity sensor, the author performed the modulation of the resistance and capacitance in the modeled equivalent circuit as a function of relative humidity. As relative humidity increased from 20% to 90%, the resistance decreased more than 450 times, while the capacitance only varied by less than a two-fold, indicating the sensing mechanism was based on the ionic current modulation. Besides, they further verified that few-layered phosphorene thin film obtained a more substantial channel resistance than bulk phosphorene, and the ionic conduction through the absorbed moist layer would be detectable.

5.3.3. Electrochemical Sensing

Since 2D TMDs have advantageous characteristics such as an excellent electrical conductivity, hydrophilicity and that they are environmentally benign, they have been experimented for numerous applications. The most studied electrochemical sensors are based on MXenes. To date, approximately 20 different MXenes have been explored for sensing applications. The surfaces of MXenes are terminated by oxygen-containing groups, such as hydroxyl, oxygen and can be functionalized with fluorine as well. The functionalized surface of Ti_2C based MXene in a study showed a strong anchoring effect to polysulfides in lithium-sulfur batteries application [99]. Due to the negatively charged Ti_3C_2 surface, MXenes have shown preferential adsorption of cationic dye [100]. Thus, tailoring the surface functionalities can be extremely feasible to manipulate the properties and characteristics of MXenes. The inter-layer spaces in MXene sheets beget spontaneous intercalation of numerous alkaline and alkaline earth cations (Li^+, K^+, Na^+, Mg^{2+}) [101–103]. These cationic intercalations often lead to the changed surface chemistry of MXenes and based on this, many applications of MXenes have been explored [43]. Studies have shown that surface modification in MXenes by alkalization interaction treatment affects the binding energies and adsorption capacities of heavy metal ions such as cadmium (Cd II), lead (Pb

II), copper (Cu II), mercury (Hg II), etc. [104]. In a study, the pristine Ti_3C_2 modified electrode has shown sensitivity towards electrochemical sensing and also towards the ultra-sensitive detection of hydrogen peroxide(H_2O_2) [105]. A theoretical study has revealed a strong anchoring effect of carbendazim on the surface of fluorine functionalized $Ti_3C_2T_x$ [106]. From various studies, it is clear that the fantastic electrochemical sensing behavior of MXenes is the result of its adsorption ability [107–109].

As the extensive ongoing research on 2D TMDs suggests, the structure of TMDs plays a vital role in defining the properties. For instance, the hexagonal lattice structure of VS_2 has been reported to demonstrate enhanced catalytic properties due to the presence of more number of active sites [110]. Interestingly the numerous active sites occurring on the surface of the ReS_2 has also been beneficial in the production of the hydrogen molecules from the electrocatalysis of water. Various studies have shown that the Gibbs free energy of the hydrogen adsorption on ReS_2 is tuned by the Re-Re bond properties [111, 112]. In contrast to TMDs like MoS_2, ReS_2 properties do not depend solely on the layered structures, thereby making it a suitable catalyst [73].

5.3.4. Biosensing Applications

In the previous sections, we have seen how 2D TMD's material properties have been advantageous for developing numerous functional devices as required. Also, most of the 2D TMDs are naturally stable in various ambiance, be it air or water, making them viable for biosensing applications. Different from graphene, electrons in 2D TMDs are massive. Additionally, the 2D TMDs based semiconductors have bandgaps mostly appropriate for biosensing applications. These bandgaps provide comparable optical properties with photoluminescence beneficial for bio-imaging. Yet another advantage of 2D TMDs over other materials is their low toxicity and environmentally benign nature, thus making them suitable for drug storage and delivery. Additionally, they also possess high photothermal response, fluorescence, and moderate conductivity suitable for multiple biomedical applications. However, most of these 2D TMDs have been surface-functionalized in order to enhance their sensitivity as a biosensor. Since TMDs have a large surface-to-volume ratio, their functionalization is relatively easy and direct and can be either non-covalent or covalent.

To enhance the electrocatalytic ability of 2D TMDCs to make them suitable for bio-imaging sensors, their surface is often decorated by nanoparticles [27, 113]. Other such modifications include the stacking of different 2D materials and sometimes pore creations as well, to increase the sensitivity of the biosensor. Depending on these modifications 2D TMDCs respond to the environment changes and set the basis for biomedical applications

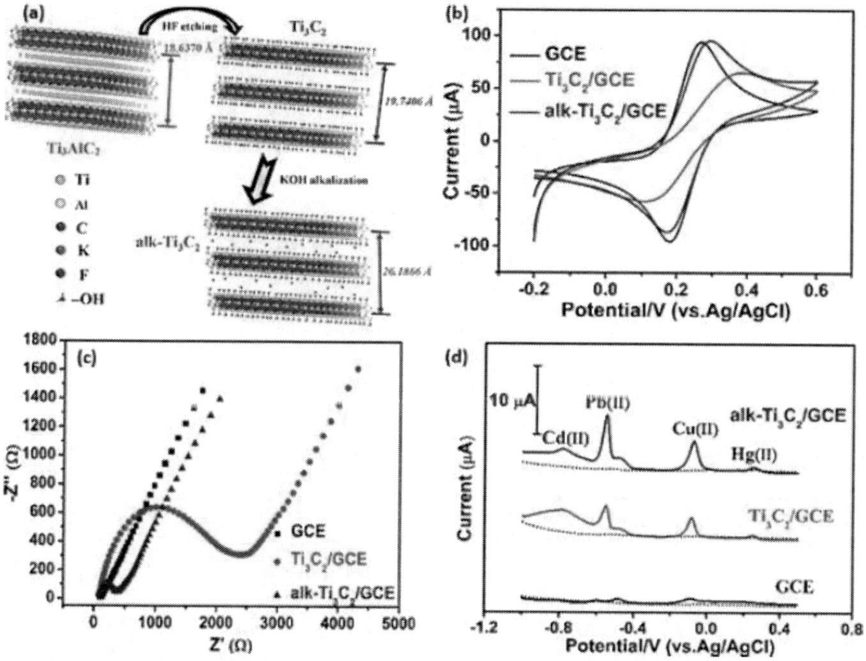

Figure 5.9. a) Schematic structural illustrations Ti_3AlC_2 and the post-etching MXene before and after alkalization, b) and c) Cyclic voltammograms and Nyquist plot comparison of glossy carbon and MXene electrodes, and d) Detection response of Cd(II), Pb(II), Cu(II) and Hg(II) at the deposition potential of 1.2 V with a deposition time of 150 s. The dotted line represents the baselines. Reproduced with permission and license from Elsevier for Zhu et al. *Electrochimica Acta* 248 (2017): 46-57.

Despite the various favorable properties of 2D TMDs for biosensing, their uses remain limited and relatively unexplored. Some reports have demonstrated an excellent electrochemical activity of MXene toward glucose [114]. A study presented a TiO_2-Ti_3C_2 MXene composite encapsulating hemoglobin having efficient H_2O_2 detection capacity with a limit of detection as low as 14 nM and a linear range of 0.1–380 mM. In a similar study on

Ti_3C_2-based sensor supplied with hemoglobin detected sodium nitrite ($NaNO_2$) in the range of 0.5–11,800 mM with the limit of detection being 0.12 mM [115]. The performance of the sensor was credited to the large surface area and high conductivity of the 2D TMD.

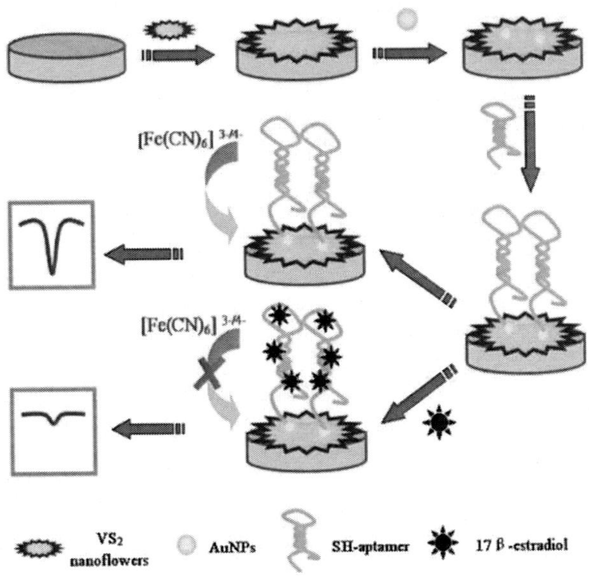

Figure 5.10. Schematics of vanadium disulphide nanoflower based electrochemical sensor. Reproduced with permission and license from Elsevier for Zhu et al. *Electrochimica*.

In a new report, a chemiluminescence biosensor with high sensitivity was developed for the detection of exosomes Ti_3C_2-MXenes nanosheets [116]. Exosomes are promising biomarkers for early cancer diagnosis. The study showed that exosomes can be adequately captured on the modified 2D TMD electrode, significantly enhancing the electrochemiluminescence response. The claimed detection of limit was 125 particles per μL, i.e., approximately 100 times efficient than the ELISA method. Further, VS_2 nanoflowers have been reported to facilitate electron transfer and enhance the number of active surfaces [117]. These nanoflowers were used as electrode materials for an aptamer biosensor with a low detection limit for 17β-estradiol detection. Another example of superior photoelectrochemical of 2D TMDs can be seen in a study on ReS_2 nanosheets for the detection of carcinoembryonic antigen (CEA) [118]. The reported sensor showed excellent stability, high selectivity, high sensitivity, and great accuracy for CEA determination. The limit of detection extremely low (0.468 $pgmL^{-1}$) with a broad linear range of 0.0005 ~ 10.0 $ngmL^{-1}$. MXenes have also been exploited for health monitoring systems and biomimetic. Therapeutic drug monitoring (TDM) has become very essential in health care system. A timely and economic TDM is the need of the hour. In this perspective, MXenes have been exploited for therapeutic monitoring. MXene-$Ti_3C_2T_x$ based screen-printed electrode has been successfully designed and tested in lab to monitor the whole blood [119]. The simplistic surface modification of the MXenes can facilitate in-vivo performance as a result of their low toxicity, superior colloidal stability, and good pharmacokinetic [120]. Further, MXene based biomimetic paper has also been reported for application in the bioelectronics [121]. Phosphorene, an MXene, was reported as a novel nanomaterial in fluorescence-based di-nucleic acid (DNA) detection [122]. Employing phosphorene as fluorophores, they proposed a detection test which was able to detect DNA concentrations with low limit of detection (LOD) and limit of quantification (LOQ) values as low as 5.9 pM and 19.7 pM, respectively, and demonstrated a large linear dynamic detection range from less than 4 pM to 4000 pM. It also showed good linearity with an r-value of 0.91. The DNA probe was also capable of realizing triple nucleotide polymorphism determination i.e., it could distinguish three-base mismatched nucleic acid sequences (mDNA-3) and non-complementary nucleic acid sequences (ncDNA) according to different fluorescence recovery. The mechanism is based on the dissimilarities in affinities of characterized single-stranded and double-stranded DNA oligonucleotides toward impulsive self-assembly with black phosphorus nanoparticles. As a result, an excellent linear dynamic range of detection extended from 4 to 4000 × 10^{-12} m of complementary nucleic acid sequences (cDNA) concentration.

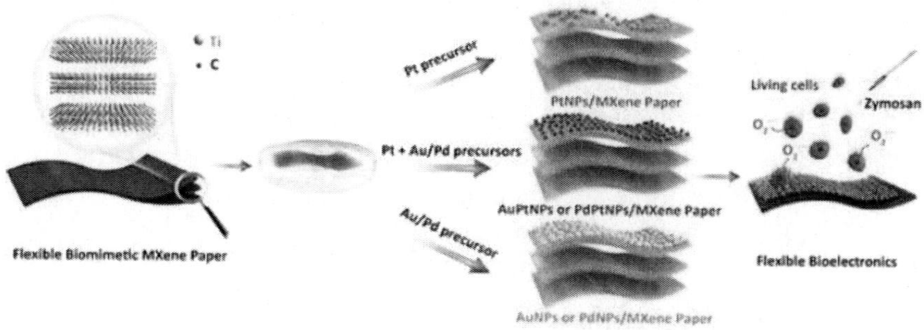

Figure 5.11. Schematic illustration of flexible biomimetic MXene paper for application in extracellular biosensing. Reproduced with permission and license from Elsevier for Yao et al. *Biosensors and Bioelectronics* 148 (2020): 111799.

Apart from fluorescent detection, plasmonic biosensing has also become the core of label-free technology. Typically, when incident wave irradiates on analyte it will cause a refractive index and even an insignificant change in the refractive index of sensing medium will cause a vivid variation in the SPR signal [123]. Having seen such excellent examples, it is encouraged to fabricate 2D TMDs based biosensors having a low limit of detection and a wide linear range.

5.4. Conclusion and Future Perspective

Most of the 2D TMDs present a very high area to the thickness ratio. Their unique structures and exceptionally tunable properties have made them an important class of materials in chemistry, physics, and materials science, having been comprehensively explored for numerous applications in various fields. The modified surfaces of 2D TMDs have seen to enhance the adsorption of gaseous molecules and favorable the surface reactions of adsorbed species due to increased active sites. Additionally, the limited to almost absent interlayer activities in these 2D materials can enable fascinating electrical properties in comparison to other nanomaterials. Furthermore, as they show metallic, semiconducting, and superconducting behavior, their properties can be further fine-tuned to improve their electrical characteristics. These properties are highly favorable for not only gas or biosensing applications but also for optoelectronic applications such as photodiodes, solar cells, and light-emitting diodes. These nanostructures are recently being exploited at a high rate for their various applications and regular publications in these areas may determine the path of the most recent progress and future direction of the research.

The revolutionary work on emerging 2D TMDs materials that have been conducted in recent years leads to understanding many of their interesting properties. It is the basis to address certain existing problems about the use of such materials for multiple applications. As pointed out in this chapter, recent advances in the two-dimensional metallic materials such as VS_2, VSe_2, NbS_2, HfS_2, ReS_2, and MXenes with their characteristic physical, chemical, and electronic properties, predict a promising future for the development of novel chemical-, gas- and biosensing applications. However, research so far suggests that extensive and significant changes are necessary before these materials can be brought into practical applications with mass production of devices. In this respect, it is imperative to discuss the preparation of significantly pronounced, reproducible, and bulk quantities of such two-dimensional materials. Certainly, other technical problems need to be overcome; however, it looks good majorly for the gas sensing applications that are difficult to accomplish with traditional materials.

References

[1] Zhang, X., L. Hou, A. Ciesielski, P. Samorì, "2D Materials Beyond Graphene for High-Performance Energy Storage Applications," *Advanced Energy Materials*, vol. 6, no. 23, pp. 1600671, 2016.

[2] Shao, Y., J. Wang, H. Wu, J. Liu, I. A. Aksay, Y. Lin, "Graphene based electrochemical sensors and biosensors: A review," *Electroanalysis*, vol. 22, no. 10, 2010.

[3] Liu, X., T. Ma, N. Pinna, J. Zhang, "Two-Dimensional Nanostructured Materials for Gas Sensing," *Advanced Functional Materials*, vol. 27, no. 37, 2017.

[4] Garcia, J. C., D. B. De Lima, L. V. C. Assali, J. F. Justo, "Group IV graphene- and graphane-like nanosheets," *Journal of Physical Chemistry* C, vol. 115, no. 27, pp. 13242–13246, 2011.

[5] Yang, W., L. Gan, H. Li, T. Zhai, "Two-dimensional layered nanomaterials for gas-sensing applications," *Inorganic Chemistry Frontiers*, vol. 3, no. 4, pp. 433–451, 2016.

[6] Barzegar, M., B. Tudu, "Two-dimensional materials for gas sensors: From first discovery to future possibilities," *Surface Innovations*, vol. 6, no. 4–5, 2018.

[7] Tricoli, A., N. Nasiri, S. De, "Wearable and Miniaturized Sensor Technologies for Personalized and Preventive Medicine," *Advanced Functional Materials*, vol. 27, no. 15, 2017.

[8] Shi, S., R. Hu, E. Wu, Q. Li, X. Chen, W. Guo, C. Sun, X. Hu, D. Zhang, J. Liu, "Highly-sensitive gas sensor based on two-dimensional material field effect transistor," *Nanotechnology*, vol. 29, no. 43, pp. 435502, 2018.

[9] Susarla, S., A. Kutana, J. A. Hachtel, V. Kochat, A. Apte, R. Vajtai, J. C. Idrobo, B. I. Yakobson, C. S. Tiwary, P. M. Ajayan, "Quaternary 2D Transition Metal Dichalcogenides (TMDs) with Tunable Bandgap," *Advanced Materials*, vol. 29, no. 35, pp. 1702457, 2017.

[10] Krasnok, A., S. Lepeshov, A. Alú, "Nanophotonics with 2D transition metal dichalcogenides [Invited]," *Optics Express*, vol. 26, no. 12, pp. 15972, 2018.

[11] Wang, Q. H., K. Kalantar-Zadeh, A. Kis, J. N. Coleman, M. S. Strano, "Electronics and optoelectronics of two-dimensional transition metal dichalcogenides," *Nature Nanotechnology*, vol. 7, no. 11, pp. 699–712, 2012.

[12] Zhou, J., F. Liu, J. Lin, X. Huang, J. Xia, B. Zhang, Q. Zeng, H. Wang, C. Zhu, L. Niu, X. Wang, W. Fu, P. Yu, T. R. Chang, C. H. Hsu, D. Wu, H. T. Jeng, Y. Huang, H. Lin, Z. Shen, C. Yang, L. Lu, K. Suenaga, W. Zhou, S. T. Pantelides, G. Liu, Z. Liu, "Large-Area and High-Quality 2D Transition Metal Telluride," *Advanced Materials*, vol. 29, no. 3, pp. 1603471, 2017.

[13] Hu Z., Z. Wu, C. Han, J. He, Z. Ni, W. Chen, "Two-dimensional transition metal dichalcogenides: interface and defect engineering," *Chemical Society Reviews*, vol. 47, no. 9, pp. 3100–3128, 2018.

[14] Ponnusamy, R., C. S. Rout, *Two Dimensional Transition Metal Dichalcogenides*, Springer Singapore, Singapore, 2019.

[15] Joshi, N., T. Hayasaka, Y. Liu, H. Liu, O. N. Oliveira, L. Lin, "A review on chemiresistive room temperature gas sensors based on metal oxide nanostructures, graphene and 2D transition metal dichalcogenides," *Microchimica Acta,* vol. 185, no. 4, pp. 213, 2018.

[16] Choi, S. J., I. D. Kim, "Recent Developments in 2D Nanomaterials for Chemiresistive-Type Gas Sensors," *Electronic Materials Letters*, vol. 14, no. 3, pp. 221–260, 2018.

[17] Sarkar, D., X. Xie, J. Kang, H. Zhang, W. Liu, J. Navarrete, M. Moskovits, K. Banerjee, "Functionalization of Transition Metal Dichalcogenides with Metallic Nanoparticles: Implications for Doping and Gas-Sensing," *Nano Letters*, vol. 15, no. 5, pp. 2852–2862, 2015.

[18] Berger, C., Z. Song, T. Li, X. Li, A. Y. Ogbazghi, R. Feng, Z. Dai, A. N. Marchenkov, E. H. Conrad, P. N. First, W. A. de Heer, "Ultrathin Epitaxial Graphite: 2D Electron Gas Properties and a Route toward Graphene-based Nanoelectronics," *The Journal of Physical Chemistry* B, vol. 108, no. 52, pp. 19912–19916, 2004.

[19] Akinwande, D., C. J. Brennan, J. S. Bunch, P. Egberts, J. R. Felts, H. Gao, R. Huang, J.-S. Kim, T. Li, Y. Li, K. M. Liechti, N. Lu, H. S. Park, E. J. Reed, P. Wang, B. I. Yakobson, T. Zhang, Y.-W. Zhang, Y. Zhou, Y. Zhu, "A review on mechanics and mechanical properties of 2D materials—Graphene and beyond," *Extreme Mechanics Letters*, vol. 13, pp. 42–77, 2017.

[20] Ouyang, Y., J. Guo, "A theoretical study on thermoelectric properties of graphene nanoribbons," *Applied Physics Letters,* vol. 94, no. 26, pp. 263107, 2009.

[21] Singh, G., A. Choudhary, D. Haranath, A. G. Joshi, N. Singh, S. Singh, R. Pasricha, "ZnO decorated luminescent graphene as a potential gas sensor at room temperature," *Carbon*, vol. 50, no. 2, pp. 385–394, 2012.

[22] Gautam, M., A. H. Jayatissa, "Detection of organic vapors by graphene films functionalized with metallic nanoparticles," *Journal of Applied Physics*, vol. 112, no. 11, pp. 114326, 2012.

[23] Cho, B., J. Yoon, M. G. Hahm, D. H. Kim, A. R. Kim, Y. H. Kahng, S.-W. Park, Y. J. Lee, S. G. Park, J. D. Kwon, C. S. Kim, M. Song, Y. Jeong, K.-S. Nam, H. C. Ko, "Graphene-based gas sensor: metal decoration effect and application to a flexible device," *J. Mater. Chem. C*, vol. 2, no. 27, pp. 5280–5285, 2014.

[24] Johnson, J. L., A. Behnam, S. J. Pearton, A. Ural, "Hydrogen Sensing Using Pd-Functionalized Multi-Layer Graphene Nanoribbon Networks," *Advanced Materials,* vol. 22, no. 43, pp. 4877–4880, 2010.

[25] Wei, W., X. Qu, "Extraordinary Physical Properties of Functionalized Graphene," *Small*, vol. 8, no. 14, pp. 2138–2151, 2012.

[26] Pumera, M., A. H. Loo, "Layered transition-metal dichalcogenides (MoS2 and WS2) for sensing and biosensing," *TrAC Trends in Analytical Chemistry*, vol. 61, pp. 49–53, 2014.

[27] Bollella, P., G. Fusco, C. Tortolini, G. Sanzò, G. Favero, L. Gorton, R. Antiochia, "Beyond graphene: Electrochemical sensors and biosensors for biomarkers detection," *Biosensors and Bioelectronics*, vol. 89, pp. 152–166, 2017.

[28] Yang, S., C. Jiang, S. Wei, "Gas sensing in 2D materials," *Applied Physics Reviews*, vol. 4, no. 2, pp. 021304, 2017.

[29] Cho, B., J. Yoon, S. K. Lim, A. R. Kim, D. H. Kim, S. G. Park, J. D. Kwon, Y. J. Lee, K. H. Lee, B. H. Lee, H. C. Ko, M. G. Hahm, "Chemical Sensing of 2D Graphene/MoS2 Heterostructure device," *ACS Applied Materials and Interfaces*, vol. 7, no. 30, pp. 16775–16780, 2015.

[30] Kuriakose, S., T. Ahmed, S. Balendhran, V. Bansal, S. Sriram, M. Bhaskaran, S. Walia, "Black phosphorus: ambient degradation and strategies for protection," *2D Materials,* vol. 5, no. 3, pp. 032001, 2018.

[31] Di Bartolomeo, A. "Emerging 2D Materials and Their Van Der Waals Heterostructures," *Nanomaterials,* vol. 10, no. 3, pp. 579, 2020.

[32] Wang, Q., Y. Lei, Y. Wang, Y. Liu, C. Song, J. Zeng, Y. Song, X. Duan, D. Wang, Y. Li, "Atomic-scale engineering of chemical-vapor-deposition-grown 2D transition metal dichalcogenides for electrocatalysis," *Energy & Environmental Science*, 2020.

[33] Bertolazzi, S., M. Gobbi, Y. Zhao, C. Backes, P. Samorì, "Molecular chemistry approaches for tuning the properties of two-dimensional transition metal dichalcogenides," *Chemical Society Reviews*, vol. 47, no. 17, pp. 6845–6888, 2018.

[34] Winkler, C., S. S. Harivyasi, E. Zojer, "Controlling the electronic properties of van der Waals heterostructures by applying electrostatic design," *2D Materials*, vol. 5, no. 3, pp. 035019, 2018.

[35] Chhowalla, M., H. S. Shin, G. Eda, L. J. Li, K. P. Loh, H. Zhang, "The chemistry of two-dimensional layered transition metal dichalcogenide nanosheets," *Nature Chemistry,* vol. 5, no. 4, pp. 263–275, 2013.

[36] Duan, X., C. Wang, A. Pan, R. Yu, X. Duan, "Two-dimensional transition metal dichalcogenides as atomically thin semiconductors: opportunities and challenges," *Chemical Society Reviews*, vol. 44, no. 24, pp. 8859–8876, 2015.

[37] Wang, H., H. Yuan, S. Sae Hong, Y. Li, Y. Cui, "Physical and chemical tuning of two-dimensional transition metal dichalcogenides," *Chemical Society Reviews*, vol. 44, no. 9, pp. 2664–2680, 2015.

[38] Shi, M., P. Mo, J. Lu, J. Liu, "Strain-enhanced electron mobility and mobility anisotropy in a two-dimensional vanadium diselenide monolayer," *Journal of Applied Physics*, vol. 126, no. 4, pp. 044305, 2019.

[39] Kanazawa, T., T. Amemiya, A. Ishikawa, V. Upadhyaya, K. Tsuruta, T. Tanaka, Y. Miyamoto, "Few-layer HfS2 transistors," *Scientific Reports*, vol. 6, no. 1, pp. 22277, 2016.

[40] Guo, Y., H. Deng, X. Sun, X. Li, J. Zhao, J. Wu, W. Chu, S. Zhang, H. Pan, X. Zheng, X. Wu, C. Jin, C. Wu, Y. Xie, "Modulation of Metal and Insulator States in

2D Ferromagnetic VS 2 by van der Waals Interaction Engineering," *Advanced Materials,* vol. 29, no. 29, pp. 1700715, 2017.

[41] Balci, E., Ü. Ö. Akkuş, S. Berber, "Band gap modification in doped MXene: Sc_2CF_2," *Journal of Materials Chemistry* C, vol. 5, no. 24, pp. 5956–5961, 2017.

[42] Sinha, A., Dhanjai, H. Zhao, Y. Huang, X. Lu, J. Chen, R. Jain, "MXene: An emerging material for sensing and biosensing," *TrAC - Trends in Analytical Chemistry,* vol. 105, pp. 424–435, 2018.

[43] Zhu, J., E. Ha, G. Zhao, Y. Zhou, D. Huang, G. Yue, L. Hu, N. Sun, Y. Wang, L. Y. S. Lee, C. Xu, K.-Y. Wong, D. Astruc, P. Zhao, "Recent advance in MXenes: A promising 2D material for catalysis, sensor and chemical adsorption," *Coordination Chemistry Reviews*, vol. 352, pp. 306–327, 2017.

[44] Aaryashree, B. Mandal, R. Bhardwaj, S. Maiti, D. S. Sharma, A. K. Das, S. Mukherjee, "Functionalized Oligo(p-Phenylenevinylene) and ZnO-Based Nanohybrid for Selective Ammonia Sensing at Room Temperature," *IEEE Sensors Journal*, vol. 19, no. 8, pp. 2847–2854, 2019.

[45] Mandal, B., S. Maiti, Aaryashree, G. Siddharth, M. Das, A. Agarwal, A. K. Das, S. Mukherjee, "Organo-di-benzoic-acidified ZnO Nanohybrids for Highly Selective Detection of CO at Low Temperature," *The Journal of Physical Chemistry C*, vol. 124, no. 13, pp. 7307–7316, 2020.

[46] Morrison, S. R. "Selectivity in semiconductor gas sensors," *Sensors and Actuators*, vol. 12, no. 4, pp. 425–440, 1987.

[47] Yuan, W., G. Shi, "Graphene-based gas sensors," *Journal of Materials Chemistry* A, vol. 1, no. 35, pp. 10078, 2013.

[48] Moseley, P. T. "Solid state gas sensors," *Measurement Science and Technology*, vol. 8, no. 3, pp. 223–237, 1997.

[49] Miasik, J. J., A. Hooper, B. C. Tofield, "Conducting polymer gas sensors," *Journal of the Chemical Society, Faraday Transactions 1: Physical Chemistry in Condensed Phases*, vol. 82, no. 4, pp. 1117, 1986.

[50] Cai, Y., Q. Ke, G. Zhang, Y. W. Zhang, "Energetics, Charge Transfer, and Magnetism of Small Molecules Physisorbed on Phosphorene," *The Journal of Physical Chemistry C,* vol. 119, no. 6, pp. 3102–3110, 2015.

[51] Cui, S., H. Pu, S. A. Wells, Z. Wen, S. Mao, J. Chang, M. C. Hersam, J. Chen, "Ultrahigh sensitivity and layer-dependent sensing performance of phosphorene-based gas sensors," *Nature Communications*, vol. 6, no. 1, pp. 8632, 2015.

[52] Abbas, A. N., B. Liu, L. Chen, Y. Ma, S. Cong, N. Aroonyadet, M. Köpf, T. Nilges, C. Zhou, "Black phosphorus gas sensors," *ACS Nano*, vol. 9, no. 5, pp. 5618–5624, 2015.

[53] Cho, S. Y. Y., H. J. J. Koh, H. W. W. Yoo, H. T. T. Jung, "Tunable Chemical Sensing Performance of Black Phosphorus by Controlled Functionalization with Noble Metals," *Chemistry of Materials*, vol. 29, no. 17, pp. 7197–7205, 2017.

[54] Arabieh, M., Y. T. Azar, "In silico insight into ammonia adsorption on pristine and X-doped phosphorene (X = B, C, N, O, Si, and Ni)," *Applied Surface Science*, vol. 396, pp. 1411–1419, 2017.

[55] Donarelli, M., L. Ottaviano, L. Giancaterini, G. Fioravanti, F. Perrozzi, C. Cantalini, "Exfoliated black phosphorus gas sensing properties at room temperature," *2D Materials*, vol. 3, no. 2, pp. 025002, 2016.

[56] Kim, S. J., H. J. J. Koh, C. E. Ren, O. Kwon, K. Maleski, S. Y. Y. Cho, B. Anasori, C. K. K. Kim, Y. K. K. Choi, J. Kim, Y. Gogotsi, H. T. T. Jung, "Metallic Ti_3C_2 Tx MXene Gas Sensors with Ultrahigh Signal-to-Noise Ratio," *ACS Nano*, vol. 12, no. 2, pp. 986–993, 2018.

[57] Lee, E., A. Vahidmohammadi, B. C. Prorok, Y. S. Yoon, M. Beidaghi, D. J. J. Kim, "Room Temperature Gas Sensing of Two-Dimensional Titanium Carbide (MXene)," *ACS Applied Materials and Interfaces*, vol. 9, no. 42, pp. 37184–37190, 2017.

[58] Yu, X., Y. Li, J. Cheng, Z. Liu, Q. Li, W. Li, X. Yang, B. Xiao, "Monolayer Ti_2CO_2 : A Promising Candidate for NH_3 Sensor or Capturer with High Sensitivity and Selectivity," *ACS Applied Materials & Interfaces*, vol. 7, no. 24, pp. 13707–13713, 2015.

[59] Lee, E., A. Vahidmohammadi, Y. S. Yoon, M. Beidaghi, D. J. Kim, "Two-Dimensional Vanadium Carbide MXene for Gas Sensors with Ultrahigh Sensitivity Toward Nonpolar Gases," *ACS Sensors*, vol. 4, no. 6, pp. 1603–1611, 2019.

[60] Golias, E., E. Xenogiannopoulou, D. Tsoutsou, P. Tsipas, S. A. Giamini, A. Dimoulas, "Surface electronic bands of submonolayer Ge on Ag(111)," *Physical Review B*, vol. 88, no. 7, pp. 075403, 2013.

[61] Dimoulas, A. "Silicene and germanene: Silicon and germanium in the "flatland,"" *Microelectronic Engineering*, vol. 131, no. september, pp. 68–78, 2015.

[62] Vogt, P., P. De Padova, C. Quaresima, J. Avila, E. Frantzeskakis, M. C. Asensio, A. Resta, B. Ealet, G. Le Lay, "Silicene: Compelling Experimental Evidence for Graphenelike Two-Dimensional Silicon," *Physical Review Letters*, vol. 108, no. 15, pp. 155501, 2012.

[63] Liu, C., C. S. Liu, X. Yan, "Arsenene as a promising candidate for NO and NO_2 sensor: A first-principles study," *Physics Letters A*, vol. 381, no. 12, pp. 1092–1096, 2017.

[64] Wang, Y., P. Huang, M. Ye, R. Quhe, Y. Pan, H. Zhang, H. Zhong, J. Shi, J. Lu, "Many-body Effect, Carrier Mobility, and Device Performance of Hexagonal Arsenene and Antimonene," *Chemistry of Materials*, vol. 29, no. 5, pp. 2191–2201, 2017.

[65] Ji, J., X. Song, J. Liu, Z. Yan, C. Huo, S. Zhang, M. Su, L. Liao, W. Wang, Z. Ni, Y. Hao, H. Zeng, "Two-dimensional antimonene single crystals grown by van der Waals epitaxy," *Nature Communications*, vol. 7, no. 1, pp. 13352, 2016.

[66] Meng, R. S., M. Cai, J. K. Jiang, Q. H. Liang, X. Sun, Q. Yang, C. J. Tan, X. P. Chen, "First Principles Investigation of Small Molecules Adsorption on Antimonene," *IEEE Electron Device Letters*, vol. 38, no. 1, pp. 134–137, 2017.

[67] Wang, D. W., A. J. Yang, J. F. Chu, P. L. Lv, Yang-Liu, X. H. Wang, M. Z. Rong, Antimonene: A promising candidate for acetone sensors with high selectivity and sensitivity, in: *2017 IEEE SENSORS, IEEE*, 2017.: pp. 1–3.

[68] Tao, L., E. Cinquanta, D. Chiappe, C. Grazianetti, M. Fanciulli, M. Dubey, A. Molle, D. Akinwande, "Silicene field-effect transistors operating at room temperature," *Nature Nanotechnology*, vol. 10, no. 3, pp. 227–231, 2015.

[69] Prasongkit, J., R. G. Amorim, S. Chakraborty, R. Ahuja, R. H. Scheicher, V. Amornkitbamrung, "Highly Sensitive and Selective Gas Detection Based on Silicene," *The Journal of Physical Chemistry* C, vol. 119, no. 29, pp. 16934–16940, 2015.

[70] Gupta, S. K., D. Singh, K. Rajput, Y. Sonvane, "Germanene: a new electronic gas sensing material," *RSC Advances*, vol. 6, no. 104, pp. 102264–102271, 2016.

[71] Zhang, X., Q. Li, "Electronic and magnetic properties of nonmetal atoms adsorbed ReS 2 monolayers," *Journal of Applied Physics*, vol. 118, no. 6, pp. 064306, 2015.

[72] Martín-García, B., D. Spirito, S. Bellani, M. Prato, V. Romano, A. Polovitsyn, R. Brescia, R. Oropesa-Nuñez, L. Najafi, A. Ansaldo, G. D'Angelo, V. Pellegrini, R. Krahne, I. Moreels, F. Bonaccorso, "Extending the Colloidal Transition Metal Dichalcogenide Library to ReS 2 Nanosheets for Application in Gas Sensing and Electrocatalysis," *Small*, vol. 15, no. 52, pp. 1904670, 2019.

[73] Rahman, M., K. Davey, S. Z. Qiao, "Advent of 2D Rhenium Disulfide (ReS2): Fundamentals to Applications," *Advanced Functional Materials*, vol. 27, no. 10, 2017.

[74] Lee, K., R. Gatensby, N. McEvoy, T. Hallam, G. S. Duesberg, "High-Performance Sensors Based on Molybdenum Disulfide Thin Films," *Advanced Materials*, vol. 25, no. 46, pp. 6699–6702, 2013.

[75] Carmalt, C., E. Peters, I. Parkin, T. Manning, A. Hector, "Chemical Vapor Deposition of Niobium Disulfide Thin Films," *European Journal of Inorganic Chemistry*, vol. 2004, no. 22, pp. 4470–4476, 2004.

[76] Kim, Y., K. C. Kwon, S. Kang, C. Kim, T. H. Kim, S. P. Hong, S. Y. Park, J. M. Suh, M. J. Choi, S. Han, H. W. Jang, "Two-Dimensional NbS2 Gas Sensors for Selective and Reversible NO_2 Detection at Room Temperature," *ACS Sensors*, vol. 4, no. 9, pp. 2395–2402, 2019.

[77] Lemmon, W., E. W., McLinden, M. O., Friend, D. G., Linstrom, P., Mallard, "*NIST Standard Reference Database Number 69*," National Institute of Standards and Technology; Gaithersburg, MD, 2018.

[78] Kruse, J. A. "Methanol poisoning," *Intensive Care Medicine*, vol. 18, no. 7, pp. 391–397, 1992.

[79] Hovda, K. E., O. H. Hunderi, A. B. Tafjord, O. Dunlop, N. Rudberg, D. Jacobsen, "Methanol outbreak in Norway 2002-2004: epidemiology, clinical features and prognostic signs," *Journal of Internal Medicine*, vol. 258, no. 2, pp. 181–190, 2005.

[80] Mandal, B., Aaryashree, R. Singh, S. Mukherjee, "Highly Selective and Sensitive Methanol Sensor Using Rose-Like ZnO Microcube and MoO 3 Micrograss-Based Composite," *IEEE Sensors Journal*, vol. 18, no. 7, pp. 2659–2666, 2018.

[81] Das, S., S. Sharma, S. K. Sharma, "Facile Synthesis of 2D-HfS2 Flakes for μ-IDE-Based Methanol Sensor: Fast Detection at Room Temperature," *IEEE Sensors Journal*, vol. 19, no. 20, pp. 9090–9096, 2019.

[82] Yan, S., B. Wang, Z. Wang, D. Hu, X. Xu, J. Wang, Y. Shi, "Supercritical carbon dioxide-assisted rapid synthesis of few-layer black phosphorus for hydrogen peroxide sensing," *Biosensors and Bioelectronics*, vol. 80, pp. 34–38, 2016.

[83] Ryder, C. R., J. D. Wood, S. A. Wells, Y. Yang, D. Jariwala, T. J. Marks, G. C. Schatz, M. C. Hersam, "Covalent functionalization and passivation of exfoliated black phosphorus via aryl diazonium chemistry," *Nature Chemistry*, vol. 8, no. 6, pp. 597–602, 2016.

[84] Guo, Z., H. Zhang, S. Lu, Z. Wang, S. Tang, J. Shao, Z. Sun, H. Xie, H. Wang, X.-F. Yu, P. K. Chu, "From Black Phosphorus to Phosphorene: Basic Solvent Exfoliation, Evolution of Raman Scattering, and Applications to Ultrafast Photonics," *Advanced Functional Materials*, vol. 25, no. 45, pp. 6996–7002, 2015.

[85] Ambrosi, A., Z. Sofer, M. Pumera, "Electrochemical Exfoliation of Layered Black Phosphorus into Phosphorene," *Angewandte Chemie International Edition*, vol. 56, no. 35, pp. 10443–10445, 2017.

[86] Kou, L., C. Chen, S. C. Smith, "Phosphorene: Fabrication, Properties, and Applications," *The Journal of Physical Chemistry Letters*, vol. 6, no. 14, pp. 2794–2805, 2015.

[87] Zhao, Y., Y. Chen, Y. H. Zhang, S. F. Liu, "Recent advance in black phosphorus: Properties and applications," *Materials Chemistry and Physics*, vol. 189, pp. 215–229, 2017.

[88] Mayorga-Martinez, C. C., Z. Sofer, M. Pumera, "Layered Black Phosphorus as a Selective Vapor Sensor," *Angewandte Chemie - International Edition*, vol. 54, no. 48, pp. 14317–14320, 2015.

[89] Pumera, M. "Phosphorene and black phosphorus for sensing and biosensing," *TrAC - Trends in Analytical Chemistry*, vol. 93, no. 2, pp. 1–6, 2017.

[90] Yang, A., J. Gao, B. Li, J. Tan, Y. Xiang, T. Gupta, L. Li, S. Suresh, J. C. Idrobo, T. M. Lu, M. Rong, N. Koratkar, "Humidity sensing using vertically oriented arrays of ReS 2 nanosheets deposited on an interdigitated gold electrode," *2D Materials*, vol. 3, no. 4, pp. 045012, 2016.

[91] Farahani, H., R. Wagiran, M. Hamidon, "Humidity Sensors Principle, Mechanism, and Fabrication Technologies: A Comprehensive Review," *Sensors,* vol. 14, no. 5, pp. 7881–7939, 2014.

[92] Gu, L., Q. A. Huang, M. Qin, "A novel capacitive-type humidity sensor using CMOS fabrication technology," *Sensors and Actuators B: Chemical*, vol. 99, no. 2–3, pp. 491–498, 2004.

[93] Li, N., Y. Jiang, C. Zhou, Y. Xiao, B. Meng, Z. Wang, D. Huang, C. Xing, Z. Peng, "High-Performance Humidity Sensor Based on Urchin-Like Composite of Ti_3C_2 MXene-Derived TiO_2 Nanowires," *ACS Applied Materials and Interfaces,* vol. 11, no. 41, pp. 38116–38125, 2019.

[94] Yasaei, P., A. Behranginia, T. Foroozan, K. Kim, F. Khalili-araghi, A. Salehi-khojin, "Stable and Selective Humidity Sensing Using Stack of Black Phosphorus Flakes Stable and Selective Humidity Sensing Using Stack of Black Phosphorus Flakes Abstract," *ACS Nano*, no. 10, pp. 9898–9905, 2015.

[95] Erande, M. B., M. S. Pawar, D. J. Late, "Humidity Sensing and Photodetection Behavior of Electrochemically Exfoliated Atomically Thin-Layered Black

Phosphorus Nanosheets," *ACS Applied Materials and Interfaces*, vol. 8, no. 18, pp. 11548–11556, 2016.

[96] Zhu, X., T. Zhang, Z. Sun, H. Chen, J. Guan, X. Chen, H. Ji, P. Du, S. Yang, "Black Phosphorus Revisited: A Missing Metal-Free Elemental Photocatalyst for Visible Light Hydrogen Evolution," *Advanced Materials*, vol. 29, no. 17, pp. 1605776, 2017.

[97] Yao, Y., H. Zhang, J. Sun, W. Ma, L. Li, W. Li, J. Du, "Novel QCM humidity sensors using stacked black phosphorus nanosheets as sensing film," *Sensors and Actuators, B: Chemical*, vol. 244, pp. 259–264, 2017.

[98] Yasaei, P., B. Kumar, T. Foroozan, C. Wang, M. Asadi, D. Tuschel, J. E. Indacochea, R.F. Klie, A. Salehi-Khojin, "High-Quality Black Phosphorus Atomic Layers by Liquid-Phase Exfoliation," *Advanced Materials*, vol. 27, no. 11, pp. 1887–1892, 2015.

[99] Liu, X., X. Shao, F. Li, M. Zhao, "Anchoring effects of S-terminated Ti_2C MXene for lithium-sulfur batteries: A first-principles study," *Applied Surface Science,* vol. 455, no. january, pp. 522–526, 2018.

[100] Mashtalir, O., K. M. Cook, V. N. Mochalin, M. Crowe, M. W. Barsoum, Y. Gogotsi, "Dye adsorption and decomposition on two-dimensional titanium carbide in aqueous media," *Journal of Materials Chemistry A*, vol. 2, no. 35, pp. 14334–14338, 2014.

[101] Mashtalir, O., M. Naguib, V. N. Mochalin, Y. Dall'Agnese, M. Heon, M. W. Barsoum, Y. Gogotsi, "Intercalation and delamination of layered carbides and carbonitrides," *Nature Communications*, vol. 4, no. 1, pp. 1716, 2013.

[102] Kajiyama, S., L. Szabova, K. Sodeyama, H. Iinuma, R. Morita, K. Gotoh, Y. Tateyama, M. Okubo, A. Yamada, "Sodium-Ion Intercalation Mechanism in MXene Nanosheets," *ACS Nano*, vol. 10, no. 3, pp. 3334–3341, 2016.

[103] Srimuk, P., F. Kaasik, B. Krüner, A. Tolosa, S. Fleischmann, N. Jäckel, M. C. Tekeli, M. Aslan, M. E. Suss, V. Presser, "MXene as a novel intercalation-type pseudocapacitive cathode and anode for capacitive deionization," *Journal of Materials Chemistry A*, vol. 4, no. 47, pp. 18265–18271, 2016.

[104] Zhu, X., B. Liu, H. Hou, Z. Huang, K. M. Zeinu, L. Huang, X. Yuan, D. Guo, J. Hu, J. Yang, "Alkaline intercalation of Ti3C2 MXene for simultaneous electrochemical detection of Cd(II), Pb(II), Cu(II) and Hg(II)," *Electrochimica Acta*, vol. 248, pp. 46–57, 2017.

[105] Lorencova, L., T. Bertok, E. Dosekova, A. Holazova, D. Paprckova, A. Vikartovska, V. Sasinkova, J. Filip, P. Kasak, M. Jerigova, D. Velic, K. A. Mahmoud, J. Tkac, "Electrochemical performance of Ti_3C_2Tx MXene in aqueous media: towards ultrasensitive H_2O_2 sensing," *Electrochimica Acta*, vol. 235, pp. 471–479, 2017.

[106] Wu, D., M. Wu, J. Yang, H. Zhang, K. Xie, C. T. Lin, A. Yu, J. Yu, L. Fu, "Delaminated Ti_3C_2Tx (MXene) for electrochemical carbendazim sensing," *Materials Letters*, vol. 236, pp. 412–415, 2019.

[107] Li, L., F. Wang, J. Zhu, W. Wu, "The facile synthesis of layered Ti_2C MXene/carbon nanotube composite paper with enhanced electrochemical properties," *Dalton Transactions*, vol. 46, no. 43, pp. 14880–14887, 2017.

[108] Zhou, J., X. Zha, X. Zhou, F. Chen, G. Gao, S. Wang, C. Shen, T. Chen, C. Zhi, P. Eklund, S. Du, J. Xue, W. Shi, Z. Chai, Q. Huang, "Synthesis and Electrochemical Properties of Two-Dimensional Hafnium Carbide," *ACS Nano*, vol. 11, no. 4, pp. 3841–3850, 2017.

[109] Kong, F., X. He, Q. Liu, X. Qi, Y. Zheng, R. Wang, Y. Bai, "Improving the electrochemical properties of MXene Ti3C2 multilayer for Li-ion batteries by vacuum calcination," *Electrochimica Acta,* vol. 265, pp. 140–150, 2018.

[110] Das, J. K., A. K. Samantara, A. K. Nayak, D. Pradhan, J. N. Behera, "VS2 an efficient catalyst for an electrochemical hydrogen evolution reaction in an acidic medium," *Dalton Transactions*, vol. 47, no. 39, pp. 13792–13799, 2018.

[111] Zhang, Q., W. Wang, J. Zhang, X. Zhu, Q. Zhang, Y. Zhang, Z. Ren, S. Song, J. Wang, Z. Ying, R. Wang, X. Qiu, T. Peng, L. Fu, "Highly Efficient Photocatalytic Hydrogen Evolution by ReS 2 via a Two-Electron Catalytic Reaction," *Advanced Materials,* vol. 30, no. 23, pp. 1707123, 2018.

[112] L. Y., Li, Y. Li, C. Tang, "Strain engineering and photocatalytic application of single-layer ReS2," *International Journal of Hydrogen Energy*, vol. 42, no. 1, pp. 161–167, 2017.

[113] Hu, Y., Y. Huang, C. Tan, X. Zhang, Q. Lu, M. Sindoro, X. Huang, W. Huang, L. Wang, H. Zhang, "Two-dimensional transition metal dichalcogenide nanomaterials for biosensing applications," *Mater. Chem. Front.*, vol. 1, no. 1, pp. 24–36, 2017.

[114] Rakhi, R. B., P. Nayuk, C. Xia, H. N. Alshareef, "Novel amperometric glucose biosensor based on MXene nanocomposite," *Scientific Reports*, vol. 6, no. november, pp. 1–10, 2016.

[115] Liu, H., C. Duan, C. Yang, W. Shen, F. Wang, Z. Zhu, "A novel nitrite biosensor based on the direct electrochemistry of hemoglobin immobilized on MXene-Ti_3C_2," *Sensors and Actuators, B: Chemical*, vol. 218, pp. 60–66, 2015.

[116] H. Zhang, Z. Wang, Q. Zhang, F. Wang, Y. Liu, "Ti_3C_2 MXenes nanosheets catalyzed highly efficient electrogenerated chemiluminescence biosensor for the detection of exosomes," *Biosensors and Bioelectronics*, vol. 124–125, no. october 2018, pp. 184–190, 2019.

[117] Huang, K. J., Y. J. Liu, G. W. Shi, X.R . Yang, Y. M. Liu, "Label-free aptamer sensor for 17β-estradiol based on vanadium disulfide nanoflowers and Au nanoparticles," *Sensors and Actuators, B: Chemical*, vol. 201, pp. 579–585, 2014.

[118] Liu, K., H. Deng, Y. Wang, S. Cheng, X. Xiong, C. Li, "A sandwich-type photoelectrochemical immunosensor based on ReS2 nanosheets for high-performance determination of carcinoembryonic antigen," *Sensors and Actuators B: Chemical,* pp. 128341, 2020.

[119] Liu, J., X. Jiang, R. Zhang, Y. Zhang, L. Wu, W. Lu, J. Li, Y. Li, H. Zhang, "MXene-Enabled Electrochemical Microfluidic Biosensor: Applications toward Multicomponent Continuous Monitoring in Whole Blood," *Advanced Functional Materials*, vol. 29, no. 6, pp. 1807326, 2019.

[120] Sinha, Dhanjai, S., A. M. Mugo, J. Chen, K. S. Lokesh, MXene-based sensors and biosensors: next-generation detection platforms, in: Handb. *Nanomater. Anal. Chem.*, Elsevier, 2020.: pp. 361–372.

[121] Yao, Y., L. Lan, X. Liu, Y. Ying, J. Ping, "Spontaneous growth and regulation of noble metal nanoparticles on flexible biomimetic MXene paper for bioelectronics," *Biosensors and Bioelectronics*, vol. 148, no. september 2019, pp. 111799, 2020.

[122] Yew, Y. T., Z. Sofer, C. C. Mayorga-Martinez, M. Pumera, "Black phosphorus nanoparticles as a novel fluorescent sensing platform for nucleic acid detection," *Materials Chemistry Frontiers*, vol. 1, no. 6, pp. 1130–1136, 2017.

[123] Wu, L., J. Guo, Q. Wang, S. Lu, X. Dai, Y. Xiang, D. Fan, "Sensitivity enhancement by using few-layer black phosphorus-graphene/TMDCs heterostructure in surface plasmon resonance biochemical sensor," *Sensors and Actuators, B: Chemical,* vol. 249, pp. 542–548, 2017.

Chapter 6

Energy Storage and Energy Conversion Applications of Transition Metal Chalcogenides

Amreetha Seetharaman[1], Manikandan Kandasamy[1] and Brahmananda Chakraborty[2,3,*]

[1]Nonlinear Optical Materials Laboratory, Department of Physics, Bharathidasan University, Tiruchirappalli, Tamil Nadu, India
[2]High Pressure and Synchrotron Radiation Physics Division, Bhabha Atomic Research Centre, Trombay, Mumbai, India
[3]Homi Bhabha National Institute, Mumbai, India

Abstract

Nowadays, renewable energy sources need efficient energy storage and conversion technologies. Supercapacitors and solar cells are the most promising energy storage and conversion devices. Metal chalcogenides attract in the supercapacitors aspect due to their physicochemical properties. As electrode materials in supercapacitors, these crystal structures, and multiple oxidation states facilitate enhanced specific capacitance, energy density, and longer charge and discharge cycles. In the case of solar cells, 2D materials are considered the better substitute for silicon-based solar cells due to their excellent electrical, mechanical, physical, and chemical properties. Among the various 2D materials, 2D transition metal chalcogenides are of better choice because they have high carrier mobility, tuneable bandgap, and high transparency. It can be prepared as single layers to a few layers through easy fabrication methods (mechanical exfoliation or chemical vapor deposition). It acts as a hole transport and an electron transport layer in perovskite solar cells by

* Corresponding Author's Email: brahma@barc.gov.in.

In: 2D Metallic Transition Metal Dichalcogenides
Editors: Chandra Sekhar Rout and Brahmananda Chakraborty
ISBN: 978-1-68507-965-9

tuning the work function. This chapter focuses recent advancement of 2D metal chalcogenides in supercapacitor and solar cell applications.

Keywords: TMDCs, supercapacitors, solar cells

6.1. Introduction

The widespread use of fossil fuels forces us to concentrate on alternative energy renewable sources (hydro, solar, and wind) to generate/store energy. Developing renewable energy-producing systems and storage technologies is critical to meet commercial/residential energy sources. The renewable energy sources are intermittent, which creates continuous energy demand. It needs an efficient energy storage device (Batteries/supercapacitors) to short out. Lithium-ion batteries (LIBs) are widely used in electronic items/vehicles. In 2019, M. Stanley Whittingham, John B. Goodenough, and Akira Yoshino were awarded the Nobel Prize for their contribution to the advancement of lithium-ion battery [1-3]. LIBs benefit from being a high-density energy source, but they also have the disadvantages of short cycle life, low power efficiency, and explosive. Supercapacitors are alternatives to LIBs because it exhibits high power density (10 times higher than LIBs) and long cyclic stability of tens of thousands. The inherently low energy density is one of the main drawbacks of supercapacitors. It is due to the electrostatic charge storage in electrical double layer (EDL) based electrodes, which can overcome through high surface area and exact pore sizes.

Moreover, another electrochemical capacitor type is pseudocapacitors that quickly store the charges on the electrode surface through faradaic redox reactions. Pseudocapacitive-based supercapacitors show greater energy density than that EDL capacitors. It is because of multiple oxidation states involved in the redox charge transfer reactions. However, it has a comparatively poor cycle life (charge/discharge) due to sluggish ion transfer rates and low electrical conductivity. It restricts their extensive industrial realization. Nonetheless, these energy storage/conversion systems have revealed a range of problems that require to be tackled along with the ever-rising power requirements. Figure 6.1 compares the power density and energy density of secondary batteries and supercapacitors [4].

The energy landscape might be changed considerably once the abovementioned boundaries are finely addressed. For example, in electrochemistry, the effectiveness of energy storage devices depends on the electrode material characteristics. Table 6.1 quantitatively compares the various vital features of the energy storage devices [2, 5].

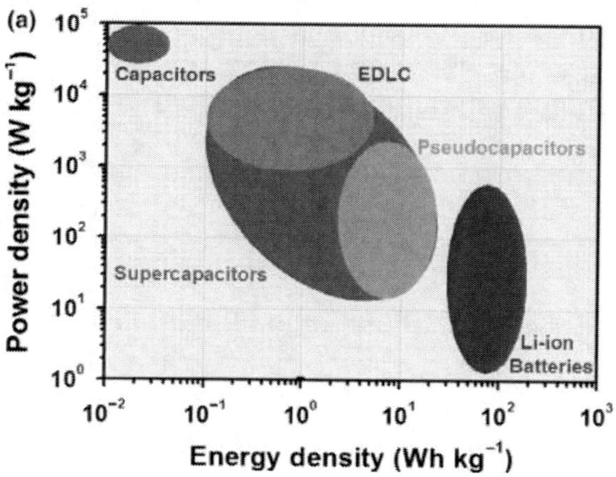

Figure 6.1. Energy density and power density comparison for the energy storage devices [4].

Table 6.1. Properties of various energy storage devices

Properties	Capacitors	Supercapacitors (SCs)			Batteries
		EDL capacitors	Pseudocapacitors	Hybrid capacitors	
Energy density (Wh/kg)	< 1	~ 10	>100		>1000
Maximum Cell Voltage (V)	~ 1000	3.8	4.0	5.0	5.0
Power density (kW/kg)	>10	0.5 - 10	~ 100	>30	<1
Duration for Charge/ Discharge	10^{-6} -10^{-3} s	s-min	s-min	s-min	0.3-10 h
Operational Life Time (cycles)	Almost infinite	>1000000	>500000	>100000	<10000
Coulombic efficiency (%)	~ 100	Up to 100	~ 99	~ 99	70-99

In energy storage systems, 2D materials (Figure 6.2) with their peculiar physical/chemical properties of the single-layered structure, a broad degree of anisotropy, and an adjustable band structure have been encouraged. With a growing number of 2D materials being discovered and synthesized using simple techniques, enormous research is investigating and applying this new family of materials. After the discovery of graphene (2004), 2D materials like transition metal dichalcogenides (TMDCs), MXenes, hexagonal boron nitride (h-BN), black phosphorous/phosphorene, borophene, silicene, germanene, graphynes, and graphdiynes have been explored. Borophene, phosphorene, germanene, and silicene structure like graphene. In contrast, TMDCs are a class of MX_2 materials (X-M-X), where M denotes the metal atom such as Mo, W, Zr, Hf, Nb, Ta, Ti, Ni, V, and Re. X states the chalcogen atom-like S, Se, and Te [6-9].

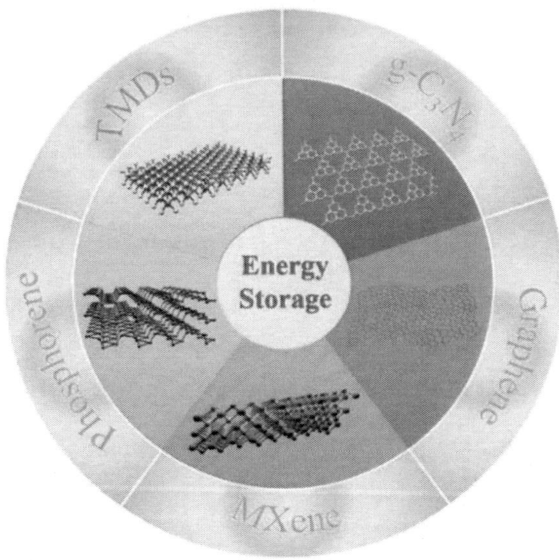

Figure 6.2. Different types of 2D materials consider for energy storage applications [10].

Individual TMDCs monolayer consists of three atomic layers. Two chalcogen layers cover the top and bottom of the transition metal layer. In TMDCs, the intralayer of X- M-X connects through a covalent bond, whereas each layer of X-M-X links via a non-covalent bond with weak van der Waals forces. TMDCs produce varieties of crystal poly-types, and as an example, MOS_2 shows four different crystal systems 1T, 1T′, 2H, and 3R. The Mo and S atom's coordination and layers packing orders are responsible for this crystal system. Bulk MoS_2 owns an indirect bandgap (1.29 eV) with semiconducting behavior. The monolayer 2H-phase of MoS_2 shows the direct bandgap of 1.9 eV, resulting in solid photoemission. Hence, benefiting from the direct bandgap of MoS_2 is more useful in photonics and electronic device fabrication. The graphene having room temperature carrier mobility is about ~ 10000 cm^2

V/s. Unfortunately, it lacks a bandgap considered inappropriate for high-performance applications. However, MoS_2, WS_2, $MoSe_2$, and WSe_2, have large bandgap (~1-2 eV) and moderate carrier mobility (a few hundred cm^2 V/s). The direct bandgap of monolayer MoS_2 shows the potential photoresponsivity (880 A/W) compared to the graphene (0.5 mA/W) at 561 nm wavelength excitation due to the increase of the high photon absorption coefficient. Bulk MoS_2 has the young's modulus value of 205 GPa, whereas the single-layer MoS_2 shows 270 GPa [11,12].

2D layered TMDCs with a higher surface area make the enormous contact zone between the anode and electrolyte. It leads to rapid Faradaic non-Faradaic reactions. The layered TMDCs contain an under-coordinated edge, which can serve as metal ion adsorption sites. It contributes to the storage ability of extra metal ions. The neighboring X-M-X interlayer layers are connected through a weak van der Waals force of attraction, which favors quick ion diffusion. 2D TMDCs nanosheets with thin and versatile characteristics make them flexible electrochemical energy storage devices. Another important class of this material is WS_2 which shows wide-bandgap and inherent crystal structures.

TMDCs frequently hybridize with other 2D materials like MXenes, metal sulfides, metal oxides, carbon materials, and conducting polymers for electrochemical energy storage. The layered TMDCs mainly hybridize with MXene and carbon materials. The high conductivity of TMDCs will speed up electron transport. Furthermore, the hybridization successfully prevents TMDCs nanosheet agglomeration due to van der Waals forces. It retains a large surface area of TMDCs [13,14]. The present chapter will focus on the 2D TMDCs for energy storage and energy conversion applications, specifically supercapacitors and solar cells.

6.2. Energy Storage Devices and Their Storage Mechanisms

Supercapacitors are categorized into electric double-layer capacitors (EDLCs), pseudocapacitors, and hybrid supercapacitors. Hybrid supercapacitors can have the ability to extend the operating voltage beyond 2 V. Whereas symmetric aqueous supercapacitor shows a limited voltage of 1.2 V. The following equation calculates the energy density of the supercapacitors,

$$E = \frac{1}{2} CV^2 \quad (6.1)$$

C and V are the capacitance and operating voltage, respectively. We can observe that a 2-fold increment of the voltage leads to a 4-fold increment of the energy density of the same capacitance. Supercapacitors' developments began with a static charge storage mechanism in the mid of the 18th century. First, von Helmholtz has constructed the electric double-layer model by examining the colloidal state of suspensions and explained the electrical charge-storage mechanism. Then, Gouy, Chapman, Stern, and Grahame, have been established the modern theory of electrical double-layer capacitance through two metal electrodes and liquid electrolytes interfaces. Another capacitor named pseudocapacitors, which worked under Faradaic processes, was established in 1971 for RuO_2 material. Hence the breakthrough of pseudocapacitance enhances the charge-storage performance of electrochemical capacitors. Supercapacitors focused on applications like heavy loading trucks or cranes (high-power energy sources) and portable electronics (energy backup).

In the 21st century, the quantity of research associated with supercapacitors has increased due to the demand for high- reliability, power, and safe energy-storage devices. The charge storage process of supercapacitor devices mainly focuses on fast electrostatic or Faradaic electrochemical processes. It consists of the electrodes (positive and negative) and is distinguished by an insulating separator. The working mechanism of the supercapacitor is related to the conventional capacitors. The supercapacitors' specific capacitance and energy density can be increased to tens of thousand-fold compared to the regular capacitors. The charge is accumulated predominantly at the electrode-electrolyte interface of the working materials, e.g., porous carbons, metal oxides, or polymers. The charging and discharging time of the supercapacitor is much faster than the batteries. Also, it shows better advantages of high-power density, long cycling life, high operating safety, high efficiency, and higher stability. EDLC is the easiest and most wanted commercial supercapacitor, and the charges are stored physically by electrostatic charge adsorption at the interface. When the electrode is submerged in an electrolyte solution, charges are separated, forming a double layer. In EDLCs, no charge transfer arises among the interfaces of electrodes and the electrolytes, i.e., no Faradaic processes happen. Hence, for the EDLC materials, the specific capacitance is reliant on the surface area of the electrode material. The specific capacitance can be determined by,

$$C= \varepsilon_r \varepsilon_0 A /d \quad (6.2)$$

Where ε_r denotes the relative permittivity of the electrolyte, ε_0 is vacuum permittivity, A is the electrode's surface area, and d is the Debye length, i.e., charge separation distance between the electrical double layers. The formation and relaxation of the electric double layer happen in ~ 10^{-8} s, while the redox reaction takes time in the range of 10^{-2} to 10^{-4} s. Hence, the EDLCs can respond more quickly to the potential changes. RuO_2 was the first reported electrode material that shows the Faradaic reaction. However, the CV curve exhibits near to a rectangular shape which is a typical capacitive nature. The pseudocapacitance mechanism is related to valence state changes of the electrode material. The name pseudocapacitance refers to the material's capacitive nature, but charges are stored in the form of Faradaic reactions. In pseudocapacitive energy storage systems, the charges are stored intermediate between EDLCs and solid-state diffusion dominated by Faradaic reactions. Various kinds of Faradaic mechanisms are involved in pseudocapacitance (Figure 6.3), such as (a) under potential deposition, (b) redox pseudocapacitance, (c) intercalation pseudocapacitance [15,16].

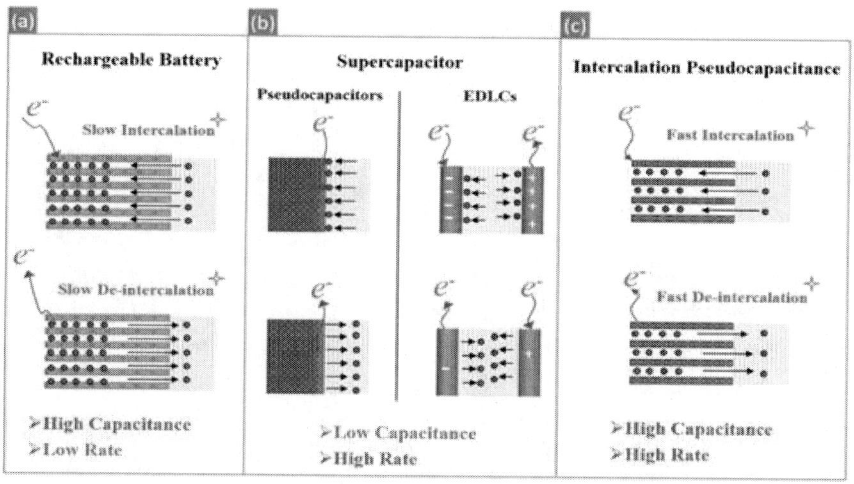

Figure 6.3. Various kinds of Faradaic mechanisms in pseudocapacitance [16].

6.2.1. 2D TMDCs for Supercapacitor

The 2D layer structured metal chalcogenides (MX_2 (M = Mo and W)) have a tuneable bandgap, layered type crystal structure, and high electron transportation, which encourages excellent electrochemical behavior for supercapacitor application. Furthermore, it has enhanced electron density states in the d subshell level, improving electronic conductivity and enhancing charge storage performance. Recently, Gupta et al. [17] have reported the rose flower-like nanostructured MoS_2 for supercapacitor applications. The electrode shows excellent specific capacitance (255 F/g) and energy density (35.5 Wh/kg) at 0.25 A/g for KOH electrolyte. Xu et al. [18] have reported the 1T phase MoS_2 micro-supercapacitors (MSCs) for future electronic devices. They have used the laser direct written (LDW) method to prepare MoS_2 (1T phase) MSCs.

The conventional nanosecond laser, femtosecond laser, and different pulse delays (5 ps, 10 ps) having temporally shaped femtosecond laser (t-fs laser) were utilized to improve the electrochemical performance. The electrolyte was H_2SO_4/PVA hydrogels which were drop-casted onto the MSCs (10 mL/cm^2) and kept overnight at atmospheric temperature. In this study they have observed excellent energy density (15.6 mWh/cm^3 at 1.12 W/cm^3) which is superior compared to s-fs MoS_2 (1T phase) MSCs (12.2 mWh/cm^3 at 0.87 w/cm^3) and traditional LDW MSCs (4.8 mWh/cm^3 at 0.34 W/cm^3). Li et al. [19] have prepared the AgNWs-MoS_2 transparent flexible electrode (TFE) (Figure 6.4a) which, showed the superior areal energy density (2.453 μWh/cm^2) at 0.2 V/s (scan rate) and power density (1.472 mW/cm^2). Ag NWs/MoS_2 hybrid surface offers additional active sites, improving the more ion intercalation and rapid electron transition. Hence, increasing electronic conductivity can make the abundant pathways for electron transport (Figure 6.4b).

From Figure 6.4c, the rate stability and the cyclic stability were plotted, which shows that the specific capacitance decreased ~ 70.3% (from 22.9 to 14.3 mF/cm^2 at 0.1 V/s) with raising scan rate (0.01V/s to 5 V/s) due to the transfer limit on the contribution of ions. On the other hand, the fabricated TFE shows excellent cyclic stability up to 20000 cycles which decreased to 3.7% at 0.05 V/s (scan rate).

Mn-doped MoS_2 nanoflowers show excellent specific capacitance (430 F/g). In addition, it shows high energy density (48.9 Wh/kg) and power density (5 kW/kg) with stable performance of 5000 life cycles at the current density of 10 A/g, according to Singha et al. [20]. The doped Mn atom in MoS_2 can prevent the restacking of the nanosheet and also increase the adsorption of the electrolyte ions. MoS_2/NiS Yolk-Shell-like structured has been tested by Qin et al. [21] for the asymmetric supercapacitor (ASC), and in this, activated carbon is used as a negative electrode. (Figure 6.5a-g) explains the possible structural formation of MoS_2/NiS yolk-shell.

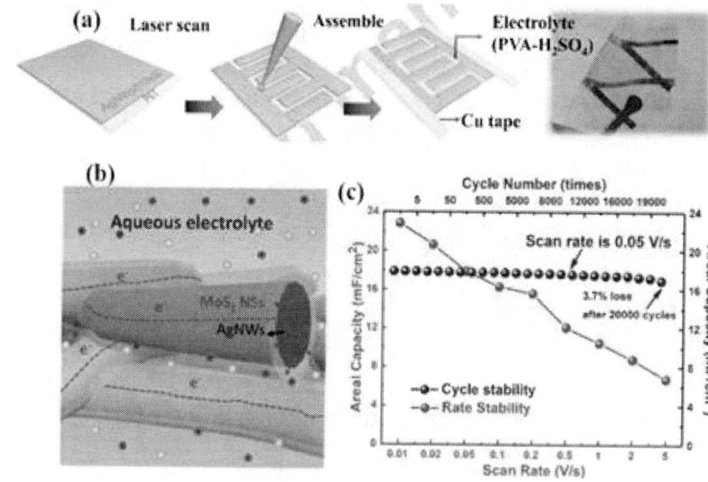

Figure 6.4. (a) laser printing method for the fabrication of transparent flexible micro-supercapacitor (b) electron transportation between MoS_2 nanosheet and Ag nanowire (c) stability for about 20000 cycles [19].

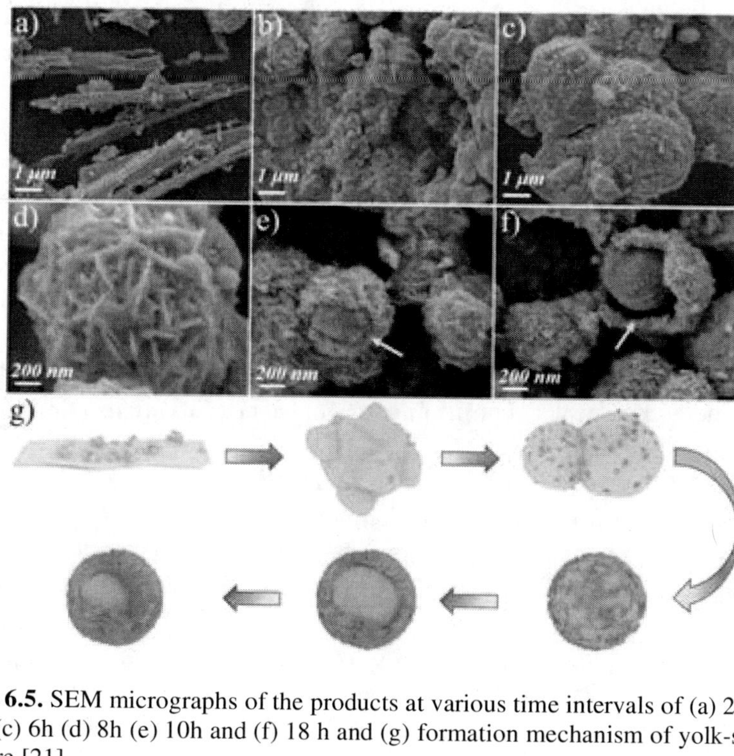

Figure 6.5. SEM micrographs of the products at various time intervals of (a) 2h (b) 4h (c) 6h (d) 8h (e) 10h and (f) 18 h and (g) formation mechanism of yolk-shell structure [21].

With the increasing reaction time from 2 to 18 hours, the amorphous crystal phase of $Ni_{2.5}Mo_6S_{6.7}$ decomposed and formed the stable phase of MoS_2/NiS, confirmed by XRD analysis. FESEM analysis reveals the structural changes from rod-like to yolk-shell structures from 2 to 18 hours. Figure 6.6a shows the excellent cyclic stability of 100% up to 10000 cycles at 0.5 A/g. Ragone plot shows (Figure 6.6b) that the excellent energy density (31 Wh/kg) with the power density (155.7 W/kg).

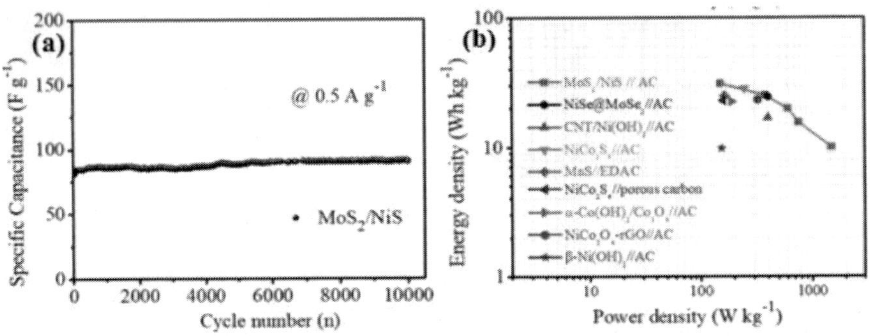

Figure 6.6. (a) Life cycle performance of MoS_2/NiS//AC device (b) comparison of energy density and power density of MoS_2/NiS//AC (asymmetric device) with other reported work [21].

The fabricated MoS_2/NiS microsphere can increase the surface volume resulting in excellent electro-active sites. The potential window reached about 1.48 V, owing to the decrease in the diffusion path lengths of the electrolyte's ions. MoS_2-$Ti_3C_2T_x$ hybrid was prepared through a hydrothermal method by Hou and his group [22]. The hybrid formation was confirmed through TEM (Figure 6.7a) HRTEM (Figure 6.7b, c)) analysis. From the CV curves (Figure 6.7d), it was observed that MoS_2-$Ti_3C_2T_x$ has rectangular-like geometry with two wide redox peaks, confirming the existence of faradaic redox-based charge storage. Compared to MoS_2 and $Ti_3C_2T_x$, the hybrid (MoS_2-$Ti_3C_2T_x$) has an enhanced area that claims excellent electrochemical performance. MoS_2-$Ti_3C_2T_x$ based electrode performs stable charge/discharge cycles and exhibits excellent capacitance retention (82%) beyond 10000 cycles. The fabricated symmetric supercapacitor exhibits outstanding capacitance (115.2 F/g) at the current density of 0.5 A/g and obtained noticeable Coulombic efficiency (90%) and cycling stability (72.3%) over 10000 charge/discharge cycles. The device (MoS_2-$Ti_3C_2T_x$) displays a high energy density of 5.1 Wh/kg at a power density of 298 W/kg. This excellent electrochemical performance results from the constructed mixed-dimensional heterostructure exhibiting the synergistic effect.

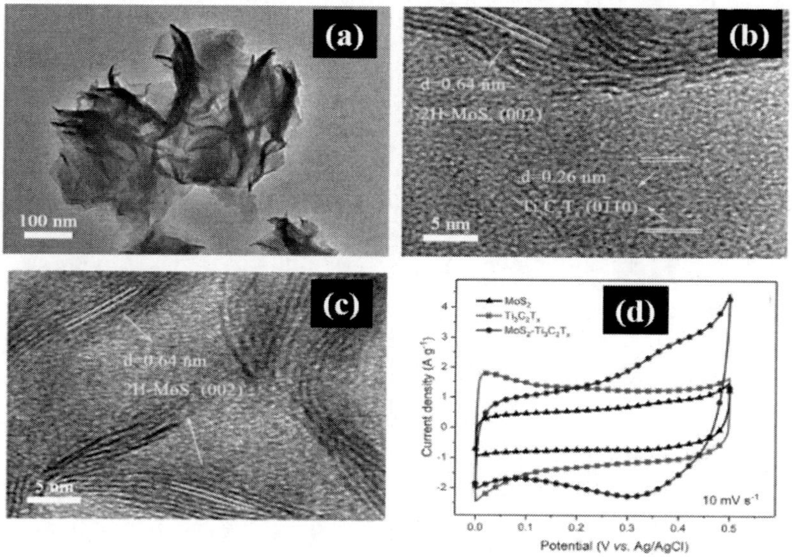

Figure 6.7. TEM (a) and HRTEM (b, c) images for MoS_2-$Ti_3C_2T_x$. CV curves (d) compare the electrochemical performance of MoS_2, $Ti_3C_2T_x$, and MoS_2-$Ti_3C_2T_x$ [22].

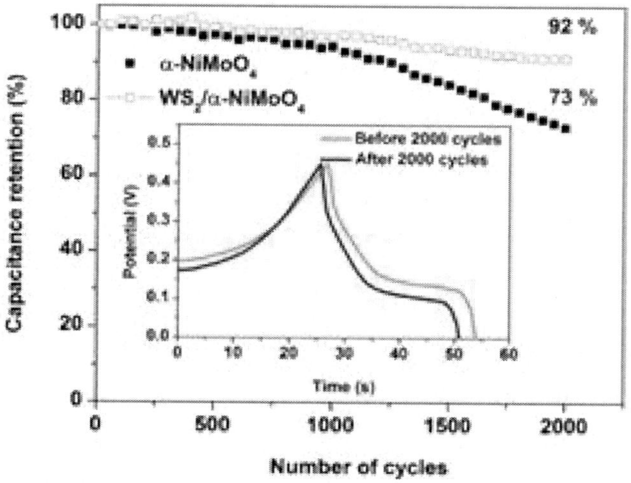

Figure 6.8. Comparison of life cycle performance of WS_2/α-$NiMoO_4$ hybrid electrode to the α-$NiMoO_4$ and inset depicts GCD curve of WS_2/α-$NiMoO_4$ before and after stability study [23].

WS_2 is another critical class of 2D metal chalcogenides for making better heterostructures. Ray et al. have fabricated the WS_2/α-$NiMoO_4$ hybrid electrode for supercapacitor application. The prepared cavity-like hierarchical structured electrode shows better cyclic stability of 92% up to 2000 cycles compared to α-$NiMoO_4$ (73%) (Figure 6.8) [23].

VS_2 has a half-filled energy band at the Fermi level. However, fabrication of a single phase of layered VS_2 is still challenging because the stoichiometric formation leads to the intermediate products of V_3S_4, V_4S_8, VS_2, etc. Ji et al. [24] have prepared VS_2/MoS_2 nanohybrid electrodes and show higher specific capacitance (8.6×10^2 F/g) for an aqueous electrolyte.

6.2.1.1. MoS_2 Hybrid

The structure of MoS_2 is similar to that of graphene. It has an enormous surface area, demonstrating the EDLC behavior. The Mo edges of MoS_2 are responsible for oxidation and reduction processes that generate extra capacitance via pseudocapacitive reactions. Ali et al. [25] have reported the impact of three key phases of 2H-MoS_2 (stable semiconductor), 3R-MoS_2 (metastable semiconductor), and 1T-MoS_2 (metastable metallic) in lithium sulfate (Li_2SO_4), sodium sulfate (Na_2SO_4), and potassium sulfate (K_2SO_4) electrolytes on the capacitance performance. Figure 6.9 depicts the formation of 1T/2H-MoS_2 (Figure 6.9a), 3R/2H-MoS_2 (Figure 6.9b), and 2H-MoS_2 (Figure 6.9c) samples. Pristine 2H-MoS_2 shows the pseudocapacitive behavior with EDLC and greater than 100% capacitance even after 2000 cycles for these electrolytes. However, it offers a low specific capacity. 3R/2H mixed-phase shows the capacitance retention for the initial 1000 life cycles; after stabilization, it displays $\geq$120% capacitance retention. The 1T/2H MoS_2 exhibited a capacitance tendency to electrolyte series (K_2SO_4> Na_2SO_4>Li_2SO_4). It shows capacitance retention drops for the complete first 50 cycles. After that, it exhibited excellent capacitance for the electrolytes of Li_2SO_4 (72%), Na_2SO_4 (50%), and K_2SO_4 (80%). 1T/2H MoS_2 was assembled in a symmetric supercapacitor and tested for K_2SO_4 electrolytes. CV curve of the device shown in Figure 6.9d, demonstrates a stable quasi-rectangular behavior up to the scan rate of 300 mV/s. GCD curves for various current densities (Figure 6.9e) clearly showed that the Coulombic efficiency was improved, and the GCD behavior is constant from 0.5 to 10 A/g. The fabricated device exhibited the specific capacitance of 79.4 and 37.2 F/g for the lower scan rate (5 mV/s) and current density (0.5 A/g), respectively. The specific capacitance was decreased to 31 and 13.44 F/g for the higher scan rate (300 mV/s) and current density (10 A/g), respectively (Figure 6.9f, g). Ragone plot (Figure 6.9h) shows the obtained energy densities (4.18 and 1.5 Wh/kg) with respect to power densities (225 and 4500 W/kg). Figure 6.9i shows that the device exhibits 82% and 72% capacitance retention beyond 1000 and 2000 cycles.

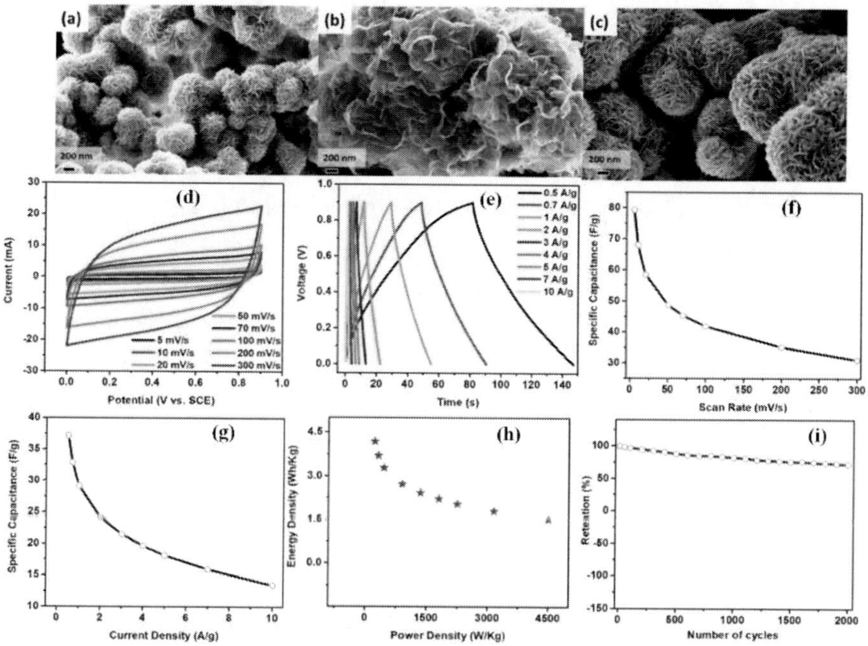

Figure 6.9. FESEM micrographs of (a) 1T/2H-MoS_2, (b) 3R/2H-MoS_2, and (c) 2H-MoS_2. (d) CV curves (e) GCD curves, (f) Rate capability (various scan rates) (g) Rate capability (current densities) (h) Ragone plot and (i) Stability of 1T/2H MoS_2 symmetric device [25].

Zhang et al. [26] have fabricated the sheared MnO_2/MoS_2 composites for symmetry all-solid-state supercapacitor (ASSS), which shows the capacitive behavior with varying energy density (0.0038 to 0.011 mWh/cm^2) and power density (0.05 to 0.5 mWh/cm^2) at 0.5mA/cm^2. It also shows excellent cyclic stability (90%) even beyond the 3000 charge/discharge cycles. Zhao et al. [27] have reported the graphene decorated $Ni(OH)_2/MoS_2$ (G-Ni $(OH)_2/MoS_2$) hybrids (Figure 6.10) for the asymmetric hybrid supercapacitors (AHSC) behavior. For the fabrication of AHSC, activated carbon and G-$Ni(OH)_2/MoS_2$ are employed as negative and positive electrodes, respectively. The faradaic redox reaction is attributed to the changes of $Ni(OH)_2$ to NiOOH (β). Graphene has been facilitated to raise the electronic conductivity and avoid the dissolution of $Ni(OH)_2$. Figure 6.11a represents the CV profiles of the ASCH electrode, which reveals battery type behavior owing to the redox behavior of Ni $(OH)_2$.

GCD curves (Figure 6.11b) were highly symmetrical, which ascribed outstanding electrochemical reversibility. Figure 6.11c indicates specific and volumetric density variation for various scan rates. The Ragone plot (Figure 6.11d) demonstrated that the device has outstanding energy density (13 Wh/kg) and power density (3500 W/kg). In addition, the device (G-Ni $(OH)_2/MoS_2$//AG AHSC) owned greater cyclic stability (Figure 6.11e). This enhancement in ASHS is due to the activated graphene, which promotes the ionic property and extends the potential window to enrich energy density and power density.

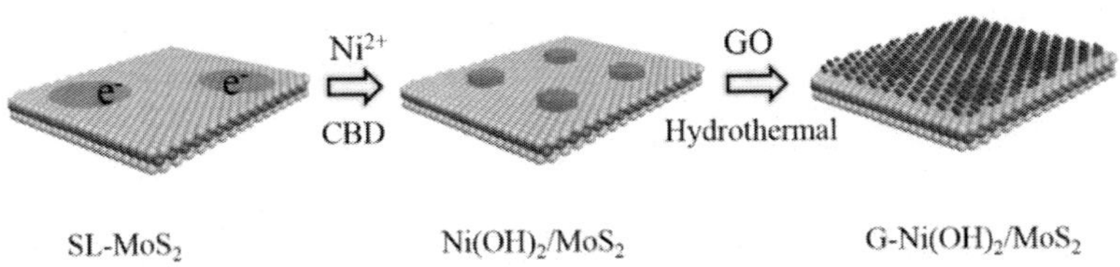

Figure 6.10. Fabrication process of $Ni(OH)_2/MOS_2$ and G-$Ni(OH)_2/MoS_2$ [27].

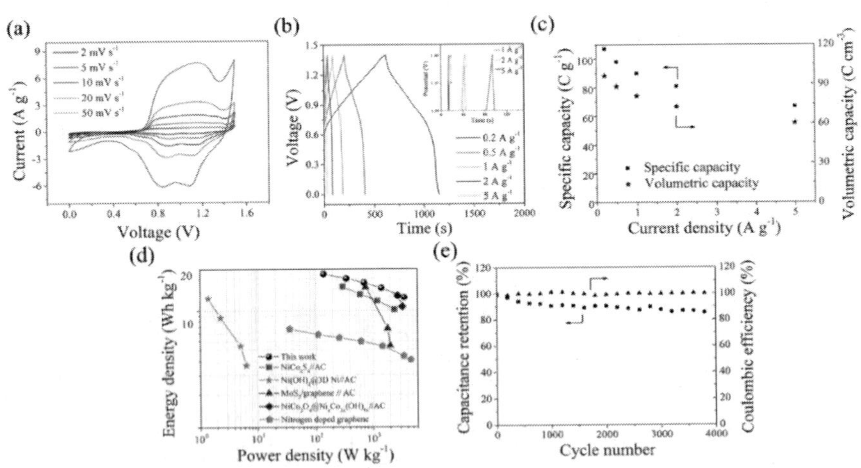

Figure 6.11. Electrochemical performance of G-$Ni(OH)_2/MoS_2$//AG AHSC for two-electrode system (a) CV profiles (b) GCD profiles (c) Specific capacity as well as volumetric capacity at different scan rates (d) Ragone plots (e) Capacitance retention versus the number of charge/discharge cycles and Coulombic efficiency [27].

Zeng et al. [28] have reported the covalently bonded PANI with 2D MoS_2 enhanced the electrical conductivity even at higher current densities. In addition, 4-aminophenyl has been functionalized on MoS_2 (MoS_2-NH_2) to hold PANI tightly onto the MoS_2 surface, and hence S - C covalent bond was formed.

This hybrid shows the Faradaic process due to the oxidation states holding Mo and PANI atoms. Figure 6.12a shows the formation of the MoS_2-NH_2/PANI hybrid. Figure 6.12b shows the Ragone plot of symmetric MoS_2-NH_2/PANI-150//MoS_2-NH_2/PANI-150 electrode with an energy density of 4 Wh/kg at a power density of 175 W/kg. Figure 6.12c shows the electrode's extraordinary stability, 96.5% up to 10000 cycles. This symmetric electrode was further examined in the practical implementation, which powdered the LED and ran the electrical fan (Figure 6.12d).

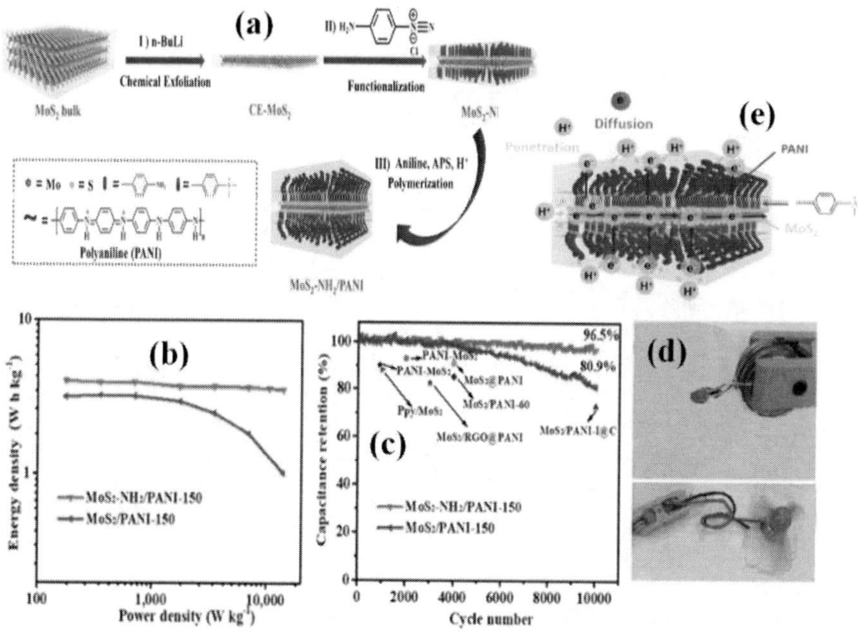

Figure 6.12. a) Schematic representation of fabrication of MoS_2-NH_2/PANI-x (b) Ragone plot (c) comparison of cyclic stability of MoS_2-NH_2/PANI -150 and MoS_2/PANI-150 constructed SSC at 5 A/g with various MoS_2 based electrode supercapacitors (d) Turning on red LED and electronic fan by the MoS_2-NH_2/PANI -150 [28]. (e) Schematic depiction for energy storage mechanism for MoS_2-NH_2/PANI-150.

Figure 6.12e expresses the charge storage mechanism of the electrode in which the linker 4-aminophenyl helped to bind the PANI in an ordered manner onto the MOS_2 nanosheet and provides the highway of ion diffusion. In MoS_2-NH_2/PANI-150 hybrids, PANI is covalently connected with MoS_2-NH_2, which provides a freeway for the diffusion of the ions during charging and discharging. Hence, the fast ion diffusion leads to the high-rate capability of the hybrids. In addition, this hybrid also enhances the high surface area resulting in enhanced electrochemical capacitance.

Chen et al. [29] have fabricated the MoS_2/PANI/CNT hybrid paper ASSS (Figure 6.13a). Figure 6.13b shows the Nyquist plot of the fabricated ASSS, which has the series resistance (R_s) of 19.09 Ω and charge transfer resistance (R_{CT}) of 13.13 Ω. It indicates the excellent electronic conductivity and the ionic diffusion rate. Figure 6.13c shows the difference in the potential window as single and three ASSS have been connected in series, which shows the enhanced potential window from 1.0 to 3.0 V, respectively. LED light has been powered with the help of series-connected ASSS, which declares the practical usage of the flexible device (Figure 6.13d).

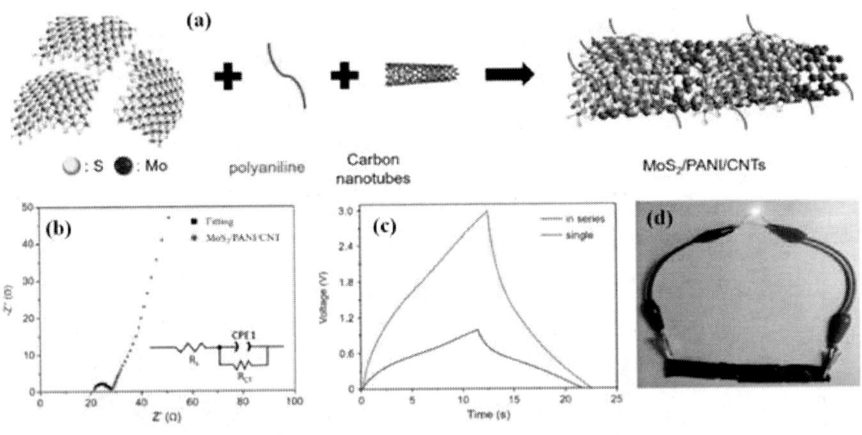

Figure 6.13. (a) Schematic process of MoS_2/PANI/CNT freestanding hybrid paper (b) Nyquist plot for MoS_2/PANI/CNT solid-state supercapacitor (c) GCD profiles for single and serially connected MoS_2/PANI/CNT solid-state supercapacitor (d) LED powered by a serially connected solid-state supercapacitor [29].

2D/2D layered hybrids greatly enhance the capacitive behavior caused by the quick ion intercalation and charge transfer along the channels. Wang et al. [30] have fabricated the SnS_2/MoS_2 hybrid for supercapacitor application (Figure 6.14a). In this, CV curves (Figure 6.14b) show no geometrical distortion for increasing scan rate, and GCD curves are depicted in Figure 6.14c. The constructed symmetric supercapacitor shows an excellent capacitance (98.8 F/g) at 1 A/g (current density) and energy density (115 Wh/kg) at 2230 Wh/kg (power density) (Figure 6.14d).

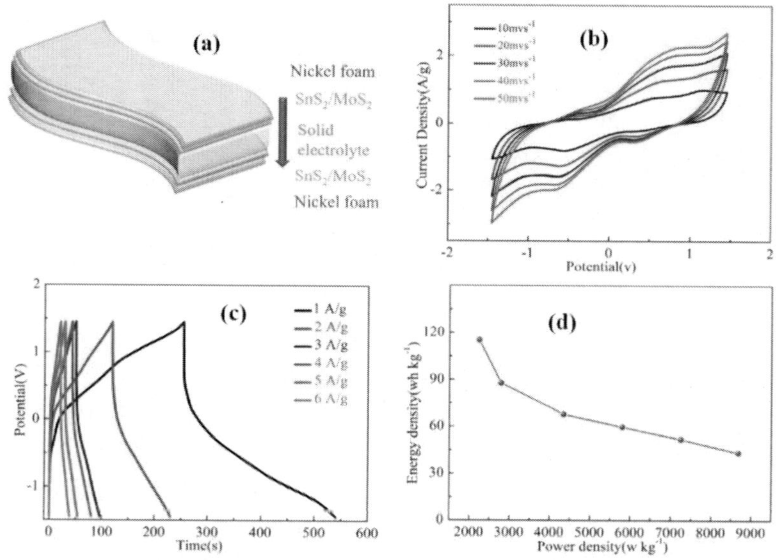

Figure 6.14. (a) Constructed symmetric supercapacitor of SnS_2/MoS_2// SnS_2/MoS_2 (b) CV profiles of the device (different scan rate) (c) Charge-discharge profiles (different current density) (d) Variation of energy density versus power density [30].

Chen et al. [31] have made the flexible supercapacitor by RGO/ MOS_2/Poly ternary electrode (Figure 6.15a) for various electronics applications. The fabricated ternary electrode shows the areal capacitance of 241.8 mF/cm^2 at 0.5 mA/cm^2 (current density). The author suggested that the improved performance of the ternary electrode is due to the synergistic effect among the three components. RGO improves the network's electrical conductivity; meanwhile, MoS_2 and poly afford pseudocapacitive behavior. Figure 6.15b shows the average cycling stability of the prepared materials with the capacitance retention of ~ 93% up to 5000 cycles. Figure 6.15c explains the various interaction mechanism among the components. Finally, Figure 6.15d shows the flexibility of the fabricated solid-state supercapacitor with no variation in capacitance while bending (0° to 180°). The enhanced capacitive behavior has been explained as the formation of RGO, forming the conductive network and providing a large active surface area. At the same time, poly nano whiskers covered the RGO surface, which affords the large pseudocapacitance. In addition to that, MoS_2 prevents the restacking of RGO and expands porous structure on poly. Hence the effective electron transmission improved the electrochemical behavior of the ternary nanohybrids.

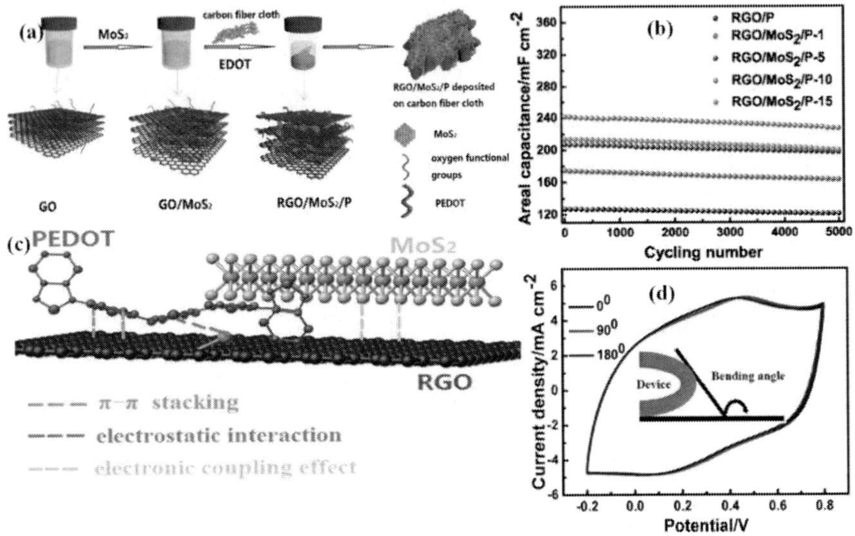

Figure 6.15. (a) Schematic representation of flexible electrode preparation of RGO/MoS_2/P (b) Stability test of the different composition of RGO/MoS_2/P (c) Pictorial illustration for possible interaction among three components (RGO/MoS_2/P) (d) CV profiles of the flexible electrode at various orientations [31].

WO_3 can easily hybrid with 1T phase MoS_2, which offers additional hexagonal tunnels and more active sites, leading to the quick transfer of electrolyte ions and higher charge storage capacity. Combining the EDLC and pseudocapacitive materials gives an asymmetric hybrid supercapacitor, enhancing the electrochemical process. Figure 6.16a shows the formation process of 1T MoS_2/hexagonal WO_3/RGO hybrid electrode. The calculated specific capacitance of the ternary hybrid is 836 F/g at 1 A/g (Figure 6.16b). The enhanced performance of the ternary hybrid than those of binary electrodes (MoS_2/Graphene and WO_3/Graphene) is attributed as follows. (i) 1T phase MoS_2, hexagonal WO_3, and graphene have sufficient electronic interaction, supplying fast ion transfer paths. (ii) MoS_2 (nanoflower structure) and hexagonal WO_3 store charges via EDLC and pseudocapacitance [32].

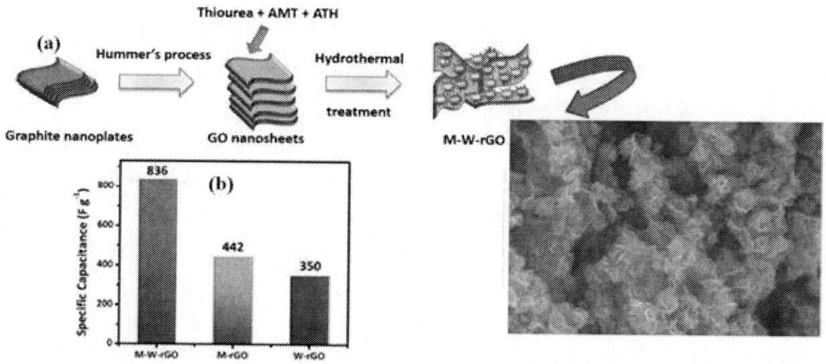

Figure 6.16. (a) Schematic representation for the process of 1T $MoS_2/WO_3/$ Graphene hybrid (b) Comparison of specific capacitance of the fabricated electrodes [32].

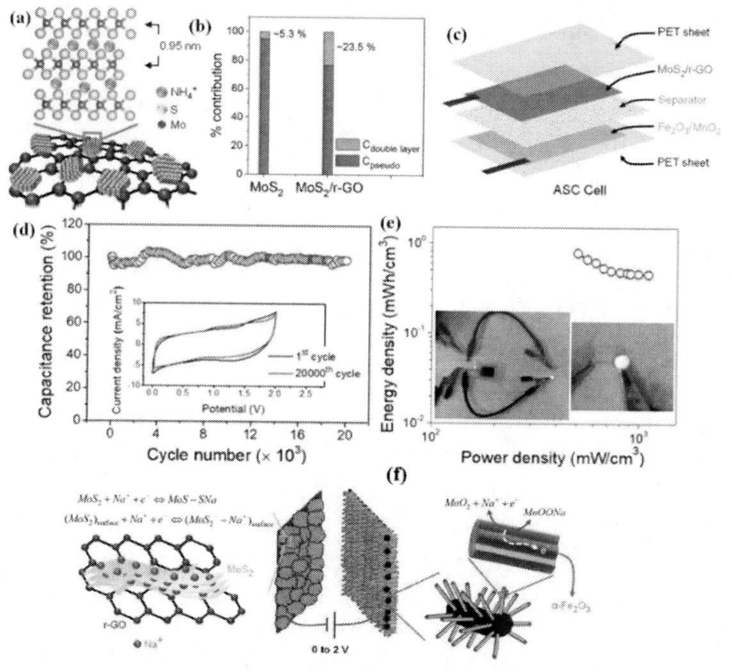

Figure 6.17. (a) Intercalation of ammonium ion into MoS_2 layer (b) contribution of EDLC and pseudocapacitance in MoS_2 and MoS_2/rGO hybrid at 0.5 V (vs. SCE) (c) formation of MoS_2/rGO //Fe_2O_3/MnO_2 asymmetric supercapacitor device (ASC) (d) cycling stability test up to 20000 cycles at the scan rate of 100 mV/s; inset primary and last CV curves (e) Ragone plot of ASC device and inset red LED powered by ASC device (f) Charge storage mechanism of an anode (MoS_2/rGO) and cathode [33].

Sarkar et al. [33] have fabricated the MoS_2/rGO//Fe_2O_3/MnO_2 asymmetric supercapacitor (ASC). In this, MoS_2/rGO hybrid has been grown directly onto the Mo foil, and thiourea is used as an exfoliating agent and a sulfur source. Ammonium ions have released from thiourea, and rGO exfoliates MoS_2 formed with the high interlayer distance (0.95 nm), which facilitates the formation of the trigonal MoS_2 phase (Figure 6.17a). This interlayer distance (0.95) nm has been reported as a higher spacing value (*d*) compared to the standard MoS_2 layer (~ 0.65 nm), which is sufficient for Na^+ ions intercalation. Hence, the MoS_2/rGO layer's expanded and opened-up structure helps improve the charge-storing process with quick insertion/exertion of Na^+ ions. The EDLC percentage increased from 5.3 to 23.5% for pristine MoS_2 to MoS_2/rGO hybrid, respectively (Figure 6.17b). It shows the effectiveness of rGO, which provides the additional active sites for the outstanding charge storage process through the EDLC process. For asymmetric capacitive behavior, the MoS_2/rGO and Fe_3O_3/MnO_2 core-shell nanorods act as anode and cathode, respectively (Figure 6.17c). Moreover, this ASC device exhibits enhanced cyclic performance of about 98% retention up to 20000 cycles (Figure 6.17d). ASC delivers volumetric energy of 0.78 mWh/cm^3 and a power density of 500 mW/cm^3. Commercial LED light is powered on using an ASC device after recharging at 2V (Figure 6.17e). Figure 6.17f explains the fast ion intercalation into the hybrids.

6.2.1.2. 2D/2D TMDCs//MXene Hybrid

MXene based hybrids are a new class of materials that show enhanced capacitive behavior. MAX phase of Ti_3C_2 was first produced in 2011. After that remarkable research, Jan et al. have prepared the few-layered Ti_3C_2/WS_2 hybrids (Figure 6.18a) for supercapacitor application. It shows the enhanced capacitive behavior (Figure 6.18b). Single/few layers of MXene have been fabricated by the delamination process, which enhances the surface area and active sites lead to the increasing of capacitive behaviour [34].

2D/2D 1T-MoS_2/Ti_3C_2 MXene hybrid electrode has been fabricated for the supercapacitor applications by Wang et al. [35]. Figure 6.19 reveals the synthesis of pure 1T phase of MoS_2 by the magneto-hydrothermal method by applying the magnetic field of 9 T. Ti_3AlC_2 was etched with hydrofluoric acid (HF) to get Ti_3C_2 MXene. Then Ti_3C_2 MXene sheets were dispersed to the solution containing citric acid monohydrate, ammonium heptamolybdate tetrahydrate, and thiourea. This solution was further subjected to the hydrothermal and magneto-hydrothermal methods. The pure MoS_2 (1T phase)/Ti_3C_2 MXene was prepared by the magneto hydrothermal method, while the mixed phase of 2H and 1T phase of MoS_2/Ti_3C_2 MXene was observed at the standard hydrothermal method. Variation of d spacings of MoS_2 with the hydrothermal (6.3 Å) and the magneto-hydrothermal (9.4 Å) method was observed, and the enhanced interlayer distance can improve the intercalation of ions and the electron conductivity.

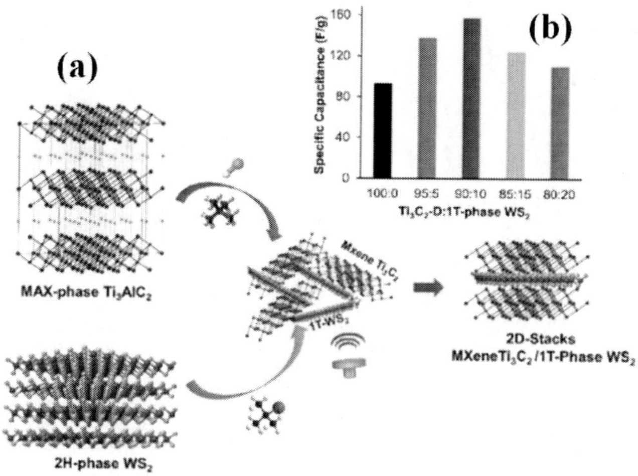

Figure 6.18. (a) Formation of MXene Ti_3C_2/1T Phase WS_2 (b) Specific capacitance value of the different ratios of Ti_3C_2:1T phase WS_2 [34].

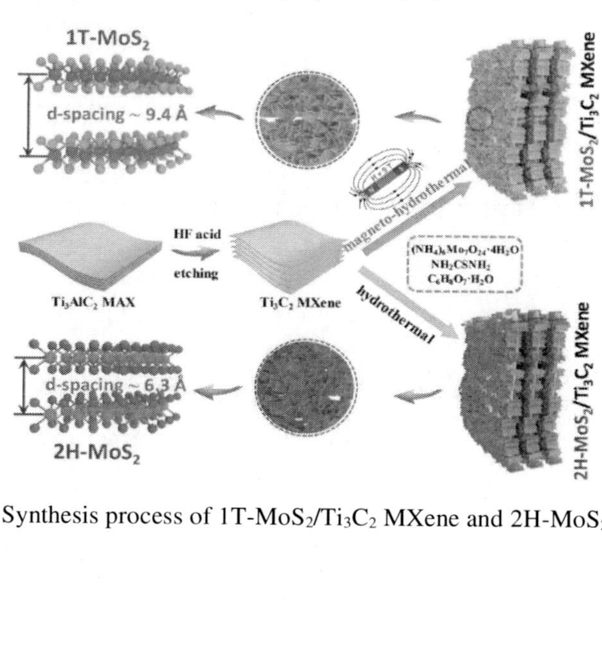

Figure 6.19. Synthesis process of 1T-MoS_2/Ti_3C_2 MXene and 2H-MoS_2/Ti_3C_2 MXene [35].

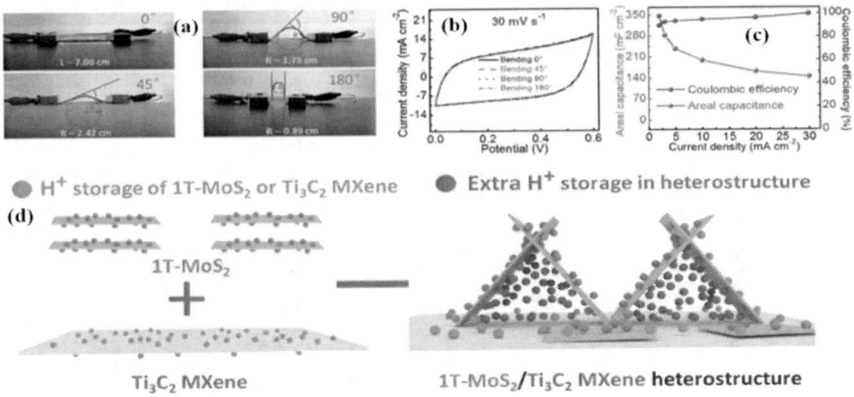

Figure 6.20. (a) 1T-MoS_2/Ti_3C_2 MXene ASSS with different bending angles (b) CV curves for ASSS (different bending angles) at 30 mV/s (c) Coulombic efficiency and areal specific capacitance at different current density (d) Extra charge storage mechanism of 1T-MoS_2/Ti_3C_2 heterostructure [35].

The fabricated 2D/2D MoS_2 (1T phase)/Ti_3C_2 ASSS consist of two 1T phase MoS_2/Ti_3C_2 MXene with PVA- H_2SO_4 electrolyte. Figure 6.20a shows the flexibility test of the fabricated ASSS device with different bending angles (45°, 90°, and 180°), which proved the admirable flexibility. Figure 6.20b is a CV curve of the ASSS device, which depicts the remarkable stability at varying bending angles. Figure 6.20c shows the maximum Coulombic efficiency (99.7%) and the capacitance of 145 mF/cm^2 (areal capacitance) at 30 mA/cm^2 (current density). The enhanced specific capacitance is due to the combined effect of 1T- MoS_2/Ti_3C_2, which can store many H^+ ions (Figure 6.20d).

Sun et al. have reported the enhanced asymmetric supercapacitive behavior of MoS_2-x@CNTs/Ni//CNT networks by creating the sulfur vacancy in the MoS_{2-x} monolayer (Figure 6.21). In this, sulfur vacancy can create an intermediate state in a deep acceptor level and trap more electrons, leading to improved electron mobility. Furthermore, this sulfur vacancy can also facilitate modifying the bandgap of the MoS_2-x@CNTs/Ni electrode. Hence, higher energy density and power density are 63 Wh/kg and 25.5 kW/kg, respectively [36]. The summary of the published works on the energy storage performance of the different metal chalcogenides with their electrochemical parameters was tabulated (Table 6.2).

Table 6.2. Electrochemical parameters of the metal chalcogenides-based electrodes

Electrode Material	Preparation Method	Electrolyte	Potential window (V)	Specific capacitance	Current density/Scan rate	Energy density of device	Ref.
2D-MoS_2	Hydrothermal	3M KOH	0-1	255 F/g	0.25 A/g	35.5 Wh/kg	Gupta et al. [17]
1T MoS_2 MSCs	Laser writing	H_2SO_4/ PVA	0-0.5	36 mF/cm^2	1 mA/cm^2	15.6×10^{-3} mWh/cm^2	Xu et al. [18]
AgNWs-MoS_2 MSCs	Spray coating Method	1 M PVA-H_2SO_4	0-1.2	27.6 mF/cm^2	0.2 V/s	2.453 μWh/cm^2	Li et al. [19]
Mn-MoS_2 MSCs	Hydrothermal	0.5 M Na_2SO_4	0-2.0	430 F/g	1 A/g	48.9 Wh/kg	Singha et al. [20]
MoS_2/NiS hybrid	Hydrothermal	1 M KOH	0-0.4	1493 F/g	0.2 A/g	31 Wh/kg	Qin et al. [21]
MoS_2-$Ti_3C_2T_x$	Hydrothermal	1 M H_2SO_4	-0.2-0.4	115.2 F/g	0.5 A/g	5.1 Wh/kg	Hou et al. [22]
WS_2/α-$NiMoO_4$	Hydrothermal	2 M KOH	0-0.45	460 F/g	1 A/g	-	Ray et al. [23]
VS_2 Nanosheets	Mechanical exfoliation	0.5 M H_2SO_4	-0.3-0.6	8.6×10^2 F/g	5 mV/s	0.01 Wh/cm^3	Ji et al. [24]
1T/2H MoS_2	Hydrothermal	0.5 K_2SO_4	-0.4 to 0.5	80 F/g	5 mV/s	4.19 Wh/kg	Ali et al. [25]
s-MnO_2/MoS_2	DC magnetron sputtering	1.0 M Na_2SO_4	0-1.0	224 mF/cm^2	0.1 mA/cm^2	0.0036 mWh/cm^2	Zhang et al. [26]
MoS_2-NH_2/PANI	Polymerization	1M H_2SO_4	0-0.7	326.4 F/g	0.5 A/g	4.0 Wh/kg	Zeng et al. [28]
MoS_2/PANI/ CNT	Polymerization	0.5 M H_2SO_4	0-1 V	245 F/cm^3	0.3 A/cm^3	0.013 Wh/cm^3	Chen et al. [29]
2D/2D SnS_2/MoS_2	Hydrothermal	0.5 M KOH	0.1-0.52	466.6 F/g	1 A/g	115 Wh/kg	Wang et al. [30]
RGO/MoS_2/P	Hydrothermal	PVA/ H_3PO_4	0-1.0 V	241.81 mF/cm^2	0.5 mA/cm^2	1.44 μWh/cm^2	Chen et al. [31]
1T-MoS_2/Hexa-WO_3/Graphene	one-pot hydrothermal	1 M H_2SO_4	-1.45 to -0.75 V	836 F/g	1 A/g	-	Naz et al. [32]

Table 6.2. (Continued)

Electrode Material	Preparation Method	Electrolyte	Potential window (V)	Specific capacitance	Current density/Scan rate	Energy density of device	Ref.
MoS_2/r-GO hybrid	Hydrothermal method	1 M Na_2SO_4	0-1.0 V	343 F/cm^3	10 mV/s	0.78 mWh/cm^3	Sarkar et al. [33]
1T-WS_2/ 2D-Ti_3C_2	Sonication-assisted procedure	1 M H_2SO_4	-0.4-0.25 V	157.0 F/g	0.11 mA/cm^2	-	Vyskočil et al. [34]
2D/2D 1T-MoS_2/Ti_3C_2	Magneto hydrothermal	PVA-H_2SO_4	-0.4-0.2	386.7 F/g	1 A/g	17.4 μWh/cm^2	Wang et al. [35]
MoS_2@CNTs/ Ni core/shell-structure	Pulsed laser deposition	1 M Na_2SO_4	-1.0-0	512 F/g	1 A/g	63 Wh/kg	Sun et al. [36]

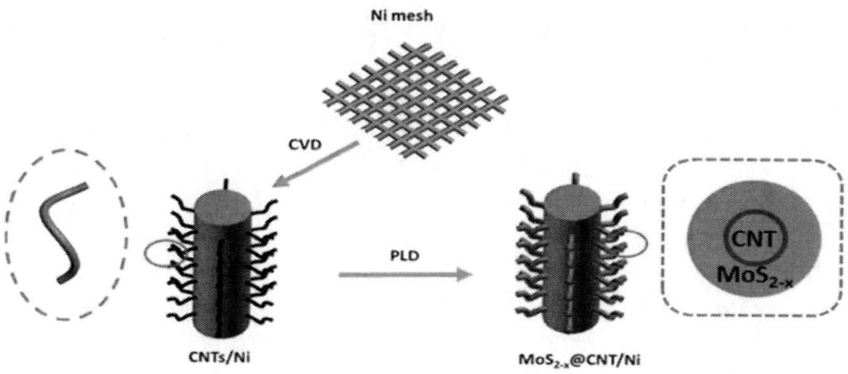

Figure 6.21. Fabrication of MoS_{2-x} @ CNT/Ni asymmetric supercapacitor (ASC) device [36].

6.3. Energy Conversion Applications

Nowadays, the energy demand has increased due to modern industrialization and the strong population growth. It is well-known that over 80% of energy consumption is only derived from fossil fuels worldwide. In reducing fossil fuels and unavoidable ecological harmfulness, an alternative energy source development is needed, which must be trustworthy, sustainable, and renewable [37]. Solar energy is the best choice to fulfill the above requirements related to other renewable energy sources, and it's highly abundant. The energy radiated by the sun in 1 hr on the earth's surface is 4.3×10^{20} J, equivalent to the one-year consumption of energy from all other sources. However, the harnessing of solar energy is another important task in the energy conversion world. The photovoltaics (PVs) approach is considered the cleanest mode for optimal solar energy conversion [38].

6.3.1. Solar Cells

PVs produce electricity by converting light energy into electrical energy. Solar cells are classified based on generation wise; silicon wafers belong to the first generation, which can deliver high power conversion efficiency. However, the cost of production is substantial. As a result, researchers concentrated on second-generation materials such as amorphous silicon, copper indium gallium selenide, and cadmium telluride. However, it shows poor performance; the researchers focused on the third-generation solar cells like dye-sensitized solar cells, organic solar cells, and perovskite solar cells.

Many efforts and development of technology are going into the PSC device due to its excellent power conversion efficiency (PCE). PSC works under the following process such as (i) Generation of free charge, (ii) transportation, and (iii) extraction. Figure 6.22 explains the schematic representation of the working mechanism of the perovskite solar cell (PEC). Light irradiation over the PSC absorbs photons and generates the excitons, followed by exciton dissociation. Then it produces the electrons and holes (charge carriers). The exciton dissociation process arises at the perovskite layer/charge-transporting layer interface. Once the electron is detached from the hole, it inserts into the electron transporting layer (ETL), and further, it moves towards the anode side (fluorine-doped tin oxide glass). At the same time, the hole was hooked on the hole transporting layer (HTL) and then shifted towards the cathode side (metal). Finally, the electrons and holes are received through working and counter electrodes. Both are moved to the external circuit to harvest current [39,40].

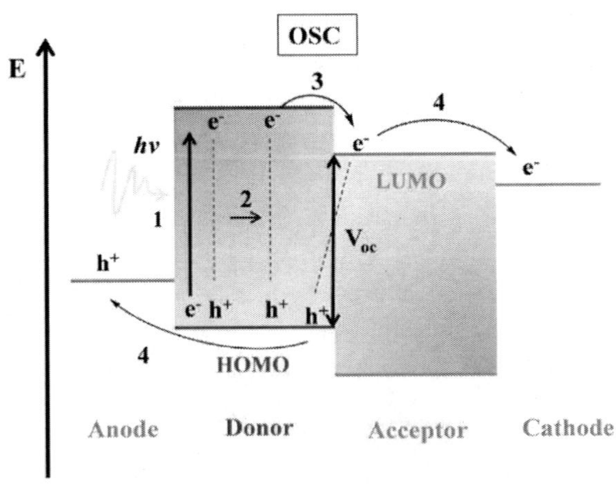

Figure 6.22. A working mechanism for PSC [39].

6.3.2. Properties of 2D TMDCs Suitable for Solar Cell Efficiency

In transparent photovoltaic devices, 2D materials serve as the active layers. After the invention of graphene, 2D materials became drawn to researchers because of their unique properties (optical, electrical, and mechanical). For example, 2D materials have atomic layer separation with weak van der Waals (vdW) forces between the layers. 2D materials also preserve their electrical properties even at the 1 nm scale of thin monolayer form.

2D semiconductors of mono or multilayers can quickly produce a p-n heterojunction that transmits visible light and displays a photovoltaic effect. The thickness of active layers plays a significant role in solar cell device performance. TMDCs with below 1 nm thickness absorb a higher percentage (5-10%) of incident solar light than monolayer CVD graphene (2-3%). 2D TMDCs have Van Hove singularities (VHS) in their density of states which enhances light absorption (visible region) and generates the electron-hole pair (exciton). But, in the case of graphene, VHS in its density of states induces photon absorption in the ultraviolet region [41].

6.3.3. 2D TMDCs for Solar Cell

Recently graphene in 2D materials has shown extensive attention in materials chemistry research. But the absence of an intrinsic bandgap limits its usage in different applications where the appropriate bandgap is needed. Compared to 2D graphene, the layered structured inorganic 2D TMDCs (MoS_2, $MoSe_2$, WS_2 and WSe_2) have paid much attention to their unique feature like 2H crystal structure arrangement makes bandgap (1-2 eV) feasibility for the optoelectronic applications [42, 43].

6.3.3.1. MoS_2

MoS_2 is a widely investigated 2D layer material among the chalcogenides. Monolayer (0.65 nm, thickness) exhibits direct bandgap (1.8 eV), a potential candidate for energy conversion and storage applications [44]. Its ambient mobility (200 cm^2/V s) and the on/off ratio of current (1×10^8) make it ideal for the transistor application [45]. The p-i-n heterojunction ($MoS_2|MAPbI_3|$ PCBM) solar cell was fabricated using UVO-treated MoS_2, CH_3NH_3PbI, and PCBM, in which the hole transport layer (HTL) is MoS_2. The device's current-voltage characteristics have been tested both the dark (Figure 6.23a) and sun illumination (Figure 6.23b) conditions for the different MoS_2 layer thicknesses (0, 3, 4, 5, and 6) [42].

In another report [46], the PSCs device was fabricated in Glass/ITO/PEDOT: PSS/MoS_2/$CH_3NH_3PbI_3$:MoS_2/PCBM/Bphen/Ag, which showed a PCE of about 18.31%. The energy level diagram and the J-V characteristics of the device were displayed in Figures 6.24a and 6.24b, respectively. MoS_2

is employed as a buffer layer between the PEDOT: PSS and $CH_3NH_3PbI_3$ layers. In addition, it serves as a dopant for the $CH_3NH_3PbI_3$ absorption layer. MoS_2 behaves as a buffer layer because its energy level is comparable (Figure 6.24a) with the $CH_3NH_3PbI_3$ energy level, which hinders the electrons transmission that can decrease the electron-hole recombination at HTL/Perovskite film interface.

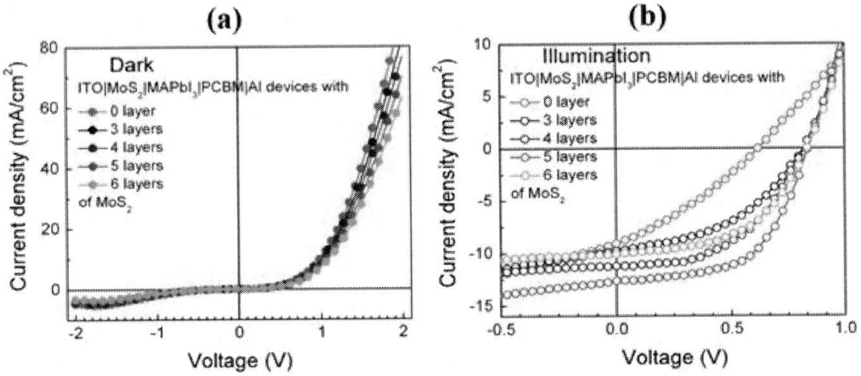

Figure 6.23. Current density-voltage (J-V) characteristics of solar cell device (MoS_2|$MAPbI_3$|PCBM) heterojunction) at (a) dark and (b) light illumination environment [42].

Moreover, the MoS_2 doping induces heterogeneous nucleation growth, improving the perovskite film quality through the tight packing of crystal grains. Finally, the charge transferred from the perovskite layer to HTL due to the lattice matching of MoS_2 and $CH_3NH_3PbI_3$. These all enhance the device performance, which is seen from the J-V curves for the buffer and without buffer layers (Figure 6.24b).

Ma et al. [47] have also checked the efficiency of Graphene/MoS_2/Si Schottky barrier-based solar cells by varying the thickness of the MoS_2 layer, and the detailed construction process was displayed in Figure 6.25a. First, the layered Graphene and MoS_2 were prepared using the chemical vapor deposition technique. After that, PMMA-supported Graphene/MoS_2 films were deposited on a silicon (n-type; Si) substrate. Then, the residual PMMA was removed using hot acetone. Furthermore, for the electrical contact purpose, Titanium (Ti, 50 nm) electrode and Ti/Ag electrode were deposited over the graphene surface and back of the n-Si substrate, respectively, through electron beam evaporation. Finally, colloidal TiO_2 nanoparticles are decorated on the graphene window to minimize the incident light reflection. Here, the number of graphene layers was constant, whereas the MoS_2 layer varied as 1, 3, and 6 and the corresponding approximate thickness is 0.7, 2, and 4 nm, respectively.

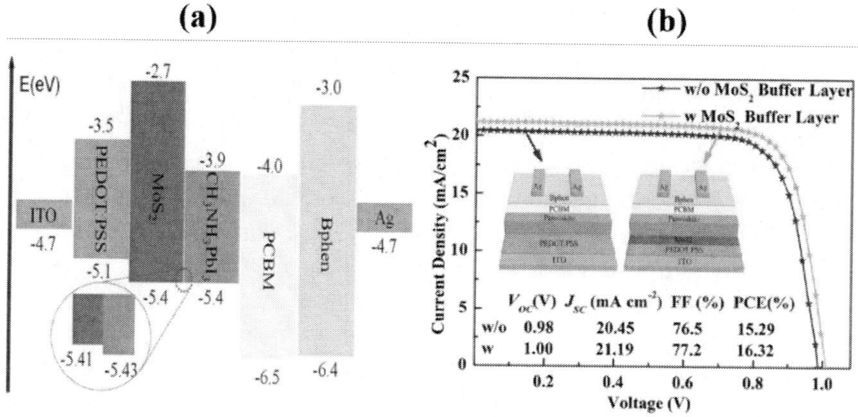

Figure 6.24. (a) Energy level diagram and (b) J-V characteristics of the perovskite solar cell (Glass/ITO/PEDOT: PSS/MoS_2/$CH_3NH_3PbI_3$:MoS_2/PCBM/Bphen/ Ag) [46].

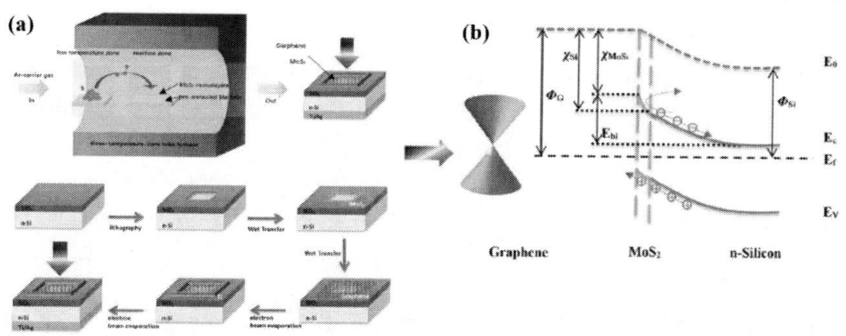

Figure 6.25. (a) Fabrication process of Graphene/MoS_2/Si device and (b) energy level diagram of Graphene/MoS_2/Si [47].

The Graphene/MoS_2/Si solar cell demonstrates enhanced conversion efficiency (8.2%) when compared with Graphene/Si configuration (6.2%). Graphene/Si has a poorer conversion efficiency because excited electrons move to the graphene electrode surface and recombine with the graphene/Si interface holes. The lower conversion efficiency of Graphene/Si is attributed to the excited electrons migrating to the surface of the graphene electrode surface and recombining with the holes at the interface (graphene/Si). But MoS_2 layer into the Graphene/MoS_2/Si solar cell configuration decreases the electron-hole recombination process, which causes improved PCE. Further, the possible charge transference is understood through Graphene/MoS_2/Si band diagram shown in Figure 6.25b. The conduction band (4.0 eV) and valence band (5.1 eV) potentials for the n-Si and the work function (4.65 eV) for graphene are indicated in the diagram. A built-in electric field (E_{bi}) appears close to the interface, separating the electron-hole pairs that shift the electron and hole toward the n-Si and graphene, respectively. After insertion of the MoS_2 layer, band alignment occurs between the MoS_2 and n-Si, wherein the conduction band maximum (E_c) of MoS_2 is larger than E_c of n-Si. It hinders the electron's transportation from n-Si to graphene, which lowers the electron-hole recombination on the graphene. Alternatively, it enhances hole transportation through the tunneling process. As the thickness of the MoS_2 layer increases (from 1 L to 6 L), the efficiency also increases from 8.2% to 13.1%. Beyond the certain thickness limit, the PCE will reduce due to the decreasing the hole transportation process. 6L thickness of MoS_2 solar cell device further enhances PCE (15.8%) after antireflection coating on the graphene surface.

6.3.3.2. $MoSe_2$

$MoSe_2$ belongs to the hexagonal crystal structure, and its density and melting point is 6.0 g/mL (25°C) and 1200°C, respectively. It has a sandwich structure similar to the MoS_2, and the bandgap converts from indirect (1.1 eV) to direct (1.55 eV) during the layer reduction, from multilayers to few layers or monolayers. The interlayer distance of $MoSe_2$ (0.646 nm) is greater than graphite and MoS_2 (0.615 nm). Moreover, monolayer $MoSe_2$ is considered a new variety of semiconductors that show an excellent photoelectric conversion efficiency so that it could be used as a component in solar cell devices [48].

The solar cell device of multi-layered Ag/AZO/C60/PMMA/$MAPbI_3$/$MoSe_2$/FTO was fabricated by change and his co-workers, which belongs to polymer/perovskite-based heterojunction (Figure 6.26a). PMMA, $MAPbI_3$, and $MoSe_2$ represent electron transport material (ETM), absorbing layer, and hole transporting material (HTM). For the optimal thickness (25 nm) of $MoSe_2$, the device shows the short-circuit current, open-circuit voltage, and fill factor are 21.11 mA/cm^2, 1.11 V, and 76.12%, respectively. Moreover, the device exhibited an excellent PCE of 18% [49]. Nowadays, the bifacial solar cell is used to enhance the outpower of the photovoltaic system. The bifacial

Cu(In, Ga)Se_2 (CIGS) with the lightweight solar cell was assembled by Nishimura et al. [50] through a lift-off process. This assembly layered a 2 μm thickness containing CIGS over Mo coated soda-lime glass. Then buffer layer (CdS) was coated over the CIGS by the chemical bath deposition technique. Further, 80 nm thick zinc oxide (i-ZnO) and Al-doped ZnO were deposited on the CdS using a sputtering method. Ni-Al grid was fabricated by the e-beam evaporation technique that is attached top and bottom of the device (Figure 6.26b). After that, a conductive wire was bonded to the Ni-Al grid using conductive epoxy glue. Fluorinated ethylene propylene (FEB) is a flexible substrate that is attached with epoxy glue at 100 °C (Figure 6.26c).

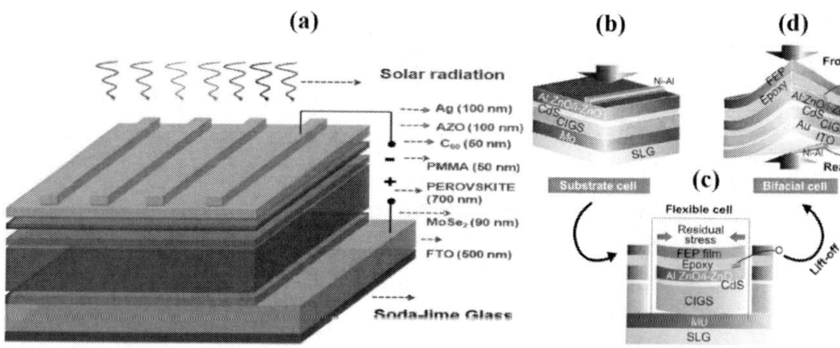

Figure 6.26. (a) Schematic representation for solar cells of $CH_3NH_3PbI_3$ ($MoSe_2$-HTM layer, C60/PMMA-ETM layer [49]. Different type of cell structures (b) substrate type, (c) flexible type, (d) flexible with bifacial type [50].

Then, layers of CIGS/CdS/i-ZnO/Al:ZnO/Ni/Al/epoxy/FEP were removed from the Mo-covered SLG (lift-off process). The bifacial solar cell process was completed after the gold and Sn-doped ITO attachment in the rear face of the CIGS through electron beam evaporation. Followed by Ni-Al grid also attached via EB method (Figure 6.26d). Moreover, the $MoSe_2$ layer with 5 nm thickness was formed at the interface of Mo/CIGS while CIGS composition grew. The authors observed that the 11.5% lift-off cell fabrication shows a high-performance ratio (95%) compared with a 12.1% efficient solar cell device.

6.3.3.3. WS_2

WS_2 with a single-layer form has a direct bandgap of 2.0 eV from the optical investigation, according to Elías et al. [51]. It has strong spin-orbit coupling due to the valence band splitting. Compared with MoS_2 (150 meV), the valance splitting is high for WS_2 (426 meV) due to the greater W atom mass. Rafiq et al. [52] have prepared the WS_2 layer through radio magnetron sputtering at 200 W deposition power. It is used as a window layer in the CdTe/WS_2 based solar cell, and its morphology was shown in Figure 6.27a. The full device form denoted as ITO/WS_2/CdTe/C/Ag which exhibited the 1.2% PCE with $V_{oc} = 0.39$ $J_{sc} = 10.45$ mA/cm^2, fill factor = 29.42%. In another work, Debbarma et al. [53] have fabricated the dual junction WS_2-graphene/n-Si heterostructure by the CVD method, and the fabrication process was displayed in Figure 6.27b. The single graphene layer was deposited over the surface of Cu foil using CVD, which was further transferred over the n-Si surface using the carrier film method. Three-zone furnace was used to synthesize WS_2-graphene/n-Si structures. The temperature of the first, second and third zone furnaces was 200, 200, and 950°C, respectively. The sulfur was taken in the quartz boat which was placed in the first zone. The graphene/n-Si substrate covered WO_3 powder containing crucible. Inside the chamber, the vacuum of 2mTorr and the steady flow (50 sccm) of H_2 were continued. After the pressure (220 mTorr) stabilization, the fixed zone temperatures were maintained for 30 min duration. The WS_2-graphene/n-Si device (Figure 6.27c) was obtained after the furnace temperature reached to room temperature. The photovoltaic parameters such as J_{sc}, V_{oc}, and η for WS_2-graphene/n-Si dual-junction are 0.52 mA/cm^2, 130 mV, and 0.015%, respectively. These values are comparatively higher than that of graphene/n-Si (J_{sc}=2.28, V_{oc}=184, η= 0.1%).

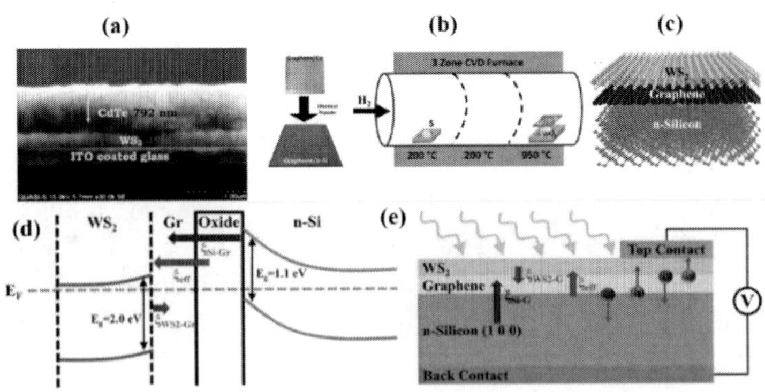

Figure 6.27. (a) SEM micrograph for ITO/WS_2/CdTe (cross-sectional view) [52], (b) WS_2-graphene/n-Si device fabrication through CVD technique, (c) Schematic view of WS_2-graphene/n-Si dual junction, (d) Energy band structure diagram of heterojunction (e) Schematic illustration for photo-generated excitons and electron and hole pair separation due to the strong built-in electric field in the direction of the WS_2 layer [53].

The enhanced performance of the WS_2-graphene/n-Si device was understood by seeing the electrostatics and the energetics of p-n junctions (Figure 6.27d). Coating WS_2 (n-type) layers formed the double junction cell over the graphene. In general, electron transit from p-type to n-type occurs at the p-n junction, but in the case of a dual junction device (WS_2-graphene/n-Si), the electrons shift from the n-type to p-type (WS_2 to graphene) and then back to the n-type (n-Si). This kind of transportation was due to graphene/n-Si ($\xi_{Si\text{-}G}$) having a stronger built-in-electric field at the interface when compared to the interface of WS_2-graphene ($\xi_{WS2\text{-}G}$), which has been shown in Figure 6.27e as schematic representation.

6.3.3.4. WSe_2

WSe_2 is a graphite-like TMDC with a layered microstructure. There are no other stoichiometries; W and Se are the only stable compounds (WSe_2). It also exhibits a robust crystal structure with varying bandgap (1.16 to 1.54 eV). It absorbs visible light and can be utilized as a photoelectric performance enhancer in solar cells as an absorption layer. Transparent 2D heterojunction with MoS_2 multilayers (n-type) and WSe_2 (p-type) was connected with indium tin oxide through ordinary lithography and lift-off technique. The fabricated device showed an 80% transparency with 4 8% PCE. In general, 2D materials can interact with moisture or gas molecules, which diminishes the device's performance so that the passivation layer like amorphous fluoropolymer (Teflon AF2400) was introduced in heterojunction of 2D-MoS_2/WSe_2. Figure 6.28a discloses the fabricated device with a chemical structure of passivation layer, and the current density-voltage (J-V) characteristics of the device at dark condition was shown in Figure 6.28b. The estimated J_0 and n for the bare device is 1.47×10^{-4} mA/cm^2 and 2.88; these values are decreased (J_0 = 1.19 mA/cm^2, n=2.23) after the passivation layer coating in the 2D device. This is because the passive layer suppresses the carrier recombination in the device. From the I-V characteristics (Figure 6.28c) under light illumination, the authors noticed that the power conversion efficiency went from 7.99% to 10% because of the passivation layer effect. This higher efficiency relates to the following causes, (i) Teflon AF2400 holds the higher amount of C-F bonds in which F belongs to the strong negative elements, which affects the charge distribution environments, (ii) The presence of electric dipole moment in the Teflon AF2400 can enhance the hole accumulation in WSe_2, during the suppression of electron arisen in $MoSe_2$, (iii) After incorporation of the passivation layer, the average reflectivity increases from 8.28% to 5.09% [54].

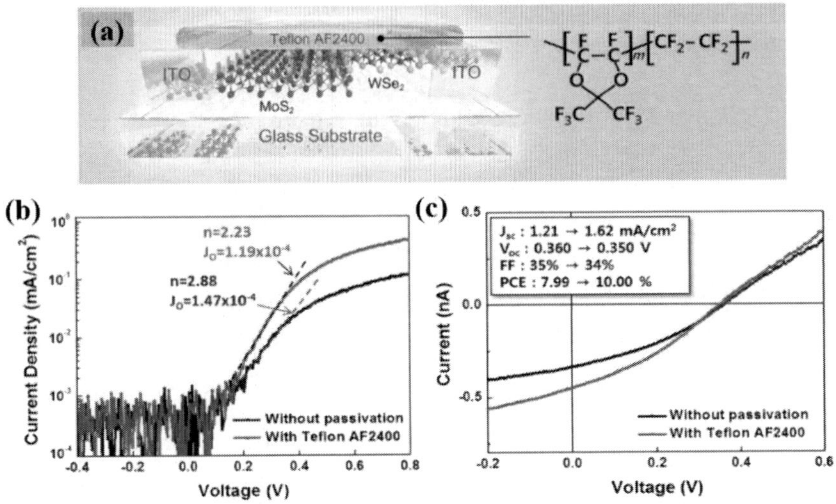

Figure 6.28. (a) WSe_2/MoS_2 solar cell with fluoropolymer (Teflon AF2400) chemical structure, (b) Current density-voltage (J-V) characteristics plot at dark condition, (c) Current-voltage (I-V) characteristics plot at light exposure (1.9 mW/cm^2) [54].

Zhang et al. [55] have assembled the Al/ITO/ZnO/CdS/CZTSSe/p-WSe_2/Mo (WSe_2-CZTSSe) structure over the Mo-deposited soda-lime glass (SLG) through pre-sputtering metal tungsten. Moreover, the estimated parameters such as short circuit current density (J_{sc}=28.11), open-circuit voltage (V_{oc}=388.84 mV), PCE (5.45%), and fill factor (FF=49.86%) are high for WSe_2-CZTSSe based solar cell compared to (Al/ITO/ZnO/CdS/CZTSSe/n-$MoSe_2$/Mo) based solar cell. In another report, Pt/WSe_2 vertical Schottky junction solar cell diodes with high work function were fabricated by McVay et al. [56]. The authors have optimized the range of WSe_2 layer thickness (80-150 nm) and studied the passivation layer (Al_2O_3) impact on the device operation. The deposited passivation layer (Al_2O_3) over the solar cell can cause the short circuit enhancement due to the enhanced external quantum efficiency (EQE) and photocurrent collection area.

This enhanced EQE due to the effects of antireflection coating, trap states passivation, and also the doping reformed carrier collection pathway. The present study claims that more than 50% EQE was obtained below the 600 nm wavelength. The surface passivation is a perfect route to tune TMDCs device performance for the suitable optoelectronic application. 2D TMDCs (MoS_2, $MoSe_2$, WS_2 and WSe_2) based perovskites of photovoltaic parameters are compared with other metal oxide-based perovskite solar cells [57-60] (Table 6.3).

Table 6.3. The performance of different 2D metal chalcogenides and metal oxide-based solar cells

Solar cells	HTL	Open circuit voltage (V)	Fill factor (%)	Short circuit current density (mA/cm^2)	Power conversion efficiency (%)	Ref
$FTO/SnO_2/Cs_2TiBr_6$ $/MoO_3/Au$	MoO_3	1.53	86.45	8.66	11.49	Ahmed et al. [57]
$FTO/SnO_2/FA_{0.83}Cs_{0.17}PbI_{1.5}Br_{1.5}/MoOx/Au$	MoO_x	1.44	81.83	17.04	20.10	Farihatun et al. [58]
$ITO/VO_X/$ Perovskite/ PCBM/ZnO/Al	VO_X	1.0	77.16	20.7	13.66	Eom et al. [59]
$ITO/CuOx/CH_3NH_3\text{-}PbI_3/$ PBCM/Ag	CuO_X	0.71	66	12.87	6.26	Makenali et al. [60]
$ITO/MoS_2|MAPbI_3|PCBM/Al$	MoS_2	0.84	69	12.60	6.01	Dasgupta et al. [42]
$FTO/MoSe_2/$ perovskite/ PMMA/C60/ AZO/Ag	$MoSe_2$	0.98	57	19.49	14.30	Chang et al. [49]
$ITO/WS_2/CdTe/$ C/Ag	WS_2	0.39	74.86	10.45	1.20	Debbarma et al. [53]
Al/ITO/ZnO/ CdS/CZTSSe/ $p\text{-}WSe_2/Mo$	WSe_2	0.38	29.42	28.11	5.45	Zhang et al. [55]

Conclusion

The current research community contributes to the energy storage and energy conversion field. In the energy storage zone, supercapacitors have paid much attention when compared to the battery due to the high-power density (1-10 kW/kg) and longer life cycles with superior cycle efficiency (98-100%). However, the main obstacle to the practical implementation is the low energy density. Therefore, the researchers concentrated on developing electrode materials to tune the specific capacitance and potential window to overcome this issue. From the electrode materials perspective, compared to metal oxides, layered type materials like metal chalcogenides attracts due to the unique crystal structures, multiple valance states, excellent redox chemistry, and higher electrical conductivity. However, it suffers from cycle stability and high-rate performance due to the phase changes during the electrochemical reactions. Therefore, the researchers pay much attention to resolving the above problems to enhance the electrochemical properties. Recent reports witnessed that the metal-doped, carbon allotropes (graphene, carbon nanotubes), and polymers (polyaniline) hybridized metal chalcogenides showed excellent electrochemical properties in the supercapacitor applications and were discussed in detail manner in the book chapter. From the commercialization point of view, our focus is on fabricating a highly conductive binder-free metal chalcogenides hybrid-based electrode that can be operated in the solid-state electrolyte environment.

In terms of the energy conversion extent, the metal chalcogenides act as an alternative light absorber in solar cell devices due to the high absorption coefficient, tunable bandgap, and cost-effectiveness. The absorber's optical band gap and thickness regulate the power conversion efficiency. The different synthesis method (CVD, RF sputtering) involves controlling the absorber's thickness (it acts as a hole transport layer). In the book chapter, we discussed the organic-in organic hybrid-based solar cells in which the role of metal chalcogenides is in the enhancement of solar cell efficiency. To achieve higher conversion efficiency without experimenting, we have to explore theoretical modeling to understand the band structures, optical properties, and physical mechanisms.

References

[1] Soares, Davi Marcelo, Santanu Mukherjee, and Gurpreet Singh. "TMDs beyond MoS_2 for electrochemical energy storage." *Chemistry-A European Journal* 26, no. 29 (2020): 6320-6341. https://doi.org/10.1002/chem.202000147.

[2] Blay, Vincent, Raquel E. Galian, Liana M. Muresan, Dmitrii Pankratov, Piyanut Pinyou, and Giorgia Zampardi. "Research Frontiers in Energy-Related Materials and Applications for 2020–2030." *Advanced Sustainable Systems* 4, no. 2 (2020): 1900145. https://doi.org/10.1002/adsu.201900145.

[3] Tao, Hengcong, Qun Fan, Tao Ma, Shizhen Liu, Henry Gysling, John Texter, Fen Guo, and Zhenyu Sun. "Two-dimensional materials for energy conversion and storage." *Progress in Materials Science* 111 (2020): 100637. https://doi.org/10.1016/j.pmatsci.2020.100637.

[4] Jiang, Yuqi, and Jinping Liu. "Definitions of pseudocapacitive materials: a brief review." *Energy & Environmental Materials* 2, no. 1 (2019): 30-37. https://doi.org/10.1002/eem2.12028.

[5] Joseph, Nikhitha, P. Muhammed Shafi, and A. Chandra Bose. "Recent advances in 2D-MoS_2 and its composite nanostructures for supercapacitor electrode application." *Energy & Fuels* 34, no. 6 (2020): 6558-6597. https://doi.org/10.1021/acs.energyfuels.0c00430.

[6] Cherusseri, Jayesh, Nitin Choudhary, Kowsik Sambath Kumar, Yeonwoong Jung, and Jayan Thomas. "Recent trends in transition metal dichalcogenide based supercapacitor electrodes." *Nanoscale Horizons* 4, no. 4 (2019): 840-858. https://doi.org/10.1039/C9NH00152B.

[7] Zhu, Zhenxiang, Zhenxing Wang, Zhaohu Ba, Xiuting Li, Jie Dong, Yuting Fang, Qinghua Zhang, and Xin Zhao. "3D MXene-holey graphene hydrogel for supercapacitor with superior energy storage." *Journal of Energy Storage* 47 (2022): 103911. https://doi.org/10.1016/j.est.2021.103911.

[8] Balasubramaniam, Bhuvaneshwari, Narendra Singh, Prasenjit Kar, Ankit Tyagi, Jai Prakash, and Raju Kumar Gupta. "Engineering of transition metal dichalcogenide-based 2D nanomaterials through doping for environmental applications." *Molecular Systems Design & Engineering* 4, no. 4 (2019): 804-827. https://doi.org/10.1039/C8ME00116B.

[9] Lv, Ruitao, Joshua A. Robinson, Raymond E. Schaak, Du Sun, Yifan Sun, Thomas E. Mallouk, and Mauricio Terrones. "Transition metal dichalcogenides and beyond: synthesis, properties, and applications of single-and few-layer nanosheets." *Accounts of chemical research* 48, no. 1 (2015): 56-64. https://doi.org/10.1021/ar5002846.

[10] Liu, Haipeng, Wen Lei, Zhaoming Tong, Xiaojian Li, Zexing Wu, Quanli Jia, Shaowei Zhang, and Haijun Zhang. "Defect engineering of 2D materials for electrochemical energy storage." *Advanced Materials Interfaces* 7, no. 15 (2020): 2000494. https://doi.org/10.1002/admi.202000494.

[11] Bang, Gyeong Sook, and Sung-Yool Choi. "Graphene and Two-Dimensional Transition Metal Dichalcogenide Materials for Energy-Related Applications." In *Nano Devices and Circuit Techniques for Low-Energy Applications and Energy Harvesting,* pp. 253-291. Springer, Dordrecht, 2016. https://doi.org/10.1007/978-94-017-9990-4_9.

[12] Jiao, Yucong, Ahmed M. Hafez, Daxian Cao, Alolika Mukhopadhyay, Yi Ma, and Hongli Zhu. "Metallic MoS_2 for high performance energy storage and energy conversion." *Small* 14, no. 36 (2018): 1800640. https://doi.org/10.1002/smll.201800640.

[13] Tan, Chaoliang, Xiehong Cao, Xue-Jun Wu, Qiyuan He, Jian Yang, Xiao Zhang, Junze Chen et al. "Recent advances in ultrathin two-dimensional nanomaterials." *Chemical reviews* 117, no. 9 (2017): 6225-6331. https://doi.org/10.1021/acs.chemrev.6b00558.

[14] Zhang, Rui, Yao Qin, Ping Liu, Chuankun Jia, Yougen Tang, and Haiyan Wang. "How does molybdenum disulfide store charge: a minireview." *ChemSusChem* 13, no. 6 (2020): 1354-1365. https://doi.org/10.1002/cssc.201903320.

[15] Shao, Yuanlong, Maher F. El-Kady, Jingyu Sun, Yaogang Li, Qinghong Zhang, Meifang Zhu, Hongzhi Wang, Bruce Dunn, and Richard B. Kaner. "Design and

mechanisms of asymmetric supercapacitors." *Chemical reviews* 118, no. 18 (2018): 9233-9280. https://doi.org/10.1021/acs.chemrev.8b00252.

[16] Wang, Yonggang, Yanfang Song, and Yongyao Xia. "Electrochemical capacitors: mechanism, materials, systems, characterization and applications." *Chemical Society Reviews* 45, no. 21 (2016): 5925-5950. https://doi.org/10.1039/C5CS00580A.

[17] Gupta, Honey, S. Chakrabarti, Sagar Mothkuri, Balaji Padya, T. N. Rao, and P. K. Jain. "High-performance supercapacitor based on 2D-MoS_2 nanostructures." *Materials Today: Proceedings* 26 (2020): 20-24. https://doi.org/10.1016/j.matpr.2019.04.198.

[18] Xu, Chenyang, Lan Jiang, Xin Li, Chen Li, Changxiang Shao, Pei Zuo, Minsheng Liang, Liangti Qu, and Tianhong Cui. "Miniaturized high-performance metallic 1T-Phase MoS_2 micro-supercapacitors fabricated by temporally shaped femtosecond pulses." *Nano Energy* 67 (2020): 104260. https://doi.org/10.1016/j.nanoen.2019.104260.

[19] Li, Jiahui, Qiuwei Shi, Yuanlong Shao, Chengyi Hou, Yaogang Li, Qinghong Zhang, and Hongzhi Wang. "Cladding nanostructured AgNWs-MoS_2 electrode material for high-rate and long-life transparent in-plane micro-supercapacitors." *Energy Storage Materials* 16 (2019): 212-219. https://doi.org/10.1016/j.ensm.2018.05.013.

[20] Singha, Shib Shankar, Siddheswar Rudra, Suchanda Mondal, Mukul Pradhan, Arpan Kumar Nayak, Biswarup Satpati, Prabir Pal, Kaustuv Das, and Achintya Singha. "Mn incorporated MoS_2 nanoflowers : A high-performance electrode material for symmetric supercapacitor." *Electrochimica Acta* 338 (2020): 135815. https://doi.org/10.1016/j.electacta.2020.135815.

[21] Qin, Qing, Lulu Chen, Tao Wei, and Xien Liu. "MoS_2/NiS yolk-shell microsphere-based electrodes for overall water splitting and asymmetric supercapacitor." *Small* 15, no. 29 (2019): 1803639. doi.org/10.1002/smll.201803639.

[22] Hou, Wentao, Yuan Sun, Yu Zhang, Tingting Wang, Liqian Wu, Youwei Du, and Wei Zhong. "Mixed-dimensional heterostructure of few-layer MXene based vertical aligned MoS_2 nanosheets for enhanced supercapacitor performance." *Journal of Alloys and Compounds* 859 (2021): 157797. https://doi.org/10.1016/j.jallcom.2020.157797.

[23] Ray, Schindra Kumar, Bishweshwar Pant, Mira Park, Jin Hur, and Soo Wohn Lee. "Cavity-like hierarchical architecture of WS_2/α-$NiMoO_4$ electrodes for supercapacitor application." *Ceramics International* 46, no. 11 (2020): 19022-19027. https://doi.org/10.1016/j.ceramint.2020.04.232.

[24] Ji, Qingqing, Cong Li, Jingli Wang, Jingjing Niu, Yue Gong, Zhepeng Zhang, Qiyi Fang et al. "Metallic vanadium disulfide nanosheets as a platform material for multifunctional electrode applications." *Nano letters* 17, no. 8 (2017): 4908-4916. https://doi.org/10.1021/acs.nanolett.7b01914.

[25] Ali, Basant A., Asmaa MA Omar, Ahmed SG Khalil, and Nageh K. Allam. "Untapped potential of polymorph MoS_2: tuned cationic intercalation for high-performance symmetric supercapacitors." *ACS applied materials & interfaces* 11, no. 37 (2019): 33955-33965. https://doi.org/10.1021/acsami.9b11444.

[26] Zhang, Honghao, Jun Wei, Yu Yan, Qianjin Guo, Liqiang Xie, Zhengchun Yang, Jie He et al. "Facile and scalable fabrication of MnO_2 nanocrystallines and enhanced electrochemical performance of MnO_2/MoS_2 inner heterojunction structure for supercapacitor application." *Journal of Power Sources* 450 (2020): 227616. https://doi.org/10.1016/j.jpowsour.2019.227616.

[27] Zhao, Chenyang, Renheng Wang, Youfang Zhang, Lunfeng Chen, Tingya Li, Xin Deng, Peixin Zhang, and Xuehong Lu. "Electrostatic force-driven anchoring of Ni $(OH)_2$ nanocrystallites on single-layer MoS_2 for high-performance asymmetric hybrid supercapacitors." *Electrochimica Acta* 320 (2019): 134591. https://doi.org/10.1016/j.electacta.2019.134591.

[28] Zeng, Rong, Zongcai Li, Longbin Li, Yizhe Li, Jun Huang, Yingbo Xiao, Kai Yuan, and Yiwang Chen. "Covalent connection of polyaniline with MoS_2 nanosheets toward ultrahigh rate capability supercapacitors." *ACS Sustainable Chemistry & Engineering* 7, no. 13 (2019): 11540-11549. https://doi.org/10.1021/acssuschemeng.9b01442.

[29] Chen, I-Wen Peter, Yu-Chen Chou, and Po-Yuan Wang. "Integration of ultrathin MoS_2/PANI/CNT composite paper in producing All-Solid-State flexible supercapacitors with exceptional volumetric energy density." *The Journal of Physical Chemistry C* 123, no. 29 (2019): 17864-17872. https://doi.org/10.1021/acs.jpcc.9b04046.

[30] Wang, Bo, Rong Hu, Jun Zhang, Zongyu Huang, Hui Qiao, Lunjun Gong, and Xiang Qi. "2D/2D SnS_2/MoS_2 layered heterojunction for enhanced supercapacitor performance." *Journal of the American Ceramic Society* 103, no. 2 (2020): 1088-1096. https://doi.org/10.1111/jace.16778.

[31] Chen, Yan, Jingge Bai, Dingyu Yang, Ping Sun, and Xu Li. "Excellent performance of flexible supercapacitor based on the ternary composites of reduced graphene oxide/molybdenum disulfide/poly (3, 4-ethylene dioxythiophene)." *Electrochimica Acta* 330 (2020): 135205. https://doi.org/10.1016/j.electacta.2019.135205.

[32] Naz, Raheela, Qinglei Liu, Waseem Abbas, Muhammad Imtiaz, Imran Zada, Javed Ahmad, Tengfei Li, and Jiajun Gu. "One-Pot Hydrothermal Synthesis of Ternary 1T-MoS_2/Hexa-WO_3/Graphene Composites for High-Performance Super capacitors." *Chemistry-A European Journal* 70 (2019): 16054-16062.: http://doi.org/10.1002/chem.201903336.

[33] Sarkar, Debasish, Debanjan Das, Shyamashis Das, Ankit Kumar, Satish Patil, Karuna Kar Nanda, D. D. Sarma, and Ashok Shukla. "Expanding interlayer spacing in MoS_2 for realizing an advanced supercapacitor." *ACS Energy Letters* 4, no. 7 (2019): 1602-1609. https://doi.org/10.1021/acsenergylett.9b00983.

[34] Vyskočil, Jan, Carmen C. Mayorga-Martinez, Kateřina Szőkölová, Apurv Dash, Jesus Gonzalez-Julian, Zdeněk Sofer, and Martin Pumera. "2D Stacks of MXene Ti_3C_2 and 1T-Phase WS_2 with Enhanced Capacitive Behavior." *ChemElectroChem* 6, no. 15 (2019): 3982-3986. https://doi.org/10.1002/celc.201900643.

[35] Wang, Xin, Han Li, Hui Li, Shuai Lin, Wei Ding, Xiaoguang Zhu, Zhigao Sheng, Hai Wang, Xuebin Zhu, and Yuping Sun. "2D/2D 1T-MoS_2/Ti_3C_2 MXene heterostructure with excellent supercapacitor performance." *Advanced Functional Materials* 30, no. 15 (2020): 0190302. https://doi.org/10.1002/adfm.201910302.

[36] Sun, Peng, Ruijing Wang, Qiang Wang, Huanwen Wang, and Xuefeng Wang. "Uniform MoS_2 nanolayer with a sulfur vacancy on carbon nanotube networks as binder-free electrodes for asymmetrical supercapacitor." *Applied Surface Science* 475 (2019): 793-802. https://doi.org/10.1016/j.apsusc.2019.01.007.

[37] Liu, Siqi, Zi-Rong Tang, Yugang Sun, Juan Carlos Colmenares, and Yi-Jun Xu. "One-dimension-based spatially ordered architectures for solar energy conversion." *Chemical Society Reviews* 44, no. 15 (2015): 5053-5075. https://doi.org/10.1039/C4CS00408F.

[38] Li, Changli, Qi Cao, Faze Wang, Yequan Xiao, Yanbo Li, Jean-Jacques Delaunay, and Hongwei Zhu. "Engineering graphene and TMDs based van der Waals heterostructures for photovoltaic and photoelectrochemical solar energy conversion." *Chemical Society Reviews* 47, no. 13 (2018): 4981-5037. https://doi.org/10.1039/C8CS00067K.

[39] Hussain, Istiak, Hoang Phong Tran, Jared Jaksik, Justin Moore, Nazmul Islam, and M. Jasim Uddin. "Functional materials, device architecture, and flexibility of perovskite solar cell." *Emergent Materials* 1, no. 3 (2018): 133-154. https://doi.org/10.1007/s42247-018-0013-1.

[40] Marinova, Nevena, Silvia Valero, and Juan Luis Delgado. "Organic and perovskite solar cells: Working principles, materials and interfaces." *Journal of colloid and interface science* 488 (2017): 373-389. https://doi.org/10.1016/j.jcis.2016.11.021.

[41] Debbarma, Rousan, Sanjay K. Behura, Yu Wen, Songwei Che, and Vikas Berry. "WS_2-induced enhanced optical absorption and efficiency in graphene/silicon heterojunction photovoltaic cells." *Nanoscale* 10, no. 43 (2018): 20218-20225. https://doi.org/10.1039/C8NR03194K.

[42] Dasgupta, Uttiya, Soumyo Chatterjee, and Amlan J. Pal. "Thin-film formation of 2D MoS_2 and its application as a hole-transport layer in planar perovskite solar cells." *Solar Energy Materials and Solar Cells* 172 (2017): 353-360. https://doi.org/10.1016/j.solmat.2017.08.012.

[43] Ma, Jun, He Bai, Wei Zhao, Yujie Yuan, and Kailiang Zhang. "High-efficiency graphene/MoS_2/Si Schottky barrier solar cells using layer-controlled MoS_2 films." *Solar Energy* 160 (2018): 76-84. https://doi.org/10.1016/j.solener.2017.11.066.

[44] Li, Xiao, and Hongwei Zhu. "Two-dimensional MoS_2: Properties, preparation, and applications." *Journal of Materiomics* 1, no. 1 (2015): 33-44. https://doi.org/10.1016/j.jmat.2015.03.003.

[45] Radisavljevic, Branimir, Aleksandra Radenovic, Jacopo Brivio, Valentina Giacometti, and Andras Kis. "Single-layer MoS_2 transistors." *Nature nanotechnology* 6, no. 3 (2011): 147-150. https://doi.org/10.1038/nnano.2010.279.

[46] Liu, Zhiyong, Kaikai Liu, Feipeng Zhang, Sagar M. Jain, Tingwei He, Yuanzhi Jiang, Pengfei Liu, Jien Yang, Hairui Liu, and Mingjian Yuan. "$CH_3NH_3PbI_3$: MoS_2 heterostructure for stable and efficient inverted perovskite solar cell." *Solar Energy* 195 (2020): 436-445. https://doi.org/10.1016/j.solener.2019.11.030.

[47] Ma, Jun, He Bai, Wei Zhao, Yujie Yuan, and Kailiang Zhang. "High-efficiency graphene/MoS_2/Si Schottky barrier solar cells using layer-controlled MoS_2 films." *Solar Energy* 160 (2018): 76-84. https://doi.org/10.1016/j.solener.2017.11.066.

[48] Wu, Xin, Yu-hua Wang, Pei-ling Li, and Zu-zhao Xiong. "Research status of $MoSe_2$ and its composites: A review." *Superlattices and Microstructures* 139 (2020): 106388. https://doi.org/10.1016/j.spmi.2020.106388.

[49] Chang, Liann-Be, Chzu-Chiang Tseng, Jia-Hsun Lee, Gwo-Mei Wu, Ming-Jer Jeng, Wu-Shiung Feng, Dave W. Chen et al. "Preparation and characterization of $MoSe_2/CH_3NH_3PbI_3$/PMMA perovskite solar cells using polyethylene glycol solution." *Vacuum* 178 (2020): 109441. https://doi.org/10.1016/j.vacuum.2020.109441.

[50] Nishimura, Takahito, Naoto Hamada, Jakapan Chantana, Abdurashid Mavlonov, Yu Kawano, Taizo Masuda, and Takashi Minemoto. "Application of Two-Dimensional $MoSe_2$ Atomic Layers to the Lift-Off Process for Producing Light-Weight and Flexible Bifacial Cu(In, Ga)Se_2 Solar Cells." *ACS Applied Energy Materials* 3, no. 10 (2020): 9504-9508. https://doi.org/10.1021/acsaem.0c01557.

[51] Elías, Ana Laura, Néstor Perea-López, Andrés Castro-Beltrán, Ayse Berkdemir, Ruitao Lv, Simin Feng, Aaron D. Long et al. "Controlled synthesis and transfer of large-area WS_2 sheets : from single layer to few layers." *ACS nano* 7, no. 6 (2013): 5235-5242. https://doi.org/10.1021/nn400971k.

[52] Rafiq, Md Khan Sobayel Bin, N. Amin, Hamad F. Alharbi, Monis Luqman, Afida Ayob, Yahya S. Alharthi, Nabeel H. Alharthi, Badariah Bais, and Md Akhtaruzzaman. "WS_2: a new window layer material for solar cell application." *Scientific reports* 10, no. 1 (2020): 1-11. https://doi.org/10.1038/s41598-020-57596-5.

[53] Debbarma, Rousan, Sanjay K. Behura, Yu Wen, Songwei Che, and Vikas Berry. "WS_2-induced enhanced optical absorption and efficiency in graphene/silicon heterojunction photovoltaic cells." *Nanoscale* 10, no. 43 (2018): 20218-20225. https://doi.org/10.1039/C8NR03194K.

[54] Cho, Ah-Jin, Min-Kyu Song, Dong-Won Kang, and Jang-Yeon Kwon. "Two-dimensional WSe_2/MoS_2 p–n heterojunction-based transparent photovoltaic cell and its performance enhancement by fluoropolymer passivation." *ACS applied materials & interfaces* 10, no. 42 (2018) : 35972-35977. https://doi.org/10.1021/acsami.8b12250.

[55] Zhang, Xiaohui, Bin Yao, Yongfeng Li, Zhanhui Ding, Haifeng Zhao, Ligong Zhang, and Zhenzhong Zhang. "Influence of WSe_2 buffer layer at back electrode on performance of Cu_2ZnSn (S, Se)$_4$ solar cells." *Solar Energy* 199 (2020): 128-135. https://doi.org/10.1016/j.solener.2020.02.033.

[56] McVay, Elaine, Ahmad Zubair, Yuxuan Lin, Amirhasan Nourbakhsh, and Tomás Palacios. "Impact of Al_2O_3 Passivation on the Photovoltaic Performance of Vertical WSe_2 Schottky Junction Solar Cells." *ACS Applied Materials & Interfaces* 12, no. 52 (2020): 57987-57995. https://doi.org/10.1021/acsami.0c15573.

[57] Ahmed, Saif, Farihatun Jannat, Md Abdul Kaium Khan, and Mohammad Abdul Alim. "Numerical development of eco-friendly Cs_2TiBr_6 based perovskite solar cell with all-inorganic charge transport materials via SCAPS-1D." *Optik* 225 (2021): 165765. https://doi.org/10.1016/j.ijleo.2020.165765.

[58] Jannat, Farihatun, Saif Ahmed, and Mohammad Abdul Alim. "Performance analysis of cesium formamidinium lead mixed halide-based perovskite solar cell with MoOx

as hole transport material via SCAPS-1D." *Optik* 228 (2021): 166202. https://doi.org/10.1016/j.ijleo.2020.166202.

[59] Eom, Kiryung, Il-han Yoo, Hoseong Han, and Hyungtak Seo. "Correlation of Band Electronic Structure to Efficiency in Perovskite Solar Cells with Vanadium Oxide Buffers." In *ECS Meeting Abstracts,* no. 11, p. 903. IOP Publishing, 2018. https://doi.org/10.1149/MA2018-01/11/903.

[60] Makenali, Marzieh, Iraj Kazeminezhad, Vahid Ahmadi, and Farzaneh Arabpour Roghabadi. "Charge transfer balancing of planar perovskite solar cell based on a low cost and facile solution-processed CuOx as an efficient hole transporting layer." *Journal of Materials Science: Materials in Electronics* 32, no. 2 (2021): 2312-2325. https://doi.org/10.1007/s10854-020-04996-y.

Chapter 7

Fuel Cell and Catalysis Applications

Antara Vaidyanathan[1], Avinash Macchindra Pardhi[1], Vaibhav Wagh[1] and Brahmananda Chakraborty[2,3,*]

[1]Department of Chemistry, Ramnarain Ruia Autonomous College, Mumbai, India
[2]High Pressure and Synchrotron Radiation Physics Division,
Bhabha Atomic Research Centre, Trombay, Mumbai, India
[3]Homi Bhabha National Institute, Mumbai, India

Abstract

This chapter provides an overview and background of catalysis, mainly electrocatalysis and photocatalysis for important reactions such as the water splitting reaction. Since the advent of graphene, there has been great interest in exploring 2D materials for their applications as catalysts. 2D Transition Metal Dichalcogenides (TMDCs) are high-performance materials, readily synthesizable and easily available, with high specific surface area and tunable electronic and magnetic properties. Their ability to be functionalized opens up opportunities for highly selective and sensitive catalysts. The water-splitting reaction specifically becomes relevant as a hydrogen production source, for which if efficient catalyst solutions are devised, it can be a promising replacement for fossil fuels for transport applications via fuel cell technology. The requirement to improve current technologies in fuel cell vehicles include replacements for noble metal catalysts, in which 2D TMDCs show ample promise. In particular, metallic 2D TMDCs are known to show excellent electronic properties suitable for catalytic applications. Phase engineering for fabrication of metallic 2D TMDCs, hybridizing nanosheets of TMDCs, defect engineering, doping, and so on, present a plethora of possibilities

[*] Corresponding Author's Email: brahma@barc.gov.in.

In: 2D Metallic Transition Metal Dichalcogenides
Editors: Chandra Sekhar Rout and Brahmananda Chakraborty
ISBN: 978-1-68507-965-9

for adapting this material towards specific applications. This chapter mentions the figures of merit for catalysis-based applications of materials, and summarizes the recent progress of 2D Metallic TMDCs.

Keywords: TMDCs, fuel cells, catalysis, solar cells

7.1. Introduction to Catalysis and Fuel Cells

Catalysis is an active field of scientific research with the main focus being to develop facile methods to prepare catalysts that are stable and highly efficient. In this regard, it has long been observed that precious metals, especially platinum group metals (like Pt, Pd, Au and their alloys) are ideally suited as catalysts [1] in numerous chemical reactions like heterogenous catalysis [2], photocatalysis [3], electrocatalysis [4], etc. However, these precious metals suffer from the drawbacks of high cost and sparse availability. Consequently, a lot of scientific research focusses on developing good alternatives to precious metals for catalyzing reactions [5]. In recent times, many 2D materials have been in the limelight for their superior structural properties, tunable electronic structure, and high-efficiency catalytic behavior [6].

With the advent of the first truly two-dimensional atomic crystal i.e., graphene [7], interest in 2D materials as catalysts has surged. Materials in bulk (or three dimension) lacked the superior surface density (surface area per unit volume) offered by planar 2D materials. Further, these 2D materials show marked differences in electronic, magnetic, mechanical, etc., characteristics with the added ability to functionalize and tune these properties [8]. Thereafter, many 2D materials have been explored in terms of synthesis and possible applications. While graphene is still exceptional in its mechanical stability and electronic characteristics, it is fairly cumbersome to produce and therefore expensive. In contrast, transition metal dichalcogenides (TMDCs) are easily available, and also show many similar properties to graphene in terms of tunability.

Going from bulk to 2D TMDCs, there are several changes in the properties of 2D TMDCs that aid their prevalent use in catalysis. The layers in bulk TMDCs which are held together by van der Waals' forces are conveniently exfoliated to monolayers. 2D materials in general have a higher specific surface area and therefore, higher fraction of active sites as compared to their bulk counterparts [9]. The lack of interlayer coupling leads to a redistribution of the charge carriers [10]. Further, the atoms present at the edges or grain boundaries serve as the catalytic active sites owing to their low degree of coordination [11]. Additionally, the type of stacking gives rise to polymorphs: primarily 2H (semi conducting) which is mostly a stable configuration, and 1T (metallic) which is metastable [12]. Metallic 2D TMDCs fare better than their 2H counterparts due to improved electric conducting properties. Similarly, many properties of TMDCs are reliant on their characteristics, such as crystallinity, number and density of active sites, defects, etc., which dictate their efficiency as a catalyst.

The invention of fuel cells by Walter Grove in 1839 is considered to be a landmark in the field of energy storage [13]. The invention came through before the advent of electric grids for the distribution of electricity, when the need for energy storage systems like batteries was high. This process is the exact opposite of that observed in electrolysis, wherein an electric current passed through a solution of an electrolyte causes it to split into its constituent ions, i.e., there is a conversion from electrical energy into chemical energy. The fuel cell on the other hand, provides the coveted chemical-to-electrical energy conversion [14]. Conventional methods, mainly combustion, follow the more energy-expensive route of chemical-to-heat energy, followed by heat-to-electrical energy. The bypass presented in fuel cell technology makes it one of the most energy efficient processes known [15].

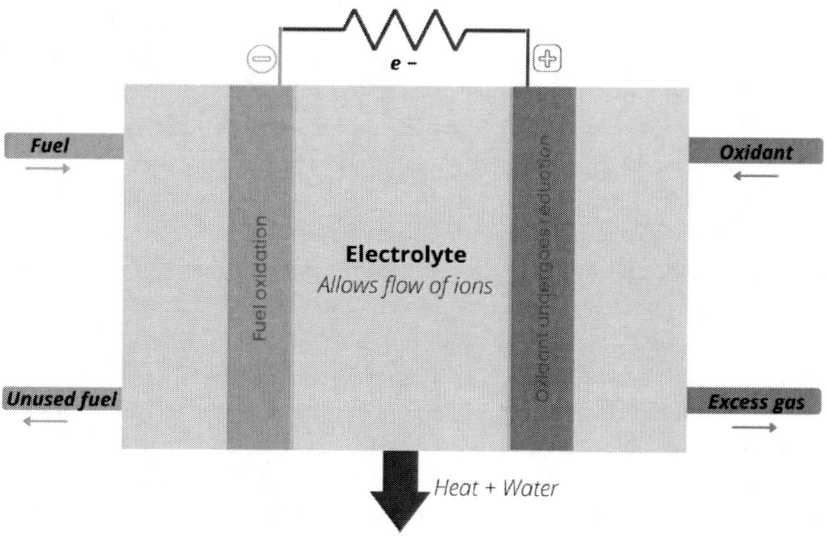

Figure 7.1. A schematic diagram of a hydrogen-oxygen fuel cell.

A fuel cell combines a fuel with an oxidant, producing energy and some by-products. For instance, a hydrogen-oxygen fuel cell combines hydrogen as the fuel with oxygen as the oxidant, producing energy and water as the by-product. The fuel is supplied to the anode, where it undergoes oxidation. The electrons thus produced are transported via the external circuit to the cathode, which is supplied with the oxidant. The movement of ions occurs through the electrolyte, similar to the set-up of a battery. At the cathode, the oxidant is acted upon by the electrons released in the external circuit, and undergoes reduction. The electrolyte serves another purpose of keeping the fuel and oxidant gas streams separate. A three-phase boundary is formed at the junction of electrodes and electrolyte, which is the site for electrochemical reactions to occur [16]. It is vital for this interface to simultaneously maintain electrical contact with the electrodes, ionic contact with the electrolyte, and be exposed to the reacting gases. Figure 7.1 depicts a schematic diagram of a fuel cell where the anodic and cathodic compartments are clearly separated by the electrolyte layer.

The reaction has an elevated energy barrier, requiring either a higher operating temperature, or an alternative reaction pathway presented by a catalyst to go to completion. The reason for the high energy barrier can be attributed to the inherent strength of the H-H bond, as well as that of the double bond between oxygen atoms in molecular oxygen [17]. Fuel cells operating at higher operating temperatures have been developed, namely: Solid Oxide fuel cells, Molten Carbonate fuel cells as suggested in Table 7.1. These are called as High Temperature Fuel Cells (HTFCs) as they function at upwards of 600°C. The engineering of a device that can be used for transport applications operating at these high temperatures remains a challenge [18]. The drawbacks of a high operating temperature include decreased longevity of the fuel cell parts, difficulties in maintaining the high temperature, and so on. For a high temperature fuel cell to be reasonably applied to transport applications, the required temperature should be rapidly attainable, and the system should be designed so as to withstand the operating temperatures for prolonged periods of time.

For the fuel cells operating at lower temperatures, the use of catalyzed electrode surfaces has been explored, and it is concluded that noble metals like platinum are good catalysts on both the anode and cathode [19]. Catalysts create an alternate reaction pathway with a lower energy barrier and thus eliminate the need for high operating temperatures and its accompanying problems. However, platinum catalysts are easily poisoned by carbon monoxide and carbon dioxide, both of which are abundantly available in the hydrogen that is obtained by the catalytic reformation of hydrocarbons. In order to use platinized electrodes successfully in a fuel cell, the hydrogen supply contain less than 10 ppm of carbon monoxide and carbon dioxide impurities, in the case of Proton Exchance Membrane Fuel Cells (PEMFCs) [20]. To achieve this, the hydrogen stream is subjected to extensive fuel processing, making the process more expensive. Furthermore, platinum catalysts significantly increase the cost of the device.

Table 7.1. Different fuel cell systems and their operating temperatures [18]

System	Operating temperature	Electrolyte
Solid Polymer Fuel Cells (SPFCS) or Proton Exchange Membrane Fuel Cells (PEMFCS)	~ 80°C	Porfluorosulfonic acid
Alkaline Fuel Cells (AFCs)	~ 100°C	Aqueous KOH
Phosphoric Acid Fuel Cells (PAFCs)	~ 200°C	Phosphoric acid
Molten Carbonate Fuel Cells (MCFCs)	~ 650°C	Solutions of lithium, sodium or potassium carbonates
Solid Oxide Fuel Cells (SOFCs)	~ 800°C – 1100°C	Yttria stabilized Zirconia

In this respect, 2D TMDCs are potential catalysts for the production of hydrogen gas to be used as a fuel in fuel cell applications, as well as on the electrodes of fuel cells. TMDCs have been used as catalysts in their 3D structure as crystals. However, as mentioned earlier, 2D TMDCs offer unique advantages for the purpose of catalysis. Two-dimensional layered materials have been researched for their catalytic properties extensively. They can also be intercalated with suitable materials to further improve their surface area. Functionalization of these materials can bring about tunability as well as specificity, both of which are desired properties in a catalyst [21]. This chapter summarizes mainly the application of 2D TMDCs for catalysis of the Hydrogen Evolution Reaction (HER), Oxygen Evolution Reaction (OER), and various photo-assisted reactions; as well as applications to fuel cells.

7.2. Applications of 2D TMDCs in Catalysis

As expounded earlier, the unique properties of 2D materials and those of 2D TMDCs in particular, make them highly suitable for the catalysis of various chemical reactions. Numerous reactions have been shown to be reasonably catalyzed by 2D TMDCs, both experimentally and computationally. Due to such promising results, there is interest among the scientific community about the potential of 2D TMDCs to replace some of the existing catalysts. Much of the work in this field focusses on the development of facile methods for their synthesis and explores their catalytic efficiency. Further, the possibility of introducing strain, defects, and the consequent effect on catalytic properties or the production of efficient composites with suitable materials as support has also been explored. The following sections elaborate on some such work.

7.2.1. Transition Metal Dichalcogenides in Hydrogen Evolution Reaction (HER)

The hydrogen evolution reaction (HER) is of profound importance for the production of hydrogen. If the notion of using hydrogen as a fuel replacement for conventional fossil fuels is to be realized, then a clean process for its production must be developed [22]. Among all existing fuels, hydrogen has the highest known energy density by weight of up to 142 MJ/kg [23]. The most popular process for hydrogen production is by reformation of hydrocarbon-based fuels such as petroleum derivatives or natural gas [24]. However, this process is financially as well as energetically expensive, while also producing hydrogen with carbon monoxide and carbon dioxide contamination. Moreover, the production of greenhouse gases and pollutants further make this process highly unsustainable [25]. Both CO and CO_2 are catalyst poisons in fuel cells that use platinum catalysts [26]. Furthermore, the sources of hydrocarbons in the world are limited, and prone to depletion. A more permanent solution could be to harvest hydrogen by splitting water into its constituent parts by electrolysis or photolysis. Water is abundant on the earth – the process produces pure hydrogen with no carbon contamination, and is free from emissions making the whole process eco-friendly [27].

The water-splitting reaction produces gaseous molecular hydrogen at the cathode [28]. Popularly, the electrolyte i.e., water is acidified to promote the discharge of hydrogen ions on the electrodes. The cell potential can be determined by calculations from the Nernst equation [29] against the reversible hydrogen electrode (RHE).

$$E_{HER} = -0.059\, pH \text{ v/s RHE} \tag{7.1}$$

The potential is pH-dependent, and is found to vary linearly by -0.059 V per unit change in pH. However, when considering RHE as the reference electrode, the pH dependence is nullified and the cell potential, E_{HER} is determined to be 0 V v/s RHE. In reality, some onset potential is required at the cathode to begin the process of hydrogen evolution. This can be thought of as a reaction energy barrier that must be crossed in order for the reaction to proceed [30]. The Nernst equation provides the equilibrium thermodynamic potential for HER. The difference between this Nernst potential and the actually applied potential is called the overpotential, η [31].

The driving potential for HER can be given as:

$$E_{Driving} = E_{HER} + iR + \eta \tag{7.2}$$

Here, iR is the potential drop across the electrolyte. It is clear that a lower value of overpotential would favor HER, and for this purpose electrocatalysts are a necessity. The reaction overpotential is one of the basic parameters which dictate whether a given material could be a suitable electrocatalyst for HER. The effect of overpotential on HER-kinetics is quantified by the Tafel equation:

$$\eta = -\frac{2.303\,RT}{\alpha nF} log\, j_0 + \frac{2.303RT}{\alpha nF} log\, j \tag{7.3}$$

Here, α represents the charge transfer coefficient, j_0 the exchange current density which describes the rate of reaction at equilibrium potential, j the current density, n the number of electrons transferred, R the ideal gas constant, T the temperature, and F the Faraday constant. The Tafel equation predicts a linear dependence of η on the current density, with a slope of $\frac{2.303RT}{\alpha nF}$ called as the Tafel slope. A large value of Tafel slope implies a steep rise of overpotential with rise in the current density. Generally, a small Tafel slope is a figure of merit to quantify the electrocatalytic activity of a given material for the HER.

The (acidic) HER is a two-step process, involving first, the adsorption of charge-less hydrogen from the electrolyte by means of a discharge reaction, and second, the evolution process of molecular hydrogen [32]. The initial discharge step produces an intermediate adsorbed species of hydrogen, H* and is named as the Volmer reaction:

$$H^+ + e^- \rightarrow H^*$$

The evolution of H_2 can proceed by one of two mechanisms, one of which involves the reaction of H* with another proton and electron from the electrolyte, called as the electrochemical desorption or Heyrovsky reaction, or by the combination of two H* at the catalyst surface called the recombination reaction or Tafel reaction.

$H^* + H^+ + e^- \rightarrow H_2$ Electrochemical desorption or Heyrovsky reaction

$H^* + H^* \rightarrow H_2$ Recombination reaction or Tafel reaction

Irrespective of whether the reaction follows the Volmer-Heyrovsky or Volmer-Tafel mechanism, an important factor affecting reaction kinetics is the intensity of the catalyst interaction with hydrogen [33]. If the interaction is very weak, the Volmer reaction will become rate-determining, making the concentration of adsorbed hydrogen much lesser, and decreasing the efficiency of the process. Alternatively, a strong catalyst-hydrogen interaction results in the a larger binding energy of molecular hydrogen and the desorption process will become rate determining, once again reducing the efficiency of the process. As per the Sabatier principle [34], for efficient HER electrocatalysis, one requires the catalyst-hydrogen interaction to be between these two extremes [35]. Accordingly, the hydrogen adsorption free energy can be another figure of merit to quantify the efficiency of an electrocatalyst, with the best results observed when the value is close to thermoneutral i.e., $\Delta G_H \approx 0$ [36].

So considering a low Tafel slope, high exchange current density, and free energy of hydrogen adsorption close to thermoneutral, as parameters for efficient electrocatalysts, a comparative study is made among the electrocatalysts that have been developed both experimentally and theoretically for the purpose of HER with respect to various factors like availability of active sites, crystallinity, phase, concentration of defects, and so on. Bulk TMDCs have been studied for their HER electrocatalytic activities, and it was observed that bulk MoS_2 gave a large Tafel slope of 1037.51 mV/decade [37] with large internal resistances hindering efficient electrocatalysis. Bulk materials as heterogeneous catalysts suffer from the fundamental drawbacks of low surface area and fewer active sites. Layered materials like graphite are of much interest to the scientific community due to their inherent structural anisotropy. Layered TMDCs also have a tunable band gap which offers opportunities for a host of exciting applications.

Table 7.2. Comparison of HER electrocatalytic performance by different materials proposed and their method of synthesis

Material	Preparation by	Tafel slope	Overpotential	Onset potential	Stability	Reference
Graphene-like WSe_2	Solvothermal method	78 mV/dec		~150 mV	1500 cycles	Wang et al. [39]
2D terrace MoS_2/CoMoSe lateral and vertical heterostructures	Chemical Vapour Deposition	95.2 mV/dec	305 mV at 10 mA/cm^2		1000 cycles	Chen et al. [40]
Metallic 2D Au foil/TaS_2	Two-zone chemical vapour deposition	53	101 mV at 10 mA/cm^2		99% current density remaining after 12 h testing	Yu et al. [41]
Metallic 2D Mo foil/TaS_2		58	110 mV at 10 mA/cm^2			
Carbon fibres/TaS_2		73	186 mV at 10 mA/cm^2			
Glassy Carbon/TaS_2		89	267 mV at 10 mA/cm^2			
MoS_x@NbS_2/Glassy Carbon	CVD + thermal annealing	43.2	164 mV at 10 mA/cm^2		20% decay in current density after 12 h	Zhou et al. [42]
Semimetallic $MoTe_2$/$Tsi_3C_2T_x$	Hydrothermal	65	293 mV at 10 mA/cm^2			Shinde et al. [43]
1T-2H MoS_2 heterostructure	One-pot annealing	65	320 mV at 20 mA/cm^2		1000 cycles	Wang et al. [44]
$MoSe_2$/SWCNT	Sonication-centrifugation assisted Liquid Phase Exfoliation	77	300 mV at 250 mA/cm^2	41 mV for 2 mA/cm^2	10% decay in current density after 500 cycles	Gholamvand et al. [45]
MoS_2	Sonication	94.91		100 for 1 mA/cm^2		Nguyen et al. [37]
WS_2		120.22		150 for 1 mA/cm^2		

Table 7.2. (Continued)

Material	Preparation by	Tafel slope	Overpotential	Onset potential	Stability	Reference
TaS_2		114.66		175 for 1 mA/cm^2		
TiS_2		91		135 for 1 mA/cm^2		
MoS_2 monolayers on Au foil	Chemical vapour deposition	60-80	100 mV	100 mV for 10 mA/cm^2	Reproducible up to 1000 cycles	Ji et al. [46]
Ultra-thin S-doped $MoSe_2$ nanosheets	Chemical synthesis route	58	156 mV	90 mV for 1 mA/cm^2	Reproducible up to 10000 cycles	Xu et al. [47]
Vertically oriented MoS_2 or WS_2	Electron beam lithography + sulfurization	~127				Woods et al. [48]
Horizontally oriented MoS_2 or WS_2		~161				
Mo-rich hierarchical ultrathin $MoSe_{2-x}$ nanosheets	Bottom-up colloidal route	98	~170 mV		Reproducible up to 1000 cycles at 400 mV overpotential	Zhou et al. [49]
MoOx/MoS_2 core-shell architecture	Chemical vapour deposition + sulfurization	90		200 mV for 2 mA/cm^2		Cummins et al. [50]
Hydrazine exposed MoOx/MoS_2 core-shell architecture	Chemical vapour deposition + sulfurization + Hydrazine exposure	50		100 mV for 22 mA/cm^2		
Nanocrystalline $MoSe_{1.3}$/Mo thin-film	Pulsed laser ablation on $MoSe_2$ target for Mo enrichment	86	270 mV for 10 mA/cm^2			Grigoriev et al. [51]
Nanocrystalline $MoSe_{1.6}$/Mo thin-film		106	220 mV for 10 mA/cm^2			

Material	Preparation by	Tafel slope	Overpotential	Onset potential	Stability	Reference
MoS_2	Theoretical modelling	98				Li et al. [52]
MoS_2 with 12.5% strain		90				
1.35% S-vacancy MoS_2		82				
1.35% S-vacancy MoS_2 + 12.5% strain		60				
Pt-C (10% Pt on activated carbon)	Commercially available	30				Huang et al. [53]
30% Pt grown epitaxially on MoS_2 layer	Wet-chemical synthesis	40				
Pure MoS_2 layer	Li-intercalation exfoliation	94				
ZnO functionalized MoS_2	Microwave-assisted chemical exfoliation	62	239 mV for 10 mA/cm^2	168 mV for 10 mA/cm^2	1000 cycles	Sumesh [54]
Au nanoparticle (39.5 mol%) decorated MoS_2	Li-intercalation exfoliation reacted with $HAuCl_4$	56.97	205 mV for 10 mA/cm^2			Kim et al. [55]
MoS_2/Ni interspersed with Acetylene black	Hydrothermal route	55		80 mV		Liu et al. [56]
$rGO/W_xMo_{(1-x)}S_2$	Wet chemical route	50.6		233 mV for 10 mA/cm^2		Lei et al. [57]
rGO/(alloyed)$W_xMo_{(1-x)}S_2$	Wet chemical route + thermal annealing for alloying	38.7		96 mV for 10 mA/cm^2		
3D carbon fibre paper/Li-MoS_2	Pyrolysis and sulfurization	62	200 mV		7000	Wang et al. [58]
WS_2/rGO nanosheets	Hydrothermal synthesis	58	150-200 mV			Yang et al. [59]

When doping is done selectively, these materials can be tuned for the catalysis of very specific reactions [38]. Table 7.2 presents a comparison of the HER electrocatalytic performance by different materials that have been proposed in the literature.

7.2.1.1. Active Sites for HER Electrocatalysis

The edge sites of TMDCs, particularly the S-edges are pivotal to the catalytic activity of TMDCs [60]. The reason for this may be coordinative unsaturation or electronic unsaturation provided by the S atoms at the edge [61]. The Mo-edge sites are also capable of HER electrocatalytic activity. Both the edges give a lower binding energy for hydrogen adsorption than that of the basal plane [62]. Further, layered 2D materials have more edges than a bulk material, and are better HER electrocatalysts. To improve the catalytic activity of HER electrocatalysts, most research groups use one of two strategies, either to increase the extent of exposure of the edge sites, or to enhance the intrinsic activity of the basal plane [63]. Layered TMDCs consist of alternating layers of the transition metals 'T' and the chalcogenide atom 'X', with layers held together by weak van der Waals' forces, similar to that in graphite. These layers can be exfoliated and the resulting nanosheet of the TMDC can be used as an electrocatalyst for HER [64].

Zhang et al. (2017) used an interesting method of electron-beam lithography to measure the electrochemical properties of their synthesized 2D TMDC MoS_2 flakes. Monolayers of MoS_2 in the stable 2H phase were produced using a chemical vapor deposition method (CVD). The flakes were then transferred to an SiO_2/Si substrate using a wet-transfer method assisted by polymethylmethacrylate (PMMA). This technique allowed them to expose specific sites of the TMDC flake individually, and perform a comparative study of their electrocatalytic activity. The catalytic activity of the basal plane was compared to that of S-terminated edge sites for MoS_2. The edge sites contributed the bulk of the activity offered by the material, with a Tafel slope of 61 mV/dec for the edge sites while that of the basal plane was 116 mV/dec [65].

Besides 2H and 1T, the distorted octahedral arrangement abbreviated as 1T' has also been investigated for its HER activity. In an ab initio study, Fan et al. predicted that the 1T' phase was most active towards HER electrocatalysis, via both, the basal planes and edges. Their calculations suggested that 50% of the basal plane sulfur atoms in monolayer MoS_2 were active in the 1T' metastable phase, with HER activity not suffering due to distortions as compared to that noted in the 1H or 1T phases. Figure 7.2 (a-d) illustrates the active sites studied (a and b), the computed Gibb's energy profile for the different active sites under study, and the relative energy profile for hydrogen evolution for the different phases, respectively.

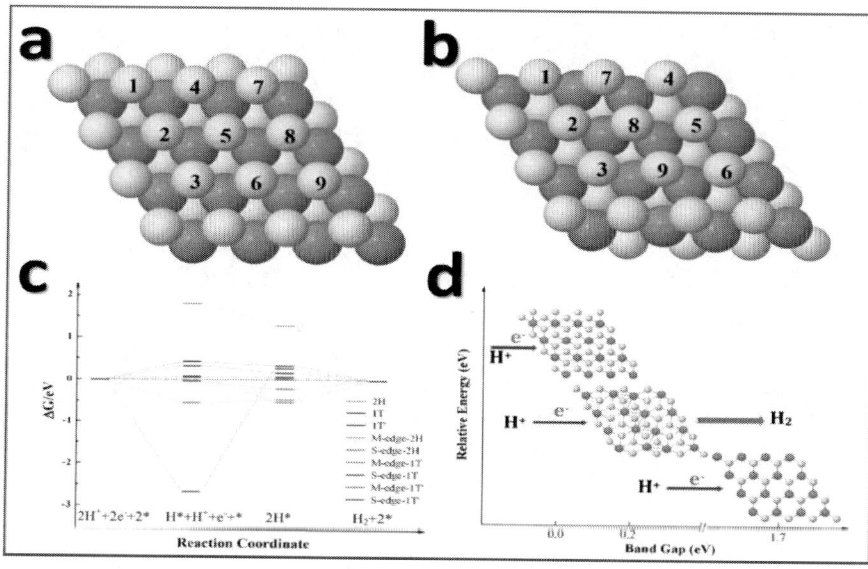

Figure 7.2. The structural plots of supercells of single layer MoS_2 in surface of (a) 1H and 1T MoS_2, and in (b) 1T' MoS_2 monolayer with probable hydrogen binding sites mentioned. (c) The Gibbs Energy Profile diagram for the various active sites studied mentioned inset (d) Energy for hydrogen production compared between the three polytypes studied by Fan et al. [65] with permission of the Royal Society of Chemistry.

A good quantitative measure of the efficiency of any given catalyst is its turnover number (TON), which is defined as the number of moles of the reactant consumed per unit moles of catalyst consumed per unit time. In the case of electrocatalysis, this formula can be modified as the ratio of the exchange current density (j_0) to the product of the density of active sites (n), the unit or elementary charge (c) and the number of electrons transferred (z) in the given electrochemical reaction, which for hydrogen is two. Based on this formula, the TON for the basal plane was found to be $1.9 \pm 0.8 \times 10^{-4}\ s^{-1}$, which is drastically less than the $1.6 \pm 1.2\ s^{-1}$ obtained for the edge sites. TMDCs have two types of edge sites, namely Mo-terminated and S-terminated. Between these two sites, the Mo-terminated edge is found to have a lower hydrogen binding energy than that of the S-terminated edge. The volcano plot depicted in Figure 7.2 (c) for the four permutations (T-edge, T-plane, H-edge and H-plane) clearly shows that the T-edge sites are most favorable and close in activity to Pt for electrocatalysis than the other sites [66].

Exfoliation also allows for the production of the metastable phase of the same material, which is otherwise difficult to produce, such as the metastable metallic 1T phase as opposed to the stable semiconducting 2H phase. The properties of these two phases differ significantly, allowing for their tunability in the synthesis protocol [68]. The possibilities in phase engineering for HER electrocatalysis have been explored in Section 7.2.1.4. Yuanyue Liu et al. have reported an improved catalytic activity in layered TMDCs with group 5 transition metals, where the enhanced performance has been attributed to basal plane activity. Their density functional theory (DFT)-based first principles investigation which made use of an energy descriptor E_{LUS}, or energy of the lowest unoccupied state, concluded that the group 5 metal disulphides, VS_2, NbS_2 and TaS_2 show promising results with their E_{LUS} value in the desired range. Their work was experimentally verified for NbS_2 and TaS_2 with Tafel slopes in the desired range [163].

Numerous methods of exfoliation of layered TMDCs have been explored, with lithium intercalation being one of the popular methods. However, the process often produces metastable phase or 1T phase TMDC nanosheets, that revert to their more stable phase over time. Zahra Gholamvand et al. have developed a method of liquid exfoliation to produce 2D TMDC nanosheets in their stable 2H phase, and have compared the catalytic activity of group 6 elements (Mo and W) for three chalcogenides (sulphides, selenides, and tellurides). They have determined that for a given metal, the HER catalytic activity decreases in the order: selenide, sulphide, telluride. Between the two metals, they conclude that $MoSe_2$ offers the best electrocatalytic activity with a Tafel slope of 76 mV/dec and an exchange current density of 1.9×10^{-3} mA/cm^2. The other TMDCs were found to show Tafel slopes in the range of 98-110 mV/dec at exchange current densities of $0.3\text{-}2.9 \times 10^{-3}$ mA/cm^2 [70].

In another study involving liquid phase exfoliation, a sonication method was used to produce atomically thin 2D TMDC nanosheets of 3-4 nm thickness with the TMDC dispersed in N-vinylpyrrolidone, and then deposited on Au/Ti electrode. MoS_2, WS_2, TaS_2, and TiS_2 nanosheets were produced by this method, and electrocatalytic activity was studied using a three-electrode setup with saturated calomel electrode as the reference in 0.5M H_2SO_4 electrolyte. 2D MoS_2 and TiS_2 synthesized by this method gave promising results with onset potentials of 100 and 135 mV respectively to achieve a current density of 1 mA cm^{-2} and Tafel slope of 94.91 and 91 mV/dec respectively. These results are a significant improvement over those of bulk TMDCs, and the authors attribute this improvement to the large number of edge sites and a greater surface-to-volume ratio in the atomically thick TMDCs [71]. Chen et al. have proposed a facile, solvothermal method for the synthesis of few-layer nanosheets of WSe_2, and have studied the electrochemical properties in 0.5 M H_2SO_4 using a Pt counter electrode and SCE reference electrode. With an onset potential of 150 mV to reach 1mA/cm^2, the graphene-like WSe_2 nanosheets give a Tafel slope of 78 mV/dec making this material a suitable candidate for HER electrocatalysts [72]. Another group has prepared nanostructured, edge-dense, triangular and dendritic MoS_2 monolayers by an optimized, etching-based CVD process. The MoS_2 triangular flakes were deposited on Au foil, and a three-electrode system was set-up. The morphology of the TMDC monolayers produced allowed for a greater number of active sites, giving a Tafel slope in the range of 60-80 mV/dec, and current densities of 24-38 μA cm^{-2}. The group suggests improved performance in HER electrocatalysis for a 1T phase flake, which would be inherently metallic and therefore more efficient [73].

Xu et al. [74] produced S-doped $MoSe_2$ nanosheets making use of a surfactant in its preparation. The surfactant reduced the surface energy of the edge sites giving greater exposure to the edge sites compared to the basal plane, thereby improving the electrocatalytic activity. $MoSe_2$ is a superior electrocatalyst than MoS_2 owing to the metallic nature of selenium. Further, the S-doping introduced defects in the crystal lattice, enhancing the electrical conductivity and reducing the bandgap. The resulting material has shown a low Tafel slope of 58 mV/dec at an onset potential of 90 mV to reach a current density of 1 mA/cm^2 [75]. The role of defects in enhancing the electrocatalytic activity is explored in the next section (Section 7.2.1.2).

7.2.1.2. Role of Defects in HER Electrocatalysis

Defects in the crystal structure of a given material has a marked effect on its electronic structure and properties [76]. Any form of dissymmetry in the crystal tends to improve the electronic conductivity, which encourages the attempt to intentionally introduce defects in the crystal lattice, such as by using dopants. A defect-rich structure may also improve the electronic conductivity of the crystal owing to their atomic arrangement which can be said to be 'quasi-periodic' and hence encourages electronic transport. This is in contrast to the polycrystalline arrangement of a defect-free crystal which is internally composed of numerous domains, each marked by domain boundaries. Each domain boundary impedes the flow of electrons and the resulting structure has poor electronic conductivity [77]. An increase in the degree of disorder in the crystal may also improve electrocatalytic activity, but will simultaneously decrease the electronic conductivity across the domain boundaries. As demonstrated by Xie et al., a hydrothermal treatment of O-MoS_2 ultrathin nanosheets showed that with decrease in temperature, the degree of disorder in the crystal increased systematically. The key to achieve best electrocatalytic results was to optimize the degree of disorder only to such extents that the electronic conductivity was still sufficient [78].

The marked effect of defects on the electronic and by extension, the HER electrocatalytic activity is evident from the comparative work carried out by Woods et al. Primarily, horizontally-oriented TMDC films are studied for their HER electrocatalytic activity given that they are rich in edge sites which are supposed to be majorly responsible for its activity. Woods et al. grew vertically-oriented TMDC sheets of about 1 nm thickness for the purpose of building van der Waal heterostructures between two layers of dissimilar TMDCs. Despite lacking in edge sites, the vertically oriented sheets were found to show some HER electrocatalytic activity. Albeit poor at the HER electrocatalysis given the large Tafel slopes (about 161 mV/dec for MoS_2) [48], the very fact that they show electrocatalytic activity is indicative of the role of defects such as grain boundaries, lattice mismatch, etc., in altering electronic properties of TMDCs.

Zhou et al. prepared Mo-rich $MoSe_{(2-x)}$ nanosheets. Although the edge sites are catalytically active in $MoSe_2$, the excess Mo sites improved the electrocatalytic activity of the basal plane, and the resulting structure gave very good HER electrocatalytic activity with an overpotential of 170 mV and Tafel slope of 98 mV/dec [49]. Cummins et al. have reported that MoS_2 structures such as nanowires, nanosheets, etc., show a marked improvement in its electrocatalytic activity when exposed to 1% hydrazine. The exposure of hydrazine on MoO_x/MoS_2 core-shell nanowire did not disrupt the crystal structure much. The onset potential to achieve 2 mA/cm^2 dropped from 200 mV for the nanowire itself to 100 mV at 22 mA/cm^2, and an improved Tafel slope from 90 to 50 mV/decade after exposure to hydrazine. They have compared their results with pure 2H MoS_2 particles, which in their bulk form are not suitable for HER with Tafel slope of 134 mV/dec. However, the exposure to hydrazine improved its activity with a Tafel slope of 108 mV/dec, and improved the onset potential from -0.25 V v/s RHE for 0.75 mA/cm^2 to -0.2V for 2 mA/cm^2 [81]. Grigoriev et al. used laser ablation on an $MoSe_2$ target, and the resulting thin films were deposited on Si, glassy carbon, and

NaCl crystal targets. It is suggested that the pulsed Q-switched yttrium aluminum garnet (YAG) laser caused local heating, which would in turn cause the Se to evaporate more exclusively resulting in the target to become Mo-rich. The resulting nanosheets would also be metal-rich in nature, which owing to its inherent defects would be a better electrocatalyst than $MoSe_2$ itself [82].

Li et al. proposed introducing a combination of strain and S-vacancies into the structure of 2H MoS_2 in a bid to improve its otherwise inert basal plane activity. Their DFT-based study showed that creating S-vacancies improved the hydrogen binding, whereas introducing axial strain was accompanied by the appearance of some new states in the band structure, and ultimately led to improved free energy change of hydrogen adsorption, i.e., it was closer to thermoneutral. Vacancies beyond the threshold of 11% were found to destabilize the structure due to an increased surface energy. Optimization could be attained by varying both strain and vacancies independently. Their experimental research indicates improvement in the HER electrocatalytic activity on straining the structure up to about 12.5%, evident in the reduction in Tafel slope from 98 mV/dec to 90 mV/dec. On the other hand, when vacancies of about 1.35% were introduced into pristine 2H MoS_2, the Tafel slope alleviated to 82 mV/dec. Still better results were obtained when the effect of strain and vacancies were observed in the same system which gave a Tafel slope of 60 mV/dec. The key point to be noted here is that these results were obtained by straining the structure without loss of stability or a transition into the metastable phase [83].

The inherent asymmetry present in Janus TMDCs can be of advantage for HER electrocatalysis applications. Janus TMDCs have the general formula MXY i.e., dissimilar chalcogenides are combined with the metal atom, as represented in Figure 7.3 (a) [69]. A first principles DFT investigation proposed by Er et al. studies the electrocatalytic activity in Janus TMDCs, i.e., introducing asymmetry in the structure by using a different layer of chalcogenide. The group studied a variety of TMDCs, and found WSSe to be most favorable for HER electrocatalysis as shown in the current density curves and volcano plot respectively depicted in Figure 7.3 (c and d). In the Janus structure, there is a net compressive and tensile strain induced on the lighter and heavier chalcogenide atom respectively. The effect of vacancies was studied for the Janus structures, and it was found that for a reasonable defect concentration of about 3.12%, the integrity of the structure was not compromised. While pristine WSSe gave a ΔG_H of -1.82 eV, on creating defects of 3.12% the ΔG_H lowered to -32 meV. Pristine WSSe has an intrinsic strain of nearly 4% due to its asymmetry, but added compressive strain further improved its electrocatalytic activity, wherein a 2% induced strain resulted in ΔG_H of -50 to -70 meV. As strain is increased, the ΔG_H decreases systematically as depicted in Figure 7.3 (b) [86]. The volcano plot depicted in Figure 7.3 (d) places these Janus structures very close to the ideal precious metal.

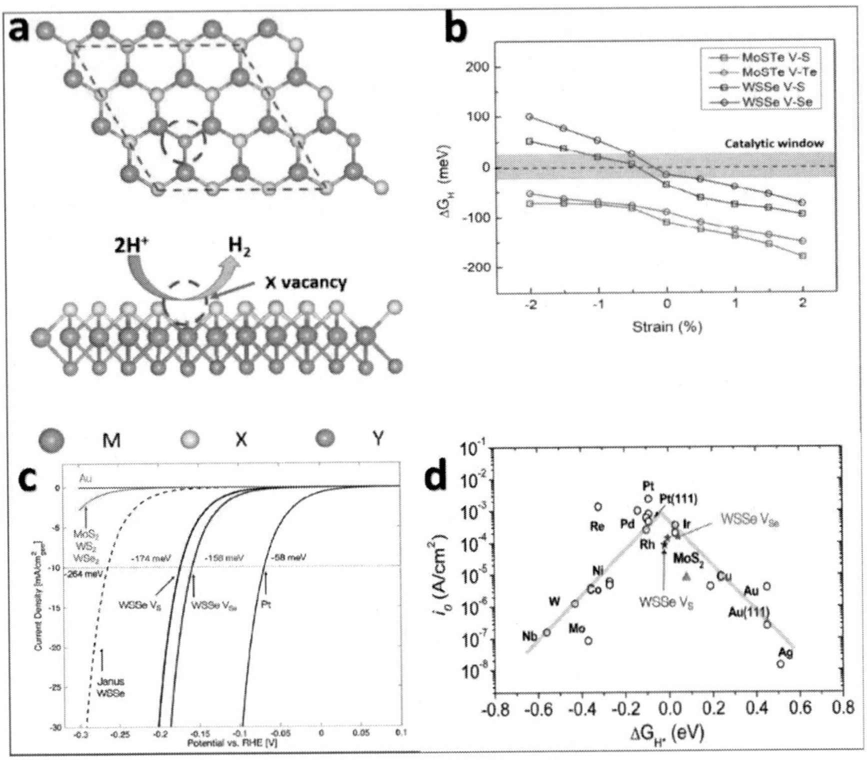

Figure 7.3. Janus structures in HER electrocatalysis (a) A schematic diagram showing HER activity at a vacancy site (b) Free energy change with variation in strain (c) Current density curves for various Janus TMDC structures compared to pristine TMDCs (d) Volcano plot for Janus WSSe with vacancy defect structure with respect to other materials. Reprinted with permission from Er, Dequan, et al. "Prediction of enhanced catalytic activity for hydrogen evolution reaction in Janus transition metal dichalcogenides." *Nano letters* 18.6 (2018): 3943-3949. Copyright © 2018, American Chemical Society [69].

7.2.1.3. Hybrid Structures as Excellent HER Electrocatalysts

One of the fundamental difficulties in using TMDCs as HER electrocatalysts appears to be the lack of electrical contact between the highly conducting electrode surface and the high-impedance TMDC surface. This problem is alleviated in hybrid structures which provide reasonable conductance between the electrode and catalyst. Additionally, hybrid structures can be viewed as having some major 'defects', presenting the same benefits discussed in the previous section. There have been numerous attempts at creating efficient TMDC-based hybrids for the purpose of HER electrocatalysis. The different hybrid structures can be categorized as under a) TMDC-metal composites b) TMDC-carbon materials composites and c) TMDC-metal oxide composites [87].

TMDC-metal composites have shown some significant improvement over pure TMDC at HER electrocatalysis [88]. When Pt was grown epitaxially on single layer nanosheets of the TMDC MoS_2, the resultant hybrid gave nearly zero overpotential and a very favorable Tafel slope of 40 mV/decade [89]. Another study of Pt on Li-intercalation exfoliated TiS_2 and TaS_2 nanosheets also gave a similar Tafel slope of 41 mV/decade proving to be as good, if not better than the pristine versions of the TMDC [90]. A detailed study on the effects of Au nanoparticles on TMDC nanosheets with respect to the relationship between loading and electron transport was made by Kim et al. [55]. The study revealed that an optimum of 21.1 mol% Au on MoS_2 nanosheet gave a hybrid with highly desirable characteristics i.e., 230 mV overpotential with a Tafel slope 57 mV/dec.

Concerning TMDC-carbon material composites, it is believed that not only will the exposure of more edge sites enhance the electrocatalytic activity, but also attaching the TMDC to a framework could avert the aggregation of TMDCs in the electrocatalytic reaction [92]. Yang et al. have proposed a temperature-sensitive hydrothermal method to prepare WS_2 nanosheets, then layered the structure with rGO forming WS_2/rGO hybrid nanosheets. The WS_2 nanosheets were prepared at 265 °C via WO_3 intermediates. They have compared HER electrocatalytic activity before and after annealing followed by freeze drying, and most desirable results were obtained when the hybrid nanostructures had undergone the additional step of annealing. The hybrid structures deposited on the glassy carbon electrode gave an overpotential of -150 to -200 mV v/s RHE, and when annealed had a Tafel slope of 58 mV/dec, a significant improvement over the 100 mV/dec of the non-annealed version. The improvement in Tafel slope was attributed to the highly conducting network formed by the underlying rGO layer which considerably improved the conductivity of the catalyst. The hybrid structures proved to be more efficient catalysts as compared to pristine WS_2 which was found to have a Tafel slope of 140 mV/dec, proving that hybrid structures have some intrinsic advantages over pristine structures [93]. Liu et al. have developed a facile hydrothermal method for the preparation of $MoSe_2$ nanosheets on a Ni substrate interspersed with acetylene black (AB), giving very good results for HER electrocatalysis with an onset potential of about -80 mV v/s RHE and a Tafel slope of 55 mV/decade [94]. Acetylene black presents a low-cost alternative over graphene, and the resulting hybrid shows similar results to graphene-based hybrids as illustrated in Table 7.2.

Another method to improve the conductivity of TMDCs by forming rGO/TMDC heterostructures has been explored by Lei et al. in which the alloying effect of Mo with W-based TMDC may also have contributed to an enhanced HER electrocatalytic activity. Their wet-chemical route for the production of rGO/$W_xMo_{(1-x)}S_2$ using graphene oxide (GO) as the template (and surfactant) to build mono- or few-layered XS_2 (X = Mo or W) involved dispersion of ammoniumtetrathio -tungstate or -molybdate in an aqueous medium containing GO, followed by sonication forming a homogenous mixture, and subsequently spin-coated onto a glassy carbon substrate. The resultant heterostructure was thermally annealed at 400 °C reducing GO to rGO and was found to develop multiple dendritic regions with ample exposed edge sites. When comparing the HER electrocatalytic activity of rGO/MoS_2, rGO/WS_2 and rGO/$W_xMo_{(1-x)}S_2$ it has been found that rGO/$W_xMo_{(1-x)}S_2$ shows the best results with a Tafel slope of 50.6 mV/dec and an onset potential of 233 mV for a current density of 10 mA/cm^2. When the same heterostructure was annealed at 300 °C, still better results were obtained with a Tafel slope of 38.7 mV/dec and a low onset potential of 96 mV for 10 mA/cm^2 current density. These results indicate that an alloying effect is also at play which may be activating the otherwise inactive basal plane towards HER electrocatalytic activity. The improved electrocatalytic activity can be similarly attributed to more exposed edges in the dendritic morphology, an activated basal plane of the TMDC layer due to W-Mo alloying, and interlayer coupling between rGO and the TMDC layer as illustrated in the density of states plot in Figure 7.4.

The band gap of pristine WS_2 or MoS_2 which is around 2.5 eV is reduced to about 1.3 eV in the rGO hybrid indicating a major change of the electronic properties in the heterostructures. There appears to be a reduction in the reaction barriers by Volmer or Heyrovsky mechanism when a hybrid is formed explaining the improvement in the HER electrocatalytic activity in the hybrids over pristine TMDCs as predicted by DFT studies [96]. Some degree of phase engineering has been made possible in the work of Wang and coworkers [97] who prepared 3D carbon fiber paper (CFP)-based composites using MoS_2 nanoparticles. Their synthesis method allowed for exposing edge sites preferentially; and the resulting 1T phase MoS_2-CFP nanocomposite showed a superior stability with a lasting electrocatalytic activity of 200 mV overpotential and 62 mV/dec for up to 7000 cycles.

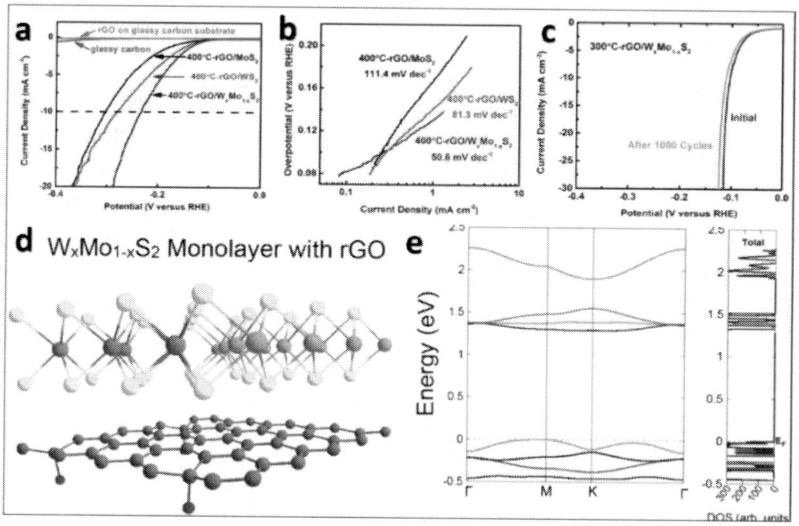

Figure 7.4. Current density curves (a), and their corresponding Tafel plots (b) for various materials mentioned inset, (c) stability of the $W_xMo_{(1-x)}S_2$ monolayer alloy over 1000 cycles (d)Structural plot of $W_xMo_{(1-x)}S_2$ monolayer hybridized with rGO layer and (e) corresponding Density of States and Band structure plots. Reprinted with permission from Lei, Y.; Pakhira, S.; Fujisawa, K.; Wang, X.; Iyiola, O. O.; Perea López, N.; Laura Elías, A.; Pulickal Rajukumar, L.; Zhou, C.; Kabius, B.; Alem, N.; Endo, M.; Lv, R.; Mendoza-Cortes, J. L.; Terrones, M. Low-Temperature Synthesis of Heterostructures of Transition Metal Dichalcogenide Alloys ($W_xMo_{1-x}S_2$) and Graphene with Superior Catalytic Performance for Hydrogen Evolution. *ACS Nano* 2017, 11 (5), 5103–5112. https://doi.org/10.1021/acsnano.7b02060, Copyright © 2017, American Chemical Society [95].

It can be anticipated that with graphene or CNTs, the costs of production skyrocket. An alternative to using high-porosity, carbon-based materials would be to use double-gyroid structures rich in nanopores having a large number of catalytically active edges [98]. Ni-foam fits in this context being relatively inexpensive, readily available, and with the desired structural characteristics. The acid-sensitivity of Ni-foam can be resolved by converting it to $NiSe_2$ foam via a direct selenization in an inert argon atmosphere [99]. The resulting $NiSe_2$ foam when hybridized with MoS_2- or MoSSe-type TMDCs show excellent HER electrocatalytic activity with a dramatic reduction in overpotential at a current density of 10 mA/cm^2, from -153 mV for pure $NiSe_2$ foam to -118 and -69 mV for $MoS_2/NiSe_2$ and $MoS_{2(1-}$

$_{x)}Se_{2x}/NiSe_2$ respectively. The $MoS_{2(1-x)}Se_{2x}/NiSe_2$ gave a low Tafel slope 42.1 mV/decade and high exchange current density 0.299 mA/cm^2. The results were substantiated computationally using DFT simulations, with the hydrogen binding energy on Mo atom as the indicator. The $MoS_{2(1-x)}Se_{2x}/NiSe_2$ hybrid gave a low ΔG_H of 2.7 kcal/mol as compared to $MoS_{2(1-x)}Se_{2x}$ (8.4 kcal/mol) and MoS_2 (10.6 kcal/mol) [100].

TMDC-metal oxide nanocomposites such as MoS_2 coated with SnO_2 nanotubes prepared by a one-step solvothermal method combined with electrospinning gave an overpotential of 150 mV and Tafel slope of 59 mV/decade [101]. The excellent HER electrocatalytic features could be attributed to the structural stability of these composites, as well as improved electronic conductivity due to the metal oxide. A core-shell assembly with MoO_3 at the core and MoS_2 forming the shell, gave a low overpotential of about 150 mV and a Tafel slope in the range of 50-60 mV/decade. In this case, the MoS_2 shell formed a protective layer around the highly conducting core of reduced MoO_3 [102]. The previously discussed acid hydrolysis differs considerably from alkaline HER which is riddled with sluggish reaction kinetics [103]. Another study used composites made from hydroxides of Ni and Co in combination with MoS_2 and WS_2 for the purpose of alkaline HER electrocatalysis [104]. The hybrid produced revealed good characteristics for HER electrocatalysis, with 2D-$MoS_2/Ni(OH)_2$ and 2D-$MoS_2/Co(OH)_2$ showing marked reduction in overpotential at 10 mA/cm^2 (from 351 mV for bare 2D MoS_2) to 185 mV and 128 mV respectively in 1M KOH. The catalysts showed a negligible depreciation in electrocatalytic activity for 1000 cycles, with ~166% increase in the exchange current densities in the hybrid structures as compared to their pristine counterparts. The computationally-determined free energy change for water dissociation (ΔG_W) was found to decrease from 2.48 eV for bare 2D MoS_2 to 1.83 and 1.18 eV respectively in 2D-$MoS_2/Ni(OH)_2$ and 2D-$MoS_2/Co(OH)_2$, respectively [105].

7.2.1.4. Importance of Phases

The phase in which the TMDC exists, whether the stable, semi-conducting 2H phase or the metastable, metallic 1T phase is also found to affect its catalytic efficiency. The 1T metallic phase has superior electrical conductivity as compared to the 2H phase, and would be the obvious choice for developing TMDC electrocatalysts. However, its lack of structural stability hinders its ready use as compared to the 2H phase TMDCs. The challenge in this case is to develop predominantly 1T phase TMDCs with the same longevity as its equivalent 2H phase structures. Some facile methods have been proposed to this end, with the incentive being in the longevity of the catalyst materials.

Tan et al. have developed TMDC nanodots which manage to remain in the metallic phase when suspended in water. The nanodots were prepared by ball milling the 2H bulk materials with Li-intercalation, followed by sonication in water to produce extremely small sized nanodots with majority of the TMDC existing in the 1T phase. The surface being charged, was found to form stable and homogenous suspensions in water. Best results for HER electrocatalysis were observed for the Janus type MoSSe nanodot with a Tafel slope of 40 mV/dec, and an onset potential of -49 mV to attain a current density of about 10 mA/cm^2. The improved electrocatalytic activity has been attributed to the alloying effect where the H adsorbed over S and Se have ΔG_H of about -0.74 to -0.95 eV, a substantial improvement over that of 1T-MoS_2 i.e., -2.8 eV [84].

The 1T-MoS_2-CFP mentioned in the previous section [107] allowed for the catalyst to achieve high levels of stability with no appreciable loss in electrocatalytic activity for up to 7000 cycles. The phase engineering was made possible by Li-intercalation giving a high cathodic current density of 200 mA/cm^2 at an overpotential of 200 mV with a Tafel slope of 62 mV/decade. Overall, the material presents a facile method for phase-tuned, highly efficient HER electrocatalyst synthesis. In another study, it was revealed that a reversible transition between 2H and 1T phases can be achieved for MoS_2-Au composites due to a resonant excitation in the plasmon modes of Au [108]. The reversible transition could facilitate production of highly tuned electrocatalytic materials that hop between the more efficient 1T phase to the more stable 2H phase and vice versa by means of some reaction conditions, allowing for a nifty method for improving longevity of the catalyst materials. In the 2D-MoS_2/$Ni(OH)_2$ and 2D-MoS_2/$Co(OH)_2$ hybrids for alkaline HER electrocatalysts mentioned in the previous section, the Li-intercalation step involved for its production induced a 2H to 1T phase transition, with a possibility of maintaining the phase transformation [109].

7.2.2. Applications of 2D-TMDCs as Electrocatalysts in Oxygen Evolution Reactions (OER)

7.2.2.1. Introduction to Oxygen Evolution Reaction

In the current energy crisis scenario, the alternatives to conventional energy sources are playing a primal role in the human fight against depleting resources. One such alternative is the water-splitting reaction, a reversible reaction producing hydrogen at the cathode (HER) which was investigated in Section 7.2.1; and oxygen at the anode in a process called as the oxygen evolution reaction (OER) [110]. The relevance of this reaction lies in the fact that along with HER, it makes the water-splitting reaction reversible, allowing for the continuous production of hydrogen [111]. In sharp contrast to HER which involves a two-electron transfer, the OER involves a transfer of four electrons in four individual steps. The kinetic energy barrier for these reactions to occur are higher than those of the cathodic half [112]. The reaction is also pH-dependent, much like HER.

To overcome the large energy barrier for OER to occur, catalytic intervention is necessary. The ideal electrocatalyst for OER should not only display a high efficiency and turnover, but it must also withstand the harsh reaction conditions (which may be acidic or alkaline), retain catalytic efficiency over time, be earth-abundant and as much as possible, inexpensive [113]. Currently the best-known catalysts for electrocatalytic OER is found to be rutile phase structures of RuO_2 and IrO_2, both of which are sparsely available and expensive [114]. While RuO_2 shows high catalytic efficiency over a versatile pH range, it is affected by higher anodic potentials, which cause it to dissolve. Whereas IrO_2 can sustain more anodic potential than RuO_2, but it is also susceptible to dissolution at still higher potentials [115]. Further, since they are sourced from noble metals, the hunt for an ideal OER electro-catalyst is far from over.

Non-noble metal based OER electrocatalysts are being explored thoroughly by researchers. Perovskite [116] and spinel-based oxides [117] have found some success in this regard, being low-cost, more available, and showing reasonable electrocatalytic activities while maintaining structural integrity. Similarly, non-oxide-based transition metal chalcogenides have also been reported as potential candidates for OER electrocatalysis. The advantage that 2D materials offer to catalysis with their large specific surface area and high degree of tunability has been illustrated earlier. 2D transition metal dichalcogenides have been explored for their possible applications in this field.

7.2.2.2. Background and Figures of Merit

The mechanism of OER varies with the media under consideration, i.e., whether alkaline or acidic. At a pH of 0, an external current is required to drive the potential difference of 1.23 V for the reaction to occur. For every unit change in pH, a 59 mV shift from reaction potential is obtained theoretically as per the Nernst equation [118]. The influence of pH on reaction potential can be avoided by using the RHE as the reference, as was the case in HER. There are some advantages of using an acidic media over alkaline in proton exchange membrane (PEM) electrolysers, mainly a quick start up, potential to reach high current densities with high efficiency [119]. It may be noted that if OER is to be carried out as a part of a water-splitting electrocatalytic reaction, then both the HER at cathode and OER at anode will undergo same electrolytic conditions, i.e., same conditions of pH. To overcome the energy barriers of both the electrode reactions, either strongly alkaline or strongly acidic electrolyte solutions will have to be employed [120]. Nevertheless, the four-electron transfer occurs in four individual steps that raise the kinetic energy barrier. There is still some ambiguity in the exact reaction mechanism for OER. The reaction mechanism under both alkaline and acidic reaction conditions are well explained in the schematic diagram Figure 7.5 (a).

The figures of merit to judge a good OER electrocatalyst are similar to those observed in HER such as overpotential (η), the exchange current density (j_0) and Tafel slope (b). These parameters have been discussed in Section 7.2.1. For OER, the ideal electrocatalyst serves to provide the highest possible exchange current density at least overpotential, and least Tafel slope, as was the case for HER. The overpotential is generally measured for an exchange current density of 10 mA/cm^2, with a range of 300-400 mV considered to be highly favorable for efficient OER electrocatalysis. In addition to these parameters, the stability of the catalyst is paramount to its subsequent use. An interesting protocol to judge whether a given catalyst passes the test of stability is mentioned in a comprehensive review article by Tahir et al. where the overpotential at a current density of 10 mA/cm^2 is compared at t=0 and t=10 h [89]. A corresponding graph of the same can be plotted where the catalyst efficiency can be divided into sections based on its stability. A volcano plot can be used to compare the catalyst efficiencies.

In a thermodynamic-kinetic model devised by Zhang et al. [98] the four reaction steps and the steps in between them depicted in Figure 7.5 (A), are considered as the rate determining step (RDS) one-by-one, and the rate expression is subsequently modified. Model fitting with a wide range of OER catalysts ensure more relatability of theoretical calculations to experimental results, to a higher degree than that promised by DFT calculations. In Figure 7.5 (B), considering the third electron transfer reaction in the reaction scheme as the rate-determining step of the OER process, the shift in Tafel plot with variation in the equilibrium potentials of each elementary step is depicted. The presence of multiple Tafel slopes in a single plot has been attributed to the possible saturation of the surface by reaction intermediates. The increase in Tafel slope as the equilibrium potential increases is prominent from the plots, specifically in the top left panel of Figure 7.6 (B), it is observed that the slope decreases in the linear region highlighted by a down arrow, however the slope of other linear regions is not affected, indicating that an increase in the energy barrier of this step will result in decreasing current at the Tafel slope. Such fascinating results can be obtained by a model-fitting approach to kinetic and thermodynamic parameters governing the OER reaction [125].

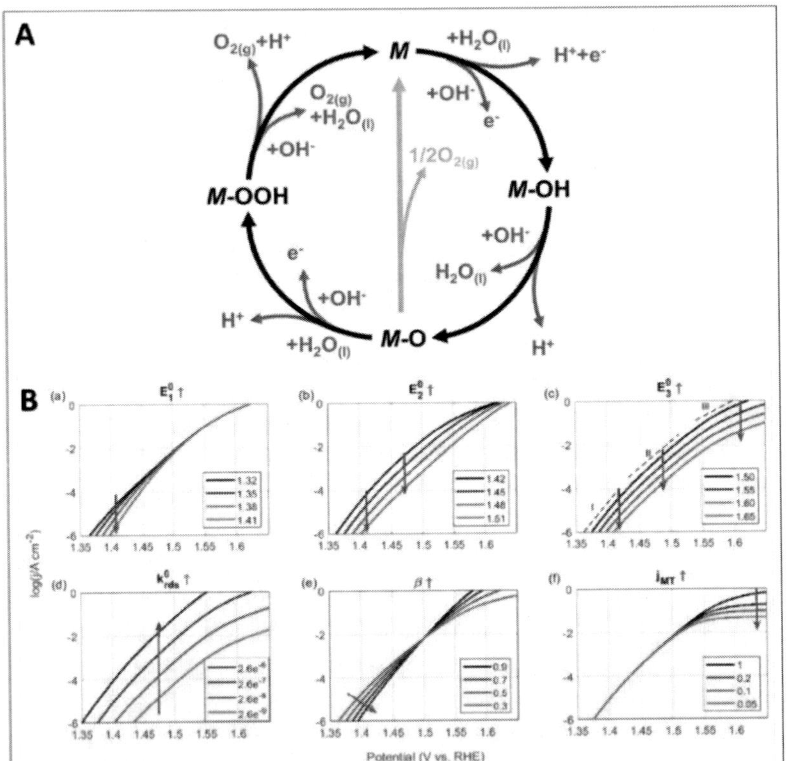

Figure 7.5. (a) Schematic flow diagram depicting the OER reaction mechanism in acidic (blue line) and alkaline conditions (red line). Reproduced with permission from Suen, Nian-Tzu, et al. "Electrocatalysis for the oxygen evolution reaction: recent development and future perspectives." *Chemical Society Reviews* 46.2 (2017): 337-365, Copyright © Royal Society of Chemistry 2017 [91]. (b) Effect of variation in equilibrium potentials on the electrocatalytic activity using model-fitting methods. Reproduced with permission from Zhang, Junming, et al. "Advances in thermodynamic-kinetic model for analyzing the oxygen evolution reaction." *ACS Catalysis* 10.15 (2020): 8597-8610, Copyright © American Chemical Society 2020 [98].

7.2.2.3. Evaluation of Electrocatalytic Performance of Some 2D TMDC Catalysts for OER

Noble metals and their corresponding oxides and alloys (in the order of Ru>Ir>Pd>Pt) are by far, considered to be the most efficient OER electrocatalysts [126]. Among other materials studied for their replacement, Co- and Ni-based chalcogenides have shown some success in water-splitting reactions where there is a need for functional bi-catalysts for both, the HER and the OER half-reactions for complete reversibility [127]. In case of TMDCs, since their layers are held together by weak van der Waals' forces, they can be exfoliated into mono-layers of 2D TMDCs via facile methods, which significantly alter their electronic properties as well as their structural properties [128]. Gao et al. have reported a $CoSe_2$ catalyst composite with N-doped graphene (NG) wherein nanobelts of $CoSe_2$ were anchored on the graphene substrate [104]. Graphene offers several desirable properties like mechanical durability, increased conductivity, etc. [130]. These advantages are clearly observed with the marked improvement in Tafel slope from 66 mV/decade for pure $CoSe_2$ nanosheets to 40 mV/decade for the NG-$CoSe_2$ composite. It is worth noting that the pure $CoSe_2$ nanosheets perform slightly better than commercial RuO_2 electrocatalysts which gave a Tafel slope of 69 mV/dec. The composite also gave a low overpotential of 366 mV at 10 mA/cm^2 and suffered negligible loss of catalytic efficiency for up to 2000 cycles. It also gave a high turnover efficiency (TOF) of 0.036, improving over the 0.017 for commercial RuO_2 [131]. It has been observed that metals with eg^1 electronic configuration may give good OER electrocatalytic results [132]. Although Co(IV) fulfils this particular criteria, it still fares poorly as an electrocatalyst mainly due to a low exposure of its active sites [133].

Transition metal d states near the Fermi level interact with oxygen atoms initiating catalytic activity [134]. Any modulation of the electronic structure will affect the d electron filling in the transition metal and this relation can be used to optimize the catalytic efficiency. The nanocomposite formed with N-doped Graphene is accompanied by a change in the electronic structure of the TM, which may be the reason for the enhanced electrocatalytic activity [135]. Another approach by Liu et al. on $CoSe_2$ catalysts negates the use of a hybridizing structure, instead relying on ultra-thin nanosheets of $CoSe_2$ to improve conductivity. 2D materials have been shown to provide significant advantages over bulk materials in the previous section, and similar advantages are prevalent here as well. The improved contact to electrode surface (in this case, glassy carbon), and the increased occurrence of disorder on the catalyst surface, both enhance the catalytic efficiency of ultrathin CoS_2 as compared to its bulk counterpart. They reported excellent results with a Tafel slope of 44 mV/decade, large turnover frequencies of 0.33 sec^{-1} at overpotentials of 500 mV for a current density of 10 mA/cm^2 [111].

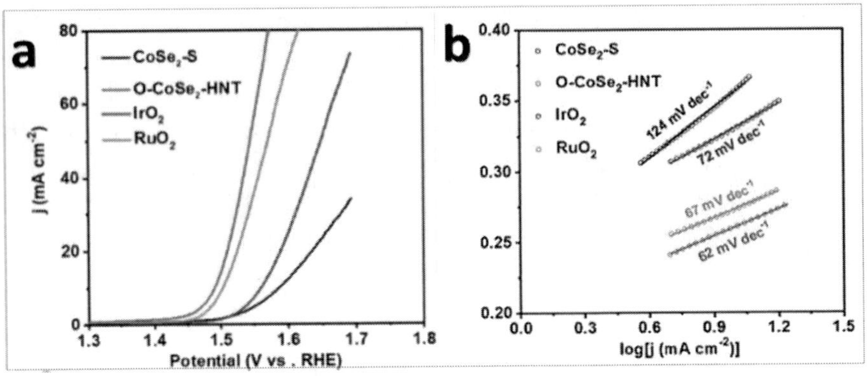

Figure 7.6. (a) Exchange current densities and (b) Tafel plots for materials under study mentioned inset, with the Tafel slopes mentioned in each case. Reproduced with permission from Jia, Baoming, et al. "Hierarchical nanotubes constructed from $CoSe_2$ nanorods with an oxygen-rich surface for an efficient oxygen evolution reaction." *Journal of Materials Chemistry* A 7.25 (2019): 15073-15078, © Royal Society of Chemistry 2019 [113].

A comprehensive study by Wu et al. compares the OER electrocatalytic activity between different phases of 2D TMDCs namely 2H (the stable, semiconducting phase) and the 1T phase (the metastable, metallic phase) in 0.5 M H_2SO_4. Their study made use of Li-exfoliated 2D nanosheets of TMDCs MoS_2 and TaS_2, with ample evidence that the crystalline state of the 1T versions were maintained even after exfoliation. Their results showed that the metallic 1T-TMDCs fared better than 2H phase, with 1T-MoS_2, 1T-TaS_2, 2H-MoS_2 and 2H-TaS_2 needing overpotentials of 420, 450, 480 and 540 mV with Tafel slopes of 322, 282, 361 and 255 mV/decade respectively. The superiority of the 1T phase was confirmed by slowly converting it to the stable 2H version by annealing the 1T-MoS_2 and recording the electrocatalytic activity at higher temperatures, which resulted in a drop in exchange current density as well as larger overpotentials. DFT simulations also revealed that the 1T-phase edges had the least activation barriers as the active site [96].

The activity of the catalyst may be increased by altering the surface morphology in a bid to enhance the number of active sites. Such specified surface engineering may be fruitful in developing efficient electrocatalysts for OER [139]. It may also be used to prevent the tendency of TMDCs to self-aggregate. Jia et al. hydrothermally synthesized Se nanowires (Se-NW) polymerized in polypyrrole (Se-NW@PPy). Zeolite Imidazole Framework-67 (ZIF-67) was grown in situ on the Se-NW@PPy and pyrolized giving the desired hierarchical nanotubes (HNT). The resulting O-$CoSe_2$-HNT electrocatalysts were found to give excellent OER electrocatalysis in 1 M KOH with Ag/AgCl electrode; and an overpotential of 252 mV for 10 mA/cm^2 and a Tafel slope of 62 mV/decade was recorded. The result of the prepared catalyst has been compared with other catalysts in the exchange current densities and Tafel plots depicted in Figure 7.6, clearly showing its superiority over the other electrocatalysts studied. The low Tafel slope indicated rapid OER kinetics, consequently resulting in a faster evolution of oxygen. The enhancement has been attributed to a modification in the electronic structure of oxygen on the surface as well as the meticulous structural advantages offered by a hierarchical nanotube array [140].

7.2.3. 2D TMDCs as Potential Photocatalysts

7.2.3.1. Introduction to Photocatalysis

Solar energy is a freely available, intense and renewable energy source and when harvested efficiently, can provide a very suitable alternative to non-renewable energy sources [141]. Photochemical reactions are useful in this respect, and are enthusiastically explored by the scientific community. Many photochemical reactions require a photocatalyst for the reaction to proceed at a more desirable rate. There is much interest in semiconductors for their role in catalysing photochemical reactions, owing to the fact that solar energy can excite electrons from the valence to conduction band of a photosensitive semiconductor material, generating electron-hole pairs or excitons which can catalyse particular reactions depending on its chemical potential [142]. Similar to electrocatalysis, 2D TMDCs possess interesting properties that render them valuable as photocatalysts for various reactions. One of these defining properties is the conversion from small indirect bandgap in bulk materials to direct bandgap when we reduce to two-dimensions. This property can be attributed to the effects of quantum confinement phenomena [143].

Enhanced surface area is also a contributing element to the efficacy of 2D TMDCs as good photocatalysts, owing to an increase the number of active sites [144]. There is a significant improvement in contact between catalyst and electrolyte, as well as catalyst and charge collectors, leading to improved reaction kinetics for 2D materials [145]. Layer-exfoliated materials improve the charge carrier separation, and also lower the conduction band minimum (CBM), meaning that electrons can be excited at lower energies [146]. This lowering of the threshold energy needed to excite electrons enables the use of a wider range of the electromagnetic spectrum. Otherwise, only UV radiation could have been utilized for the same purpose, which only accounts for about 4% of the radiation reaching the earth's surface. Another deciding factor for the development of these photocatalysts would be cost and availability. TMDCs are relevant in this respect, being earth-abundant and relatively inexpensive. 2D materials also allow for a preferential exposure of the more catalytically active facets of the material, which can help in keeping the photo-excited charge pairs separated, as the recombination of these pairs would release energy in the form of heat or light draining the absorbed energy [147].

7.2.3.2. Photocatalysis for Water-Splitting Reaction Using 2D TMDCs

In our attempt to resolve the energy crisis, we must be careful to look for alternatives that do not harm the environment. The photocatalytic splitting of water involving solar energy inputs presents an ideal solution as a virtually inexhaustible and clean energy production method [148]. In this respect, metal-based oxides [149], chalcogenides [150], and also carbides [151] have been explored thoroughly because of their excellent photocatalytic properties and availability. For this application, TMDCs are highly suitable for photocatalysing the water-splitting reaction due to its excellent inherent electronic and optical properties. 2D TMDCs tend to enhance the catalyst performance due to the improved surface-to-volume ratio as seen in the previous sections. Composite heterostructures containing TMDCs hybridized with another material are also reported to show excellent photocatalytic performance as compared to pristine TMDCs, with better separation of the photogenerated charge carriers contributing to this enhancement.

Mechanism

The mechanism for photoelectrocatalytic water splitting involves three consecutive steps, the first being capturing the energy of photons by the use of some semiconductor photoactive material. The solar energy captured in this manner will then create electron-hole pairs by excitation of an electron, provided the energy of the photon is greater than the valence-conduction band gap of the material. The generated photocarriers are prone to recombination, releasing heat or light which is undesirable for the current process. Therefore, the separation of these carriers is essential for harnessing solar energy [152]. The separation could be aided by the use of an appropriate co-catalyst, with the electrons accumulating in the photo-cathode for the reduction of hydrogen ions to molecular hydrogen, and the holes can initiate the oxidation of oxide ions to oxygen atoms at the photo-anode. The appropriate construction of the semiconductor would involve a p-type semiconductor for the photocathode, and an n-type semiconductor for the photoanode [153].

As previously discussed, the free energy change for the water-splitting reaction, when converted into potential needed by means of the Nernst equation, indicates that 1.23 V is required per electron transferred. Therefore, for a photocatalytic process, the photons must have a threshold energy of 1.23 eV or higher for the reaction to occur. This corresponds to light of wavelength of 1000 nm or lower. Each molecule of water that is split requires the generation of two and four electron-hole pairs for each molecule of H_2 and O_2 respectively. An ideal photocatalyst in this case would be a semiconductor material with a threshold band gap region as required for the water-splitting reaction, and valence and conduction band energies each straddling the electrochemical reduction potentials of oxygen (O_2/H_2O) and hydrogen (H^+/H_2). Considering the overpotential problem and other possibilities of energy loss, the minimum band gap energy for being an efficient water-splitting catalyst is considered to be in the range of 1.6 to 2.4 eV [154].

Graphene has been touted to be an excellent material for various electronic and opto-electronic applications, given its high charge carrier concentration, excellent conductivity and mechanical properties. However, its zero-band gap property is a considerable challenge for the purpose of photocatalyzing the water-splitting reaction, as seen previously for electrocatalytic applications as well. The use of defects, dopants, or hybrid materials to create a band gap by distorting the symmetry of graphene can be applied towards improving its electronic properties. Using platinum as a co-catalyst owing to its near-perfect overpotential for HER is limited by its cost and ease of poisoning by carbon monoxide. Cadmium sulphide has a desirable band gap and tuneable properties which make it useful for photoelectrocatalytic water-splitting. However, it suffers from other drawbacks such as a high toxicity and lack of photostability due to its susceptibility to photo-corrosion. Therefore, the hunt for an ideal catalyst for photoelectrocatalytic water splitting is ongoing, with promising results revealed for many non-graphene and non-noble-metal materials. Some such potent materials are 2D TMDCs [155]. An overview of the various reported 2D TMDCs and their hybrid heterostructures is summarized in the next section.

Review of Recent Work in the Field of Photocatalysis Using 2D TMDCs

Computational work can be used to establish a screening procedure to streamline more potent materials for a given application which can then be developed in a laboratory. Zhuang et al. (2013) have computationally explored potential 2D TMDC photocatalyst materials that can provide a good turnover for the water-splitting reaction via Density Functional Theory (DFT) investigations. The studied TMDC nanosheets include early transition metals like Ti, V, Zr, Nb, Mo, Hf, Ta and W, as well as Pt, combined with chalcogenide atoms such as S, Se, and Te. It was found that for a sensor material to be considered suitable for photocatalyzing the water-splitting reaction, it must have a bandgap in between 1.6 to 1.7 eV, band edges close to water's redox potentials, while remaining insoluble in water. These three criteria are fundamental requirements of a photocatalyst for this application [131]. Low formation energies of bulk TMDCs could indicate how easy it would be to exfoliate into 2D nanosheets. Further, in the case of MoS_2, WS_2, PtS_2 and $PtSe_2$, the band edges were computed to be close to redox potentials for the water splitting reaction at 0 pH, implying that these could provide efficient photocatalysis without the need for a bias voltage [157].

As expanded in Section 7.2.1, HER electrocatalysis with 2D TMDCs is found to be more efficient using metallic as compared to semiconducting TMDCs. A similar enhancement of catalytic activity is observed for photocatalysts that are in 1T phase over the bulk or 2H phases. Bai et al. (2015) evaluated the effect of phase on photocatalytic activity for MoS_2 hybridized with nanocrystals of TiO_2 applied to the photocatalytic degradation of RhB and hydrogen production reactions. They found that 1T phase TMDC was better at preventing exciton recombination than its bulk or even 2H counterparts [133]. 1T phase TMDCs have a more active basal plane, as well as improved charge transfer kinetics, both of which as seen in the previous applications, make this phase more effective for photocatalytic action.

Use of Plasmons and Phase Engineering in 2D TMDC Photocatalytic Applications

Phase engineering involve specifically producing a particular phase, for instance, the stable 1T nanosheets of the TMDC which is desired for its catalytic applications. Such methods must be facile and relatively inexpensive to compete with the noble metal catalysts available. Kang et al. (2014) suggested a method for the facile phase engineering of 1T MoS_2 using plasmon modes of nanoparticles like gold. The Au nanoparticles placed on the surface of 2D MoS_2 absorbed energy from light irradiation leading to a resonance-based excitation to a plasmon mode, which subsequently decayed, transferring the 'hot' electron to the TMDC. The electron distribution in TMDCs for the 2H phase on receiving the hot electron is destabilized as compared to that for the 1T phase, which is explained using the crystal field theory. Thus, the irradiation of light can cause desirable structural phase changes in 2D TMDCs [167]. Zhang et al. (2017) explored the possibility of using a gold nanorod-assembled monolayer of MoS_2 as a potential HER photocatalyst with the occurrence of a 2H-to-1T phase transition towards metallic MoS_2 [160]. The absorption of incident light led to the formation of 'hot electrons' in the interface between MoS_2 and Au, which approached the catalytically active sites of MoS_2 for catalysing HER. Their investigations revealed that the recombination lifetime of the photo-generated charge carriers was high, with holes formed in the Au layer capable of regeneration by some chemical reactions [161].

Further tunability can be made possible in such plasmonic photocatalysts by tweaking the wavelength and intensity of the incident beam by using a tunable laser [162]. The enhanced photocatalytic activity in such systems could also be attributed to the transition to metallic 1T character [163]. Chen et al. (2019) prepared a photoluminescent WS_2 nanosheet by a CVD methods, and then deposited islands of Au-nanofilms on its surface which shows plasmonic properties. The resulting composite displays excellent absorption properties in the visible range, with improved surface charge-carrier generation which would be excellent for photo-electrochemical catalysis [138]. An interesting variation in the effect of phase on photocatalytic activity is observed when the TMDC is hybridized with another material. As suggested by Chang et al. (2016), when MoS_2 nanosheets are hybridized with CdS, better catalytic performance is observed when the MoS_2 in 2H phase is used, which has been attributed to a better electrical contact in the latter, leading to more efficient charge separation of the charge carriers, thereby enhancing performance. This improvement is in spite of the fact that in the pristine versions, 1T MoS_2 dominates over 2H MoS_2 in terms of performance [139].

Preparation of 2D-2D Heterostructures Using 2D TMDCs for Photocatalytic Applications

2D nanosheets of TMDCs show high interfacial interaction with 2D sheets of other materials in hybrid structures. Such structures can show enhancement of certain properties which can be beneficial for photocatalytic applications depending on the combining structures, in a synergistic manner. Graphene sheets hybridised with 2D TMDC nanosheets are one such composite.

Graphene having a high carrier mobility, as well as rapid charge-transfer kinetics, can decelerate the rate of recombination of photo-generated carriers. Such properties can enhance the photocatalytic performance of TMDC materials in hybrid structures [166]. By itself, graphene being a zero-bandgap semiconductor shows poor electronic performance with poor catalytic activity [167]. Heterostructures are carefully constructed to avoid problems of structural mismatch as well as to ensure flow of charge carriers throughout. To improve the electronic contact between two hybrid materials, Zhou and co-workers (2016) employed an annealing method based on photo-reduction of MoS_2 sheets on n^+pSi photocathodes with a layer of transparent alumina protection to prevent Si – oxidation. 100 mW/cm^2 irradiation using a Xenon lamp gave excellent results of 34.5 mA photocurrent at 35 mV v/s RHE. The photocatalytic activity was also found to be stable for an extended time of up to 40 h, showing promise as practical photoelectrocatalysts [168].

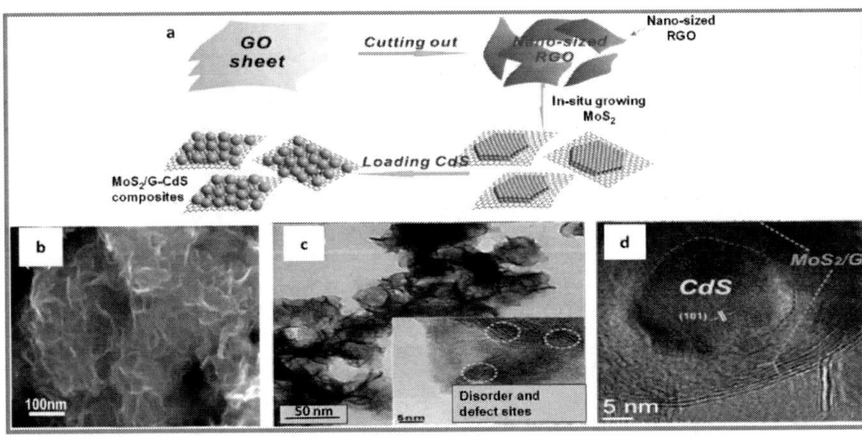

Figure 7.7. (a) Schematic flow chart for preparing the CdS/rGO hybrid with MoS_2 (b) SEM (b) TEM and (c) HRTEM images of the as-prepared the CdS/rGO hybrid with MoS_2. Reproduced with permission from Chang, Kun, et al. "MoS_2/graphene cocatalyst for efficient photocatalytic H_2 evolution under visible light irradiation." *ACS Nano* 8.7 (2014): 7078-7087. Copyright © 2014 American Chemical Society [143].

The use of graphene hybridized with TMDCs could lead to exciting optoelectronic properties. Graphene, with its high charge-carrier concentration and high conductivity can separate the photo-generated electrons and holes very efficiently, and its drawback of having a zero band gap can be overcome by the presence of the TMDC. In this respect, Yuan et al. (2017) synthesized

MoS_2 nanosheets hybridized with reduced graphene oxide (rGO) to produce a heterostructure having high electronic contact by a facile, hydrothermal method [142]. Such hybrids can also be made with a third material like CdS nanoparticles, bringing with it the advantage of good charge transport between MoS_2 and CdS as well as an improved photon absorption capacity in the hybrid structure [170]. Figure 7.7 depicts a schematic flow chart of the preparation process of the CdS/rGO hybrid with MoS_2 (a), and morphological characteristics of the hybrids (b, c, d).

Fabrication of vdW Heterostructures Using 2D TMDCs for Photocatalytic Applications

van der Waals (vdW) heterostructures are formed when 2D nanosheets are stacked into composites with interlayer van der Waals forces. The type of band structure in vdW heterostructures can dictate how efficient at photocatalysis a given heterostructure material can be. Type I, or symmetric and Type III, or broken heterostructures are not suitable for photocatalysis due to the higher exciton recombination rate in the former and inability of photo-formed electrons and holes to traverse the interface in the latter. In a Type II staggered heterostructure, the valence band maximum (VBM) of one structure and the conduction band minimum (CBM) of the other structure lie in different layers; such that both the VBM and CBM of one lies higher or lower than that of the other. Such a structure allows for ease of splitting of the excitons and therefore are suggested to be the photocatalytic heterostructures of choice. Some heterostructures constructed using 2D TMDCs as one of the combining structures have shown promise in this respect.

Such band alignment can dictate the performance of the photocatalyst. When hybridizing MoS_2 with titania or TiO_2, the CBM of 2H phase MoS_2 is lower than that of titania, making photo-electron transfer from titania to the TMDC more hindered. In case of 1T phase MoS_2, the CBM is actually higher than that of titania, facilitating photo-electron transfer and improving the overall photocatalytic efficiency. When instead of TiO_2, we consider CdS as the hybridizing structure, we observe that the CBM of both 2H and 1T MoS_2 are higher than that of CdS, therefore both offer appreciable transfer of the photogenerated carriers; and incidentally, it is 2H MoS_2 as a co-catalyst that offers superior photocatalytic activity. Thus, in case of hybrid structures, the thumb rule of metallic phase TMDCs operating at higher sensitivity or performance as compared to the semiconducting phase collapses, and deeper insight into the formation of these composites is needed, such as the relative alignment of their band structure, etc., to arrive at a judgement about a positive correlation between a phase and photocatalytic performance.

Kumar et al. (2018) combined WS_2 as the TMDC and C_2N or carbon nitride in a computational investigation, forming a Type II vdW heterostructure. This carbon nitride was shown to have a high charge-carrier mobility, similar to that of graphene, along with the heightened advantage of low exciton binding energies. The composite catalyst showed charge transfer from the TMDC to the carbon nitride, where the TMDC and C_2N were found to catalyse the OER and HER reaction respectively – suggesting that such composites can work efficiently for photocatalysing the water splitting reaction [136].

TMD/ZnO hybrids have also shown some promise in the theoretical investigations for HER photocatalysis as suggested by Wang et al. (2018). In their work, MoS_2, $MoSe_2$, WS_2, and WSe_2 were hybridized with ZnO to form vdW heterostructures, with the sulphur-based TMDCs forming Type II, and selenium-based TMDCs forming Type I heterostructures. All four materials showed good absorption properties in the visible and near IR regions which point toward applications as photocatalysts using solar energy sources [102]. Another heterostructure showing promise in photocatalysis of OER reaction is suggested by Khan A. et al. (2019) where 1T phase WS_2 is combined with graphite sheets. The composites were prepared by sonication followed by a facile solvothermal synthesis method, producing flower-like hybrid composites which showed higher efficiency than pristine TMDCs for photocatalytic HER applications [148].

7.2.3.3. Applications in Dye Sensitized Solar Cells (DSSCs)

Dye Sensitized Solar Cells form the new-age third generation solar cells, remarkably similar to photosynthesis where an electrochemical reaction occurring in light-sensitive molecules gives rise to a series of electron-transfer reactions, ultimately leading to energy storage. Solar energy is harnessed as the electrochemical energy of redox reactions, and then stored as the chemical energy of bonds in energy-dense molecules. In DSSCs, dye molecules serve the role of a photosensitive material, absorbing photons of appropriate wavelengths and undergoing molecular photoexcitation. The photoexcitation is followed by an ionization of the excited dye molecule, leading to dye-cation formation, and the subsequent redox reactions that ensue lead to the flow of electrons around an external circuit i.e., the generation of a photocurrent.

DSSCs eliminate the need for a photosensitive semiconductor, and can be constructed to achieve a relatively high efficiency which is currently in the range of 7 to 14%. The main components in a DSSC involve a working electrode or anode set-up and a counter electrode. The working electrode consists of the dye molecule, usually coated on TiO_2 nanoparticles. The use of titania nanoparticles allows the composite to be transparent to incident radiation. The molecular photoexcitation of the dye molecule, and the subsequent ionization steps generate electrons which are transferred from the dye molecule to the titania, which then pass from the anode to cathode via an external circuit, generating photocurrent. The counter electrode or cathode consists of a transparent conductive oxide (TCO) coated with an electrocatalyst, usually platinum. Here, the electrons reduce triiodide ions (I_3^-) into iodide ions (I^-), followed by regeneration of the dye molecule by a redox reaction with the iodide ions. A schematic diagram illustrating the process is depicted in Figure 7.8.

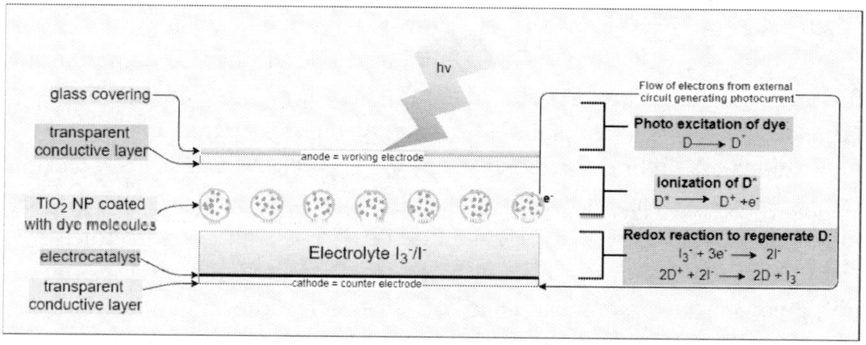

Figure 7.8. Schematic diagram of photocatalytic process by DSSC cells.

Commonly used materials for the counter electrode are platinized fluorine-doped tin oxide (FTO). Platinum as an electrocatalyst has several desirable qualities like high electrical conductivity as well as stability, but the glaring drawback of platinum is its scarcity and expenses involved. The use of suitable replacements for Pt electrocatalysts is therefore necessary for large-area fabrication of DSSCs. Graphene and graphene-based materials have been investigated as suitable counter electrode electrocatalysts on account of their high specific surface area, good optical and electronic properties, and their versatility in application to different substrates. In this regard, 2D TMDCs have shown some potential as Pt replacements due to their similarity to graphene.

Wei et al. (2016) used pristine metallic 1T MoS_2 deposited on FTO (TCO) glass prepared by a hydrothermal route as the counter electrode material. The resulting DSSC showed a power conversion efficiency of ~7.1%, which was found to be higher than that of the 2H semiconducting counterpart, and close to that of Pt-coated counter electrodes (~7.3%). Another study proposed a straightforward hydrothermal, one-pot method of deposition of thin films of MoS_2 on FTO substrates as the counter electrode. Their work reported an interesting analysis of the effect of varying the mole ratio of reactants used in the hydrothermal reaction on the power conversion efficiency. The range of mole ratios studied gave power conversion efficiencies in the range of 3.7 to 7.41% at a temperature of 423 K, with the highest power conversion efficiency slightly higher than the 7.13% of Pt-based anodes sensitized by the same dye, i.e., N719 [151].

Ahn et al. (2016) designed an edge-oriented WS_2-based counter electrode by the tape-casting method from mesoporous tungsten oxide, via a sulfurization process at high temperatures. The resulting WS_2 thin film counter electrodes showed a larger concentration of active sites translating into a better power conversion efficiency of 8.85% (compared to Pt-based counterpart with efficiency of 7.2%) in a DSSC. Titanium sulfide or TiS_2, a metallic TMDC has also been reported as a suitable counter electrode electrocatalyst for DSSC applications. A hybrid heterostructure of 2D TiS_2 nanosheets placed on graphene by mechanical ball-milling followed by an annealing step at high temperature was reported by Meng et al. (2016) as a photoanode for DSSCs. Their photoanode showed superior electrocatalytic activity as compared to Pt with a power conversion efficiency of 8.8% (0.8% more than Pt). Graphene could have contributed to the stability of the hybrid, while the metallic nature of TiS_2 and a better electrical contact between the TCO and the electrocatalyst have been attributed as the cause of the enhanced electrocatalytic activity [153].

7.3. Applications of 2D TMDCs for Fuel Cell Applications

As expanded earlier, some of the fundamental problems of hydrogen-oxygen fuel cells lie in the high energy barrier of the reaction, which can be addressed by suitable electrocatalysts or higher temperatures, both of which pose other difficulties in their development. The level of hydrogen purity needed for use as a fuel in a fuel cell depends on the electrocatalyst used, for instance, Pt electrocatalysts show excellent electrocatalytic activity for the fuel cell reaction, however it is easily poisoned in the presence of small concentrations of CO or even CO_2. Currently, the most popular method for hydrogen production involves catalytic reformation of petroleum products which invariably contributes some CO and CO_2 to the hydrogen generated. Purifying the resulting hydrogen to ensure that the concentration of CO and CO_2 in the fuel stream is lesser than the threshold limit for Pt poisoning, would require additional refining and purification steps making the whole process both economically and energetically expensive. In general, Pt is a rare and expensive element, which if replaced with another material of similar efficiency, could be extremely beneficial.

This can be addressed in many ways, one of which is to avoid catalytic reforming in general, pursuing instead the benefits of the water-splitting reaction. Electrocatalytic water splitting could generate a stream of hydrogen with very little to no carbon contamination, which would be appropriate for the fuel cell. Moreover, if the electricity for the reaction is supplied by a non-conventional energy source, such as solar energy via a photocatalytic water-splitting, the whole process would become all the more sustainable. The carbon emissions made by catalytic reforming would also be avoided, with very little environmental impact on the whole. The water-splitting reaction and various methods to catalyze it by means of 2D TMDCs, and specifically the role of metallic TMDCs in this respect have been expanded in the previous section (Section 7.2). In this section, the different 2D TMDCs that can electrocatalyse the fuel oxidation process, as well as those that can oxidize fuels other than hydrogen in modern fuel cells are reviewed.

As mentioned previously, 2D MoS_2 with its various polytypic versions shows metallic nature in its 1T form while the 2H form is thermodynamically stable. Recently, it has been reported as a good electrocatalyst or catalyst support material for oxidation of fuels in Direct Methanol Fuel Cells. Methanol is an energy-dense fuel, with the advantage of convenient transport and storage (relative to hydrogen) owing to its stability. Electrocatalyst supports for platinum can increase its Electrochemical Active Surface Area (ECSA), while decreasing the amount of Pt needed to achieve a similar catalytic activity. 2D materials have been found to fare well in this regard, due to desirable properties such as high specific surface area, good electrical conductivity, etc. The ECSA can be computed from the cyclic voltammograms, by means of the formula

$$ECSA = \frac{Q_H}{0.21 \times [Pt]} \tag{7.4}$$

where Q_H stands for the charge involved in hydrogen desorption in mA obtained from the voltammograms, [Pt] stands for milligrams of Pt in the electrode, and 0.21 accounts for the maximum surface charge transferred by a monolayer of H to the Pt catalyst.

Larger values of ECSA could indicate a higher concentration of active sites. Moreover, methanol oxidation suffers the limitation of carbon dioxide emissions which can accumulate on the catalyst surface, hampering its efficiency. Therefore, potential catalysts must be evaluated for their electro-oxidation abilities, as well as resistance to fall in efficiency due to the contamination by carbon-based by-products of the methanol oxidation reaction. The ratio of anodic current density peak in the forward scan to that in the reverse scan can be indicative of the catalyst tolerance to carbon-based residue accumulated on the catalyst surface, as the forward scan corresponds to the methanol oxidation peak, whereas the reverse scan corresponds to the removal of the carbon residue accumulated in the forward scan [160].

2D TMDCs hybridized with graphene have been reported as potential Pt electrocatalyst-support materials. Zhai et al. (2015) fabricated Pt-RGO-2D MoS_2 hybrid heterostructures by a simplistic, one-pot hydrothermal method for the generation of RGO-MoS_2 composites, with electrodeposition of Pt on the composite achieved at high dispersion rates. Their Pt-MoS_2 and Pt-RGO-MoS_2 hybrids had higher ECSA of 10.58 and 12.87 cm^2/mg respectively as compared to that of Pt/C catalysts with ECSA of 6.84 cm^2/mg. The larger ECSA values in the hybrids can be attributed to the larger concentration of active sites in these heterostructures. Further, both these heterostructures outperformed their Pt/C counterpart in terms of their current density in the forward scan between 0.6 to 0.9 V with 4.26 and 7.35 mA/cm^2 for Pt-MoS_2 and Pt-RGO-MoS_2, respectively. The authors also suggest reasonable stability of these heterostructures, indicating the possibility of preparing 2D TMDC-based electrocatalyst supports [161].

Another group, Zhang et al. (2019) suggested an Auivillius oxide, Bi_2WO_6 hybridized with 2D MoS_2 for photo-assisted catalysis of methanol oxidation. Bi_2WO_6 by itself shows promise in photocatalysis with good stability and a narrow band gap of 2.8 eV, however its efficiency is limited due to the requirement of photons of wavelengths 450 nm or shorter, and also high recombination rates of the charge carriers. When the Auivillius oxide is hybridized with 2D MoS_2, the charge separation of photo-generated carriers is facilitated to a greater extent, while the range of photosensitivity is increased as 2D MoS_2 has a lower band gap of 1.2-1.9 eV. Under the same conditions, the composite with MoS_2 showed a current density of 1781 mA/mg_{Pt}, which is much higher than 522 mA/mg$_{Pt}$ of Pt-Bi_2WO_6. The composite showed remarkable enhancement in current density under visible light irradiation, while also showing longevity and the ability to withstand poisoning [162].

Another variety of fuel cells, the PEMFCs normally involve hydrogen as a fuel, and oxygen as the oxidant, with the membrane capable of proton transfer. The oxygen reduction reaction (ORR) occurs at the cathode, and is usually the rate-determining step. The ORR involves reduction of oxygen molecule in the zero-oxidation state to oxide ions in the -2-oxidation state, a reaction characterized by its sluggish kinetics due to the large bond strength of the dioxygen bond [166]. There are two pathways for the ORR to occur, the two-electron pathway (which leads to formation of peroxide species) and the four-electron pathway (which leads to the production of water). The four-electron pathway is the preferable pathway because peroxide as a by-product can have a harmful effect on the membrane itself, as well as damage other parts which could lead to a decrease in efficiency. In this respect, Pt is considered to be a very effective electrocatalyst due to its ability to induce the four-electron route [167]. As seen previously, if Pt is replaced by another electrocatalyst material that is more easily available and inexpensive, it would make PEMFCs more economically viable.

Rowley-Neale et al. (2017) have designed screen-printed electrodes (SPEs) by combining 2D MoS_2 into pre-existing and commercially available graphitic inks to fabricate highly customizable SPEs for the purpose of electrocatalyzing ORR. The resulting ink mixture was deposited on a previously screen-printed graphite surface, over which the 2D MoS_2-containing ink was impregnated, followed by a curing process at 60 °C for 30 min resulting in a physically-blended composite. The process of fabricating this electrode is represented in Figure 7.9. The composite showed a variation in its ORR electrocatalytic performance with change in weight percentage of 2D MoS_2 incorporated. When varying the weight percentage of 2D MoS_2 in the hybrid SPEs from 5% to 40%, the onset potential was found to vary from -0.19 V to +0.16 V, whereas the peak current density went from -0.89 to -1.62 mA/cm^2. Beyond 40% of 2D MoS_2 by weight, the structural integrity of the screen-printing ink was compromised due to the high viscosity. The results were encouraging when compared with the performance of Pt, which has an onset potential of +0.46 V and current density peak at 2.92 mA/cm^2. The reaction was found to follow the desirable four-electron route, and also retained its current density over the course of multiple cycles [130].

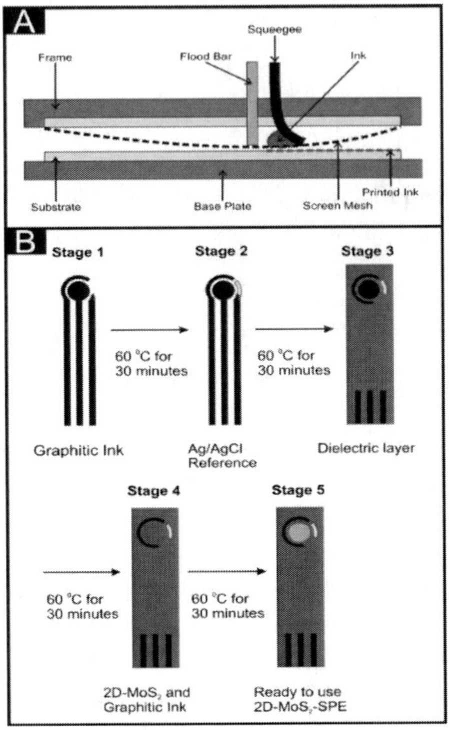

Figure 7.9. (a) A schematic diagram of the screen-printing process employed and (b) the individual stages necessary to fabricate the mass-producible 2D-MoS_2-SPEs. Reproduced with permission from Rowley-Neale, Samuel J., Graham C. Smith, and Craig E. Banks. "Mass-producible 2D-MoS_2-impregnated screen-printed electrodes that demonstrate efficient electrocatalysis toward the oxygen reduction reaction." *ACS Applied Materials & Interfaces* 9.27 (2017): 22539-22548. Copyright © 2017 American Chemical Society [130].

Huang et al. (2018) proposed a novel hybrid electrocatalyst for ORR by the use of carbon black (Vulcan XC-72) as a support for porous and laminated WS_2. The electrocatalytic properties were investigated on a rotating disc electrode (RDE) in nitrogen and oxygen saturated 0.1M KOH with the reduction reaction proceeding via the coveted four-electron pathway. The TMDC-based electrocatalyst gave an onset potential of 0.78 V v/s RHE on a rotating ring disc electrode (RRDE), and limiting current density of 4.99 mA/cm^2. It also showed stability in current density up to 1000 cycles in the range of -1.0 to 0.2 V v/s SCE at a 20 mV/second speed by cyclic voltammetry [169].

Conclusion

From this systematic review of the current research in TMDCs for catalysis and fuel cell applications, one can surmise that TMDCs show promise and potential for replacing noble metal-based catalysts. With regards to catalyzing the water splitting reaction, the design of dual catalysts for both the HER and OER process can allow for inexpensive hydrogen production by electrolysis, putting an end to the currently used hydrocarbon reformation processes. Further, they show immense potential for application in photocatalytic reactions. In light of the climate and energy crises, and the global interest in a hydrogen-based economy surging, the need to explore the properties and applications of TMDCs are well justified.

2D TMDCs are especially equipped over layered TMDCs because of their larger exposure to the active sites. Their high tunability also assures varying electronic structure which may be beneficial for the purpose of electrocatalysis. Future research should tend towards low-cost and facile alternatives to achieve the same results as that obtained with platinum as an electrocatalyst. There is a need for major improvements in the catalyst efficiency, which is mainly achieved by increasing the density of active sites or by improving activity of the basal plane. Other alternatives like introducing defects, strain and hybridizing with second-party structures may prove more effective in this regard. The catalysts presented should resist degradation for a reasonable amount of time, and most of the time catalyst efficiency is considered for up to 1000 cycles. The stability may be improved if efficient electrocatalysis is obtained in the stable 2H phase, however 1T metallic phase offers superior conductivity and catalysis. Catalysts suggested should be reproducible giving repeatable results with good efficiency.

An important problem with TMDCs as catalysts is the lack of electrical contact between the electrode surface and the catalyst reducing electronic conductivity [169]. The conductivity of pristine TMDCs can be improved drastically by the introduction of strain or defects into the system. Future research can focus on this facet of developing facile methods to synthesize TMDCs with optimized defect concentration or strain percentages for sufficient electrocatalytic activity. Hybrid structures also present some viable solutions to overcome the problem of electronic conductivity, and may provide structural stability to the catalyst. Some hybrid structures have shown promising results in this respect and have been outlined in the Section 7.2.1.3. Hybrid structures are also by-default strained, which can add to their catalytic efficiency. Generating hybrid structures on carbon materials is somewhat problematic due to the size disparity between the large TMDC nanostructures and carbon materials. One method suggested to overcome this issue is space-confined growth, wherein the space between porous materials was used to grow TMDC nano-structures in a bottom-up approach [170].

References

[1] Deng, Jiao, Dehui Deng, and Xinhe Bao. "Robust Catalysis on 2D Materials Encapsulating Metals: Concept, Application, and Perspective." *Advanced Materials* 29, no. 43 (2017): 1606967. https://doi.org/10.1002/adma.201606967.

[2] Wu, Bohua, Yinjie Kuang, Xiaohua Zhang, and Jinhua Chen. "Noble Metal Nanoparticles/Carbon Nanotubes Nanohybrids: Synthesis and Applications." *Nano Today* 6, no. 1 (February 1, 2011): 75–90. https://doi.org/10.1016/j.nantod.2010.12.008.

[3] Koh, Shirlaine, and Peter Strasser. "Electrocatalysis on Bimetallic Surfaces: Modifying Catalytic Reactivity for Oxygen Reduction by Voltammetric Surface Dealloying." *Journal of the American Chemical Society* 129, no. 42 (October 1, 2007): 12624–25. https://doi.org/10.1021/ja0742784.

[4] Wu, Binghui, and Nanfeng Zheng. "Surface and Interface Control of Noble Metal Nanocrystals for Catalytic and Electrocatalytic Applications." *Nano Today* 8, no. 2 (April 1, 2013): 168–97. https://doi.org/10.1016/j.nantod.2013.02.006.

[5] Tian, Na, Zhi-You Zhou, Shi-Gang Sun, Yong Ding, and Zhong Lin Wang. "Synthesis of Tetrahexahedral Platinum Nanocrystals with High-Index Facets and High Electro-Oxidation Activity." *Science* 316, no. 5825 (May 4, 2007): 732–35. https://doi.org/10.1126/science.1140484.

[6] Linic, Suljo, Phillip Christopher, and David B. Ingram. "Plasmonic-Metal Nanostructures for Efficient Conversion of Solar to Chemical Energy." *Nature Materials* 10, no. 12 (December 2011): 911–21. https://doi.org/10.1038/nmat3151.

[7] Tada, Hiroaki, Tomokazu Kiyonaga, and Shin-ichi Naya. "Rational Design and Applications of Highly Efficient Reaction Systems Photocatalyzed by Noble Metal Nanoparticle-Loaded Titanium(IV) Dioxide." *Chemical Society Reviews* 38, no. 7 (June 23, 2009): 1849–58. https://doi.org/10.1039/B822385H.

[8] Linic, Suljo, Umar Aslam, Calvin Boerigter, and Matthew Morabito. "Photochemical Transformations on Plasmonic Metal Nanoparticles." *Nature Materials* 14, no. 6 (June 2015): 567–76. https://doi.org/10.1038/nmat4281.

[9] Deng, Dehui, K. S. Novoselov, Qiang Fu, Nanfeng Zheng, Zhongqun Tian, and Xinhe Bao. "Catalysis with Two-Dimensional Materials and Their Heterostructures." *Nature Nanotechnology* 11, no. 3 (March 2016): 218–30. https://doi.org/10.1038/nnano.2015.340.

[10] Novoselov, K. S., D. Jiang, F. Schedin, T. J. Booth, V. V. Khotkevich, S. V. Morozov, and A. K. Geim. "Two-Dimensional Atomic Crystals." *Proceedings of the National Academy of Sciences* 102, no. 30 (July 26, 2005): 10451–53. https://doi.org/10.1073/pnas.0502848102.

[11] Li, Henan, Yumeng Shi, Ming-Hui Chiu, and Lain-Jong Li. "Emerging Energy Applications of Two-Dimensional Layered Transition Metal Dichalcogenides." *Nano Energy* 18 (November 1, 2015): 293–305. https://doi.org/10.1016/j.nanoen.2015.10.023.

[12] Mak, Kin Fai, Changgu Lee, James Hone, Jie Shan, and Tony F. Heinz. "Atomically Thin MoS2: A New Direct-Gap Semiconductor." *Physical Review Letters* 105, no. 13 (September 24, 2010): 136805. https://doi.org/10.1103/PhysRevLett.105.136805.

[13] Tsai, Charlie, Frank Abild-Pedersen, and Jens K. Nørskov. "Tuning the MoS2 Edge-Site Activity for Hydrogen Evolution via Support Interactions." *Nano Letters* 14, no. 3 (March 12, 2014): 1381–87. https://doi.org/10.1021/nl404444k.

[14] McDonnell, Stephen J., and Robert M. Wallace. "Atomically-Thin Layered Films for Device Applications Based upon 2D TMDC Materials." *Thin Solid Films* 616 (October 1, 2016): 482–501. https://doi.org/10.1016/j.tsf.2016.08.068.

[15] Grove, W.R. "LXXII. On a Gaseous Voltaic Battery." *The London, Edinburgh, and Dublin Philosophical Magazine and Journal of Science* 21, no. 140 (December 1, 1842): 417–20. https://doi.org/10.1080/14786444208621600.

[16] Behling, Noriko Hikosaka. "Chapter 1 - Introduction." In *Fuel Cells*, edited by Noriko Hikosaka Behling, 1–6. Elsevier, 2013. https://doi.org/10.1016/B978-0-444-56325-5.00001-6.

[17] Wang, Junye, Hualin Wang, and Yi Fan. "Techno-Economic Challenges of Fuel Cell Commercialization." *Engineering* 4, no. 3 (June 1, 2018): 352–60. https://doi.org/10.1016/j.eng.2018.05.007.

[18] Wendt, Hartmut, Estevam V. Spinacé, Almir Oliveira Neto, and Marcelo Linardi. "Electrocatalysis and Electrocatalysts for Low Temperature Fuel Cells: Fundamentals, State of the Art, Research and Development." *Química Nova* 28 (December 2005): 1066–75. https://doi.org/10.1590/S0100-40422005000600023.

[19] Alegre, Cinthia, Laura Álvarez-Manuel, Radu Mustata, Luis Valiño, Antonio Lozano, and Félix Barreras. "Assessment of the Durability of Low-Cost Al Bipolar Plates for High Temperature PEM Fuel Cells." *International Journal of Hydrogen Energy*, Special Issue on Selected Contributions from the *European Hydrogen Energy Conference* 2018. Málaga, Spain. March 14th - 16th, 44, no. 25 (May 17, 2019): 12748–59. https://doi.org/10.1016/j.ijhydene.2018.07.070.

[20] Abouzari-Lotf, Ebrahim, Masoumeh Zakeri, Mohamed Mahmoud Nasef, Mikio Miyake, Pooria Mozarmnia, Nur Anati Bazilah, Noor Fatina Emelin, and Arshad Ahmad. "Highly Durable Polybenzimidazole Composite Membranes with Phosphonated Graphene Oxide for High Temperature Polymer Electrolyte Membrane Fuel Cells." *Journal of Power Sources* 412 (February 1, 2019): 238–45. https://doi.org/10.1016/j.jpowsour.2018.11.057.

[21] Centi, Gabriele, and Siglinda Perathoner. "Catalysis by Layered Materials: A Review." *Microporous and Mesoporous Materials, Workshop on Innovative Applications of Layered Materials: From Catalysis to Nanotechnology*, 107, no. 1 (January 1, 2008): 3–15. https://doi.org/10.1016/j.micromeso.2007.03.011.

[22] Gupta, Uttam, and C. N. R. Rao. "Hydrogen Generation by Water Splitting Using MoS2 and Other Transition Metal Dichalcogenides." *Nano Energy* 41 (November 1, 2017): 49–65. https://doi.org/10.1016/j.nanoen.2017.08.021.

[23] Jain, I. P., Chhagan Lal, and Ankur Jain. "Hydrogen Storage in Mg: A Most Promising Material." *International Journal of Hydrogen Energy*, *Novel Hydrogen Production Technologies and Applications*, 35, no. 10 (May 1, 2010): 5133–44. https://doi.org/10.1016/j.ijhydene.2009.08.088.

[24] Breen, J. P, R Burch, and H. M Coleman. "Metal-Catalysed Steam Reforming of Ethanol in the Production of Hydrogen for Fuel Cell Applications." *Applied Catalysis B: Environmental* 39, no. 1 (November 8, 2002): 65–74. https://doi.org/10.1016/S0926-3373(02)00075-9.

[25] Morales-Guio, Carlos G., Lucas-Alexandre Stern, and Xile Hu. "Nanostructured Hydrotreating Catalysts for Electrochemical Hydrogen Evolution." *Chemical Society Reviews* 43, no. 18 (August 18, 2014): 6555–69. https://doi.org/10.1039/C3CS60468C.

[26] Pan, Hui. "Metal Dichalcogenides Monolayers: Novel Catalysts for Electro chemical Hydrogen Production." *Scientific Reports* 4, no. 1 (June 26, 2014): 5348. https://doi.org/10.1038/srep05348.

[27] Zhao, Guoqiang, Kun Rui, Shi Xue Dou, and Wenping Sun. "Heterostructures for Electrochemical Hydrogen Evolution Reaction: A Review." *Advanced Functional Materials* 28, no. 43 (2018): 1803291. https://doi.org/10.1002/adfm.201803291.

[28] Suffredini, H. B., J. L. Cerne, F. C. Crnkovic, S. A. S. Machado, and L. A. Avaca. "Recent Developments in Electrode Materials for Water Electrolysis." *International Journal of Hydrogen Energy* 25, no. 5 (May 1, 2000): 415–23. https://doi.org/10.1016/S0360-3199(99)00049-X.

[29] Zeng, Min, and Yanguang Li. "Recent Advances in Heterogeneous Electrocatalysts for the Hydrogen Evolution Reaction." *Journal of Materials Chemistry A* 3, no. 29 (July 14, 2015): 14942–62. https://doi.org/10.1039/C5TA02974K.

[30] Li, Yanguang. "Nanocarbon-Based Hybrid Materials for Electrocatalytical Energy Conversion: Novel Materials and Methods." *IEEE Nanotechnology Magazine* 8, no. 2 (June 2014): 22–28. https://doi.org/10.1109/MNANO.2014.2313471.

[31] Liang, Yongye, Yanguang Li, Hailiang Wang, and Hongjie Dai. "Strongly Coupled Inorganic/Nanocarbon Hybrid Materials for Advanced Electrocatalysis." *Journal of the American Chemical Society* 135, no. 6 (February 13, 2013): 2013–36. https://doi.org/10.1021/ja3089923.

[32] Lee, Chi Ho, and Sang Uck Lee. *Theoretical Basis of Electrocatalysis. Electrocatalysts for Fuel Cells and Hydrogen Evolution - Theory to Design.* IntechOpen, 2018. https://doi.org/10.5772/intechopen.77109.

[33] Parsons, Roger. "The Rate of Electrolytic Hydrogen Evolution and the Heat of Adsorption of Hydrogen." *Transactions of the Faraday Society* 54, no. 0 (January 1, 1958): 1053–63. https://doi.org/10.1039/TF9585401053.

[34] Sabatier, Paul. "Hydrogénations et Déshydrogénations Par Catalyse." *Berichte Der Deutschen Chemischen Gesellschaft* 44, no. 3 (1911): 1984–2001. https://doi.org/10.1002/cber.19110440303.

[35] Nørskov, J. K., T. Bligaard, A. Logadottir, J. R. Kitchin, J. G. Chen, S. Pandelov, and U. Stimming. "Trends in the Exchange Current for Hydrogen Evolution." *Journal of The Electrochemical Society* 152, no. 3 (January 24, 2005): J23. https://doi.org/10.1149/1.1856988.

[36] Tsai, Charlie, Karen Chan, Jens K. Nørskov, and Frank Abild-Pedersen. "Theoretical Insights into the Hydrogen Evolution Activity of Layered Transition Metal Dichalcogenides." *Surface Science, Reactivity Concepts at Surfaces: Coupling Theory with Experiment*, 640 (October 1, 2015): 133–40. https://doi.org/10.1016/j.susc.2015.01.019.

[37] Nguyen, Thang Phan, Seokhoon Choi, Jong-Myeong Jeon, Ki Chang Kwon, Ho Won Jang, and Soo Young Kim. "Transition Metal Disulfide Nanosheets Synthesized by Facile Sonication Method for the Hydrogen Evolution Reaction." *The Journal of Physical Chemistry C* 120, no. 7 (February 25, 2016): 3929–35. https://doi.org/10.1021/acs.jpcc.5b12164.

[38] Pumera, Martin, Zdeněk Sofer, and Adriano Ambrosi. "Layered Transition Metal Dichalcogenides for Electrochemical Energy Generation and Storage." *Journal of Materials Chemistry A* 2, no. 24 (May 27, 2014): 8981–87. https://doi.org/10.1039/C4TA00652F.

[39] Wang, Xinqiang, Yuanfu Chen, Binjie Zheng, Fei Qi, Jiarui He, Qian Li, Pingjian Li, and Wanli Zhang. "Graphene-like WSe2 Nanosheets for Efficient and Stable Hydrogen Evolution." *Journal of Alloys and Compounds* 691 (January 15, 2017): 698–704. https://doi.org/10.1016/j.jallcom.2016.08.305.

[40] Chen, Xiaoshuang, Yunfeng Qiu, Guangbo Liu, Wei Zheng, Wei Feng, Feng Gao, Wenwu Cao, YongQing Fu, Wenping Hu, and PingAn Hu. "Tuning Electrochemical Catalytic Activity of Defective 2D Terrace MoSe 2 Heterogeneous Catalyst via Cobalt Doping." *Journal of Materials Chemistry A* 5, no. 22 (2017): 11357–63. https://doi.org/10.1039/C7TA02327H.

[41] Yu, Qiangmin, Yuting Luo, Siyao Qiu, Qinye Li, Zhengyang Cai, Zhiyuan Zhang, Jiaman Liu, Chenghua Sun, and Bilu Liu. "Tuning the Hydrogen Evolution Performance of Metallic 2D Tantalum Disulfide by Interfacial Engineering." *ACS Nano* 13, no. 10 (October 22, 2019): 11874–81. https://doi.org/10.1021/acsnano.9b05933.

[42] Zhou, Jungui, Song Dai, Wen Dong, Xiaodong Su, Liang Fang, Fengang Zheng, Xiongdong Wang, and Mingrong Shen. "Efficient and Stable MoS2 Catalyst Integrated on Si Photocathodes by Photoreduction and Post-Annealing for Water Splitting." *Applied Physics Letters* 108, no. 21 (May 23, 2016): 213905. https://doi.org/10.1063/1.4952739.

[43] Shinde, Pratik V., Pratap Mane, Dattatray J. Late, Brahmananda Chakraborty, and Chandra Sekhar Rout. "Promising 2D/2D MoTe2/Ti3C2Tx Hybrid Materials for Boosted Hydrogen Evolution Reaction." *ACS Applied Energy Materials* 4, no. 10 (October 25, 2021): 11886–97. https://doi.org/10.1021/acsaem.1c02914.

[44] Wang, Sake, Hongyu Tian, Chongdan Ren, Jin Yu, and Minglei Sun. "Electronic and Optical Properties of Heterostructures Based on Transition Metal Dichalcogenides and Graphene-like Zinc Oxide." *Scientific Reports* 8, no. 1 (August 13, 2018): 12009. https://doi.org/10.1038/s41598-018-30614-3.

[45] Gholamvand, Zahra, David McAteer, Claudia Backes, Niall McEvoy, Andrew Harvey, Nina C. Berner, Damien Hanlon, et al. "Comparison of Liquid Exfoliated Transition Metal Dichalcogenides Reveals MoSe2 to Be the Most Effective Hydrogen Evolution Catalyst." *Nanoscale* 8, no. 10 (March 3, 2016): 5737–49. https://doi.org/10.1039/C5NR08553E.

[46] Ji, Qingqing, Yu Zhang, Jianping Shi, Jingyu Sun, Yanfeng Zhang, and Zhongfan Liu. "Morphological Engineering of CVD-Grown Transition Metal Dichalcogenides for Efficient Electrochemical Hydrogen Evolution." *Advanced Materials* 28, no. 29 (2016): 6207–12. https://doi.org/10.1002/adma.201504762.

[47] Xu, Chen, Shengjie Peng, Chaoliang Tan, Huixiang Ang, Huiteng Tan, Hua Zhang, and Qingyu Yan. "Ultrathin S-Doped MoSe 2 Nanosheets for Efficient Hydrogen Evolution." *Journal of Materials Chemistry A* 2, no. 16 (2014): 5597–5601. https://doi.org/10.1039/C4TA00458B.

[48] Woods, John M., Yeonwoong Jung, Yujun Xie, Wen Liu, Yanhui Liu, Hailiang Wang, and Judy J. Cha. "One-Step Synthesis of MoS2/WS2 Layered Heterostructures and Catalytic Activity of Defective Transition Metal Dichalcogenide Films." *ACS Nano* 10, no. 2 (February 23, 2016): 2004–9. https://doi.org/10.1021/acsnano.5b06126.

[49] Zhou, Xiaoli, Jun Jiang, Tao Ding, Jiajia Zhang, Bicai Pan, Jian Zuo, and Qing Yang. "Fast Colloidal Synthesis of Scalable Mo-Rich Hierarchical Ultrathin MoSe2−x Nanosheets for High-Performance Hydrogen Evolution." *Nanoscale* 6, no. 19 (September 12, 2014): 11046–51. https://doi.org/10.1039/C4NR02716G.

[50] Cummins, Dustin R., Ulises Martinez, Andriy Sherehiy, Rajesh Kappera, Alejandro Martinez-Garcia, Roland K. Schulze, Jacek Jasinski, et al. "Efficient Hydrogen Evolution in Transition Metal Dichalcogenides via a Simple One-Step Hydrazine Reaction." *Nature Communications* 7 (June 10, 2016): 11857. https://doi.org/10.1038/ncomms11857.

[51] Grigoriev, S. N., V. Yu. Fominski, R. I. Romanov, M. A. Volosova, and A. V. Shelyakov. "Pulsed Laser Deposition of Nanocomposite MoSex/Mo Thin-Film Catalysts for Hydrogen Evolution Reaction." *Thin Solid Films* 592 (October 1, 2015): 175–81. https://doi.org/10.1016/j.tsf.2015.09.024.

[52] Li, Hong, Charlie Tsai, Ai Leen Koh, Lili Cai, Alex W. Contryman, Alex H. Fragapane, Jiheng Zhao, et al. "Activating and Optimizing MoS2 Basal Planes for Hydrogen Evolution through the Formation of Strained Sulphur Vacancies." *Nature Materials* 15, no. 1 (January 2016): 48–53. https://doi.org/10.1038/nmat 4465.

[53] Huang, Huajie, and Xin Wang. "Recent Progress on Carbon-Based Support Materials for Electrocatalysts of Direct Methanol Fuel Cells." *Journal of Materials Chemistry A* 2, no. 18 (April 8, 2014): 6266–91. https://doi.org/10.1039/C3TA 14754A.

[54] Sumesh, C. K. "Zinc Oxide Functionalized Molybdenum Disulfide Hetero structures as Efficient Electrocatalysts for Hydrogen Evolution Reaction." *International Journal of Hydrogen Energy* 45, no. 1 (January 1, 2020): 619–28. https://doi.org/10.1016/j.ijhydene.2019.10.235.

[55] Kim, Jaemyung, Segi Byun, Alexander J. Smith, Jin Yu, and Jiaxing Huang. "Enhanced Electrocatalytic Properties of Transition-Metal Dichalcogenides Sheets by Spontaneous Gold Nanoparticle Decoration." *The Journal of Physical Chemistry Letters* 4, no. 8 (April 18, 2013): 1227–32. https://doi.org/10.1021/jz400507t.

[56] Liu, Xue, Ji-Zong Zhang, Ke-Jing Huang, and Pan Hao. "Net-like Molybdenum Selenide–Acetylene Black Supported on Ni Foam for High-Performance Supercapacitor Electrodes and Hydrogen Evolution Reaction." *Chemical Engineering Journal* 302 (October 15, 2016): 437–45. https://doi.org/10.1016/j.cej.2016.05.074.

[57] Lei, Yu, Srimanta Pakhira, Kazunori Fujisawa, Xuyang Wang, Oluwagbenga Oare Iyiola, Néstor Perea López, Ana Laura Elías, et al. "Low-Temperature Synthesis of Heterostructures of Transition Metal Dichalcogenide Alloys (WxMo1–XS2) and Graphene with Superior Catalytic Performance for Hydrogen Evolution." *ACS Nano* 11, no. 5 (May 23, 2017): 5103–12. https://doi.org/10.1021/acsnano.7b02060.

[58] Wang, Haotian, Zhiyi Lu, Desheng Kong, Jie Sun, Thomas M. Hymel, and Yi Cui. "Electrochemical Tuning of MoS2 Nanoparticles on Three-Dimensional Substrate for Efficient Hydrogen Evolution." *ACS Nano* 8, no. 5 (May 27, 2014): 4940–47. https://doi.org/10.1021/nn500959v.

[59] Yang, Jieun, Damien Voiry, Seong Joon Ahn, Dongwoo Kang, Ah Young Kim, Manish Chhowalla, and Hyeon Suk Shin. "Two-Dimensional Hybrid Nanosheets of Tungsten Disulfide and Reduced Graphene Oxide as Catalysts for Enhanced Hydrogen Evolution." *Angewandte Chemie* 125, no. 51 (2013): 13996–99. https://doi.org/10.1002/ange.201307475.

[60] Yang, Kunttal Keyshar, et al. "Self-Optimizing Layered Hydrogen Evolution Catalyst with High Basal-Plane Activity." *Nature Energy* 2, no. 9 (September 2017): 17127. https://doi.org/10.1038/nenergy.2017.127.

[61] Mayrhofer. "A Critical Review on Hydrogen Evolution Electrocatalysis: Re-Exploring the Volcano-Relationship." *Electroanalysis* 28, no. 10 (2016): 2256–69. https://doi.org/10.1002/elan.201600270.

[62] Greeley, Jeff, Thomas F. Jaramillo, Jacob Bonde, Ib Chorkendorff, and Jens K. Nørskov. "Computational High-Throughput Screening of Electrocatalytic Materials for Hydrogen Evolution." *Nature Materials* 5, no. 11 (November 2006): 909–13. https://doi.org/10.1038/nmat1752.

[63] Zhang, Jing, Jingjie Wu, Hua Guo, Weibing Chen, Jiangtan Yuan, Ulises Martinez, Gautam Gupta, Aditya Mohite, Pulickel M. Ajayan, and Jun Lou. "Unveiling Active Sites for the Hydrogen Evolution Reaction on Monolayer MoS2." *Advanced Materials* 29, no. 42 (2017): 1701955. https://doi.org/10.1002/adma.201701955.

[64] Fan, Xiao-Li, Yi Yang, Pin Xiao, and Woon-Ming Lau. "Site-Specific Catalytic Activity in Exfoliated MoS2 Single-Layer Polytypes for Hydrogen Evolution: Basal Plane and Edges." *Journal of Materials Chemistry A* 2, no. 48 (November 18, 2014): 20545–51. https://doi.org/10.1039/C4TA05257A.

[65] Fan, Ronglei, Jie Mao, Zhihao Yin, Jiansheng Jie, Wen Dong, Liang Fang, Fengang Zheng, and Mingrong Shen. "Efficient and Stable Silicon Photocathodes Coated

with Vertically Standing Nano-MoS2 Films for Solar Hydrogen Production." *ACS Applied Materials & Interfaces* 9, no. 7 (February 22, 2017): 6123–29. https://doi.org/10.1021/acsami.6b15854.

[66] Kreuter, W., and H. Hofmann. "Electrolysis: The Important Energy Transformer in a World of Sustainable Energy." *International Journal of Hydrogen Energy* 23, no. 8 (August 1, 1998): 661–66. https://doi.org/10.1016/S0360-3199(97)00109-2.

[67] Bockris, J. O'm, and E. C. Potter. "The Mechanism of the Cathodic Hydrogen Evolution Reaction." *Journal of The Electrochemical Society* 99, no. 4 (April 1, 1952): 169. https://doi.org/10.1149/1.2779692.

[68] Quaino, Paola, Fernanda Juarez, Elizabeth Santos, and Wolfgang Schmickler. "Volcano Plots in Hydrogen Electrocatalysis – Uses and Abuses." *Beilstein Journal of Nanotechnology* 5, no. 1 (June 13, 2014): 846–54. https://doi.org/10.3762/bjnano.5.96.

[69] Er, Dequan, Han Ye, Nathan C. Frey, Hemant Kumar, Jun Lou, and Vivek B. Shenoy. "Prediction of Enhanced Catalytic Activity for Hydrogen Evolution Reaction in Janus Transition Metal Dichalcogenides." *Nano Letters* 18, no. 6 (June 13, 2018): 3943–49. https://doi.org/10.1021/acs.nanolett.8b01335.

[70] Xie, Junfeng, and Yi Xie. "Structural Engineering of Electrocatalysts for the Hydrogen Evolution Reaction: Order or Disorder?" *ChemCatChem* 7, no. 17 (2015): 2568–80. https://doi.org/10.1002/cctc.201500396.

[71] Xie, Junfeng, Jiajia Zhang, Shuang Li, Fabian Grote, Xiaodong Zhang, Hao Zhang, Ruoxing Wang, Yong Lei, Bicai Pan, and Yi Xie. "Controllable Disorder Engineering in Oxygen-Incorporated MoS2 Ultrathin Nanosheets for Efficient Hydrogen Evolution." *Journal of the American Chemical Society* 135, no. 47 (November 27, 2013): 17881–88. https://doi.org/10.1021/ja408329q.

[72] Lu, Ang-Yu, Hanyu Zhu, Jun Xiao, Chih-Piao Chuu, Yimo Han, Ming-Hui Chiu, Chia-Chin Cheng, et al. "Janus Monolayers of Transition Metal Dichalcogenides." *Nature Nanotechnology* 12, no. 8 (August 2017): 744–49. https://doi.org/10.1038/nnano.2017.100.

[73] Zeng, Zhiyuan, Zongyou Yin, Xiao Huang, Hai Li, Qiyuan He, Gang Lu, Freddy Boey, and Hua Zhang. "Single-Layer Semiconducting Nanosheets: High-Yield Preparation and Device Fabrication." *Angewandte Chemie* 123, no. 47 (2011): 11289–93. https://doi.org/10.1002/ange.201106004.

[74] Lu, Qipeng, Yifu Yu, Qinglang Ma, Bo Chen, and Hua Zhang. "2D Transition-Metal-Dichalcogenide-Nanosheet-Based Composites for Photocatalytic and Electrocatalytic Hydrogen Evolution Reactions." *Advanced Materials* 28, no. 10 (2016): 1917–33. https://doi.org/10.1002/adma.201503270.

[75] Huang, Xiao, Zhiyuan Zeng, Shuyu Bao, Mengfei Wang, Xiaoying Qi, Zhanxi Fan, and Hua Zhang. "Solution-Phase Epitaxial Growth of Noble Metal Nanostructures on Dispersible Single-Layer Molybdenum Disulfide Nanosheets." *Nature Communications* 4, no. 1 (February 5, 2013): 1444. https://doi.org/10.1038/ncomms2472.

[76] Kibsgaard, Jakob, Zhebo Chen, Benjamin N. Reinecke, and Thomas F. Jaramillo. "Engineering the Surface Structure of MoS2 to Preferentially Expose Active Edge

Sites for Electrocatalysis." *Nature Materials* 11, no. 11 (November 2012): 963–69. https://doi.org/10.1038/nmat3439.

[77] Zhou, Haiqing, Yumei Wang, Ran He, Fang Yu, Jingying Sun, Feng Wang, Yucheng Lan, Zhifeng Ren, and Shuo Chen. "One-Step Synthesis of Self-Supported Porous NiSe2/Ni Hybrid Foam: An Efficient 3D Electrode for Hydrogen Evolution Reaction." *Nano Energy* 20 (February 1, 2016): 29–36. https://doi.org/10.1016/j.nanoen.2015.12.008.

[78] Zhou, Haiqing, Fang Yu, Yufeng Huang, Jingying Sun, Zhuan Zhu, Robert J. Nielsen, Ran He, et al. "Efficient Hydrogen Evolution by Ternary Molybdenum Sulfoselenide Particles on Self-Standing Porous Nickel Diselenide Foam." *Nature Communications* 7, no. 1 (September 16, 2016): 12765. https://doi.org/10.1038/ncomms12765.

[79] Huang, Yunpeng, Yue-E. Miao, Longsheng Zhang, Weng Weei Tjiu, Jisheng Pan, and Tianxi Liu. "Synthesis of Few-Layered MoS2 Nanosheet-Coated Electrospun SnO2 Nanotube Heterostructures for Enhanced Hydrogen Evolution Reaction." *Nanoscale* 6, no. 18 (August 21, 2014): 10673–79. https://doi.org/10.1039/C4NR02014F.

[80] Chen, Zhebo, Dustin Cummins, Benjamin N. Reinecke, Ezra Clark, Mahendra K. Sunkara, and Thomas F. Jaramillo. "Core–Shell MoO3–MoS2 Nanowires for Hydrogen Evolution: A Functional Design for Electrocatalytic Materials." *Nano Letters* 11, no. 10 (October 12, 2011): 4168–75. https://doi.org/10.1021/nl2020476.

[81] Seh, Zhi Wei, Jakob Kibsgaard, Colin F. Dickens, Ib Chorkendorff, Jens K. Nørskov, and Thomas F. Jaramillo. "Combining Theory and Experiment in Electrocatalysis: Insights into Materials Design." *Science* 355, no. 6321 (January 13, 2017): eaad4998. https://doi.org/10.1126/science.aad4998.

[82] Zhu, Zhengju, Huajie Yin, Chun-Ting He, Mohammad Al-Mamun, Porun Liu, Lixue Jiang, Yong Zhao, et al. "Ultrathin Transition Metal Dichalcogenide/3d Metal Hydroxide Hybridized Nanosheets to Enhance Hydrogen Evolution Activity." *Advanced Materials* 30, no. 28 (2018): 1801171. https://doi.org/10.1002/adma.201801171.

[83] Zhu, Yinlong, Wei Zhou, Zhi-Gang Chen, Yubo Chen, Chao Su, Moses O. Tadé, and Zongping Shao. "SrNb0.1Co0.7Fe0.2O3−δ Perovskite as a Next-Generation Electrocatalyst for Oxygen Evolution in Alkaline Solution." *Angewandte Chemie* 127, no. 13 (2015): 3969–73. https://doi.org/10.1002/ange.201408998.

[84] Tan, Chaoliang, Zhimin Luo, Apoorva Chaturvedi, Yongqing Cai, Yonghua Du, Yue Gong, Ying Huang, et al. "Preparation of High-Percentage 1T-Phase Transition Metal Dichalcogenide Nanodots for Electrochemical Hydrogen Evolution." *Advanced Materials* 30, no. 9 (2018): 1705509. https://doi.org/10.1002/adma.201705509.

[85] Cherevko, Serhiy, Simon Geiger, Olga Kasian, Nadiia Kulyk, Jan-Philipp Grote, Alan Savan, Buddha Ratna Shrestha, et al. "Oxygen and Hydrogen Evolution Reactions on Ru, RuO2, Ir, and IrO2 Thin Film Electrodes in Acidic and Alkaline Electrolytes: A Comparative Study on Activity and Stability." *Catalysis Today, Electrocatalysis*, 262 (March 15, 2016): 170–80. https://doi.org/10.1016/j.cattod.2015.08.014.

[86] Kötz, R., H. Neff, and S. Stucki. "Anodic Iridium Oxide Films: XPS-Studies of Oxidation State Changes and O2 Evolution." *Journal of The Electrochemical Society* 131, no. 1 (January 1, 1984): 72. https://doi.org/10.1149/1.2115548.

[87] Vuković, Marijan. "Oxygen Evolution Reaction on Thermally Treated Iridium Oxide Films." *Journal of Applied Electrochemistry* 17, no. 4 (July 1, 1987): 737–45. https://doi.org/10.1007/BF01007809.

[88] Yan, Ya, Bao Yu Xia, Bin Zhao, and Xin Wang. "A Review on Noble-Metal-Free Bifunctional Heterogeneous Catalysts for Overall Electrochemical Water Splitting." *Journal of Materials Chemistry A* 4, no. 45 (2016): 17587–603. https://doi.org/10.1039/C6TA08075H.

[89] Tahir, Muhammad, Lun Pan, Faryal Idrees, Xiangwen Zhang, Li Wang, Ji-Jun Zou, and Zhong Lin Wang. "Electrocatalytic Oxygen Evolution Reaction for Energy Conversion and Storage: A Comprehensive Review." *Nano Energy* 37 (July 1, 2017): 136–57. https://doi.org/10.1016/j.nanoen.2017.05.022.

[90] Cook, Timothy R., Dilek K. Dogutan, Steven Y. Reece, Yogesh Surendranath, Thomas S. Teets, and Daniel G. Nocera. "Solar Energy Supply and Storage for the Legacy and Nonlegacy Worlds." *Chemical Reviews* 110, no. 11 (November 10, 2010): 6474–6502. https://doi.org/10.1021/cr100246c.

[91] Suen, Nian-Tzu, Sung-Fu Hung, Quan Quan, Nan Zhang, Yi-Jun Xu, and Hao Ming Chen. "Electrocatalysis for the Oxygen Evolution Reaction: Recent Development and Future Perspectives." *Chemical Society Reviews* 46, no. 2 (January 23, 2017): 337–65. https://doi.org/10.1039/C6CS00328A.

[92] Frydendal, Rasmus, Elisa A. Paoli, Brian P. Knudsen, Björn Wickman, Paolo Malacrida, Ifan E. L. Stephens, and Ib Chorkendorff. "Benchmarking the Stability of Oxygen Evolution Reaction Catalysts: The Importance of Monitoring Mass Losses." *Chem Electro Chem* 1, no. 12 (2014): 2075–81. https://doi.org/10.1002/celc.201402262.

[93] Kötz, R., H. J. Lewerenz, and S. Stucki. "XPS Studies of Oxygen Evolution on Ru and RuO2 Anodes." *Journal of The Electrochemical Society* 130, no. 4 (April 1, 1983): 825. https://doi.org/10.1149/1.2119829.

[94] Suntivich, Jin, Kevin J. May, Hubert A. Gasteiger, John B. Goodenough, and Yang Shao-Horn. "A Perovskite Oxide Optimized for Oxygen Evolution Catalysis from Molecular Orbital Principles." *Science* 334, no. 6061 (December 9, 2011): 1383–85. https://doi.org/10.1126/science.1212858.

[95] Chen, Rong, Hsin-Yi Wang, Jianwei Miao, Hongbin Yang, and Bin Liu. "A Flexible High-Performance Oxygen Evolution Electrode with Three-Dimensional NiCo2O4 Core-Shell Nanowires." *Nano Energy* 11 (January 1, 2015): 333–40. https://doi.org/10.1016/j.nanoen.2014.11.021.

[96] Wu, Jingjie, Mingjie Liu, Kuntal Chatterjee, Ken P. Hackenberg, Jianfeng Shen, Xiaolong Zou, Yong Yan, et al. "Exfoliated 2D Transition Metal Disulfides for Enhanced Electrocatalysis of Oxygen Evolution Reaction in Acidic Medium." *Advanced Materials Interfaces* 3, no. 9 (2016): 1500669. https://doi.org/10.1002/admi.201500669.

[97] Yang, Xiao-Feng, Aiqin Wang, Botao Qiao, Jun Li, Jingyue Liu, and Tao Zhang. "Single-Atom Catalysts: A New Frontier in Heterogeneous Catalysis." *Accounts of*

Chemical Research 46, no. 8 (August 20, 2013): 1740–48. https://doi.org/10.1021/ar300361m.

[98] Zhang, Junming, Hua Bing Tao, Min Kuang, Hong Bin Yang, Weizheng Cai, Qingyu Yan, Qing Mao, and Bin Liu. "Advances in Thermodynamic-Kinetic Model for Analyzing the Oxygen Evolution Reaction." *ACS Catalysis* 10, no. 15 (August 7, 2020): 8597–8610. https://doi.org/10.1021/acscatal.0c01906.

[99] Zhou, Xiaofeng, Shi-Hsin Lin, Xiulin Yang, Henan Li, Mohamed Nejib Hedhili, Lain-Jong Li, Wenjing Zhang, and Yumeng Shi. "MoSx-Coated NbS2 Nanoflakes Grown on Glass Carbon: An Advanced Electrocatalyst for the Hydrogen Evolution Reaction." *Nanoscale* 10, no. 7 (February 15, 2018): 3444–50. https://doi.org/10.1039/C7NR09172A.

[100] Zhang, Jing, Shuai Jia, Iskandar Kholmanov, Liang Dong, Dequan Er, Weibing Chen, Hua Guo, et al. "Janus Monolayer Transition-Metal Dichalcogenides." *ACS Nano* 11, no. 8 (August 22, 2017): 8192–98. https://doi.org/10.1021/acsnano.7b03186.

[101] Zhang, Huanjun, Guohua Chen, and Detlef W. Bahnemann. "Photoelectrocatalytic Materials for Environmental Applications." *Journal of Materials Chemistry* 19, no. 29 (July 14, 2009): 5089–5121. https://doi.org/10.1039/B821991E.

[102] Wang, Shuai, Di Zhang, Bin Li, Chao Zhang, Zhiguo Du, Haoming Yin, Xiaofang Bi, and Shubin Yang. "Ultrastable In-Plane 1T–2H MoS2 Heterostructures for Enhanced Hydrogen Evolution Reaction." *Advanced Energy Materials* 8, no. 25 (2018): 1801345. https://doi.org/10.1002/aenm.201801345.

[103] Wang, Qing Hua, Kourosh Kalantar-Zadeh, Andras Kis, Jonathan N. Coleman, and Michael S. Strano. "Electronics and Optoelectronics of Two-Dimensional Transition Metal Dichalcogenides." *Nature Nanotechnology* 7, no. 11 (November 2012): 699–712. https://doi.org/10.1038/nnano.2012.193.

[104] Gao, Min-Rui, Xuan Cao, Qiang Gao, Yun-Fei Xu, Ya-Rong Zheng, Jun Jiang, and Shu-Hong Yu. "Nitrogen-Doped Graphene Supported CoSe2 Nanobelt Composite Catalyst for Efficient Water Oxidation." *ACS Nano* 8, no. 4 (April 22, 2014): 3970–78. https://doi.org/10.1021/nn500880v.

[105] Turner, John A. "Sustainable Hydrogen Production." *Science* 305, no. 5686 (August 13, 2004): 972–74. https://doi.org/10.1126/science.1103197.

[106] Gewirth, Andrew A., and Matthew S. Thorum. "Electroreduction of Dioxygen for Fuel-Cell Applications: Materials and Challenges." *Inorganic Chemistry* 49, no. 8 (April 19, 2010): 3557–66. https://doi.org/10.1021/ic9022486.

[107] Li, Guoqing, Du Zhang, Qiao Qiao, Yifei Yu, David Peterson, Abdullah Zafar, Raj Kumar, et al. "All The Catalytic Active Sites of MoS2 for Hydrogen Evolution." *Journal of the American Chemical Society* 138, no. 51 (December 28, 2016): 16632–38. https://doi.org/10.1021/jacs.6b05940.

[108] Gao, Min-Rui, Jin-Xia Liang, Ya-Rong Zheng, Yun-Fei Xu, Jun Jiang, Qiang Gao, Jun Li, and Shu-Hong Yu. "An Efficient Molybdenum Disulfide/Cobalt Diselenide Hybrid Catalyst for Electrochemical Hydrogen Generation." *Nature Communications* 6, no. 1 (January 14, 2015): 5982. https://doi.org/10.1038/ncomms6982.

[109] Hammer, B., and J. K. Nørskov. "Theoretical Surface Science and Catalysis—Calculations and Concepts." In *Advances in Catalysis*, 45:71–129. Impact of Surface Science on Catalysis. Academic Press, 2000. https://doi.org/10.1016/S0360-0564(02)45013-4.

[110] Banham, Dustin, Siyu Ye, Katie Pei, Jun-ichi Ozaki, Takeaki Kishimoto, and Yasuo Imashiro. "A Review of the Stability and Durability of Non-Precious Metal Catalysts for the Oxygen Reduction Reaction in Proton Exchange Membrane Fuel Cells." *Journal of Power Sources* 285 (July 1, 2015): 334–48. https://doi.org/10.1016/j.jpowsour.2015.03.047.

[111] Liu, Youwen, Hao Cheng, Mengjie Lyu, Shaojuan Fan, Qinghua Liu, Wenshuai Zhang, Yuduo Zhi, et al. "Low Overpotential in Vacancy-Rich Ultrathin CoSe2 Nanosheets for Water Oxidation." *Journal of the American Chemical Society* 136, no. 44 (November 5, 2014): 15670–75. https://doi.org/10.1021/ja5085157.

[112] Alfaifi, Bandar Y., Asif A. Tahir, and K. G. Upul Wijayantha. "Fabrication of Bi_2WO_6 Photoelectrodes with Enhanced Photoelectrochemical and Photocatalytic Performance." *Solar Energy Materials and Solar Cells* 195 (June 15, 2019): 134–41. https://doi.org/10.1016/j.solmat.2019.02.031.

[113] Jia, Baoming, Ziqian Xue, Qian Liu, Qinglin Liu, Kang Liu, Min Liu, Ting-Shan Chan, et al. "Hierarchical Nanotubes Constructed from CoSe2 Nanorods with an Oxygen-Rich Surface for an Efficient Oxygen Evolution Reaction." *Journal of Materials Chemistry A* 7, no. 25 (June 25, 2019): 15073–78. https://doi.org/10.1039/C9TA03606G.

[114] Yang, Yong, Mingchuan Luo, Weiyu Zhang, Yingjun Sun, Xu Chen, and Shaojun Guo. "Metal Surface and Interface Energy Electrocatalysis: Fundamentals, Performance Engineering, and Opportunities." *Chem* 4, no. 9 (September 13, 2018): 2054–83. https://doi.org/10.1016/j.chempr.2018.05.019.

[115] Huang, Yongkui, Shifei Kang, Yun Yang, Hengfei Qin, Zhijiang Ni, Shuijin Yang, and Xi Li. "Facile Synthesis of Bi/Bi_2WO_6 Nanocomposite with Enhanced Photocatalytic Activity under Visible Light." *Applied Catalysis B: Environmental* 196 (November 5, 2016): 89–99. https://doi.org/10.1016/j.apcatb.2016.05.022.

[116] Bahnemann, Detlef. "Photocatalytic Water Treatment: Solar Energy Applications." *Solar Energy*, *Photocatalysis*, 77, no. 5 (November 1, 2004): 445–59. https://doi.org/10.1016/j.solener.2004.03.031.

[117] Zhang, Zhijie, Wenzhong Wang, Lu Wang, and Songmei Sun. "Enhancement of Visible-Light Photocatalysis by Coupling with Narrow-Band-Gap Semiconductor: A Case Study on Bi2S3/Bi2WO6." *ACS Applied Materials & Interfaces* 4, no. 2 (February 22, 2012): 593–97. https://doi.org/10.1021/am2017199.

[118] Yuan, Yong-Jun, Hong-Wei Lu, Zhen-Tao Yu, and Zhi-Gang Zou. "Noble-Metal-Free Molybdenum Disulfide Cocatalyst for Photocatalytic Hydrogen Production." *Chem Sus Chem* 8, no. 24 (2015): 4113–27. https://doi.org/10.1002/cssc.201501203.

[119] Kalantar-zadeh, Kourosh, Jian Zhen Ou, Torben Daeneke, Michael S. Strano, Martin Pumera, and Sally L. Gras. "Two-Dimensional Transition Metal Dichalcogenides in Biosystems." *Advanced Functional Materials* 25, no. 32 (2015): 5086–99. https://doi.org/10.1002/adfm.201500891.

[120] Luo, Bin, Gang Liu, and Lianzhou Wang. "Recent Advances in 2D Materials for Photocatalysis." *Nanoscale* 8, no. 13 (March 24, 2016): 6904–20. https://doi.org/10.1039/C6NR00546B.

[121] Li, Zizhen, Xiangchao Meng, and Zisheng Zhang. "Recent Development on MoS2-Based Photocatalysis: A Review." *Journal of Photochemistry and Photobiology C: Photochemistry Reviews* 35 (June 1, 2018): 39–55. https://doi.org/10.1016/j.jphotochemrev.2017.12.002.

[122] Ong, Wee-Jun, Lling-Lling Tan, Siang-Piao Chai, Siek-Ting Yong, and Abdul Rahman Mohamed. "Highly Reactive {001} Facets of TiO2-Based Composites: Synthesis, Formation Mechanism and Characterization." *Nanoscale* 6, no. 4 (January 30, 2014): 1946–2008. https://doi.org/10.1039/C3NR04655A.

[123] Wan, Sijie, Jingsong Peng, Lei Jiang, and Qunfeng Cheng. "Bioinspired Graphene-Based Nanocomposites and Their Application in Flexible Energy Devices." *Advanced Materials* 28, no. 36 (2016): 7862–98. https://doi.org/10.1002/adma.201601934.

[124] Chen, Jianqing, Donghui Yang, Dan Song, Jinghua Jiang, Aibin Ma, Michael Z. Hu, and Chaoying Ni. "Recent Progress in Enhancing Solar-to-Hydrogen Efficiency." *Journal of Power Sources* 280 (April 15, 2015): 649–66. https://doi.org/10.1016/j.jpowsour.2015.01.073.

[125] Cao, Xiehong, Chaoliang Tan, Xiao Zhang, Wei Zhao, and Hua Zhang. "Solution-Processed Two-Dimensional Metal Dichalcogenide-Based Nanomaterials for Energy Storage and Conversion." *Advanced Materials* 28, no. 29 (2016): 6167–96. https://doi.org/10.1002/adma.201504833.

[126] Zhu, Jixin, Ken Sakaushi, Guylhaine Clavel, Menny Shalom, Markus Antonietti, and Tim-Patrick Fellinger. "A General Salt-Templating Method To Fabricate Vertically Aligned Graphitic Carbon Nanosheets and Their Metal Carbide Hybrids for Superior Lithium Ion Batteries and Water Splitting." *Journal of the American Chemical Society* 137, no. 16 (April 29, 2015): 5480–85. https://doi.org/10.1021/jacs.5b01072.

[127] Arifin, Mohd Nur Ikhmal Salehmin, Mohamad Azuwa Mohamed, and Mohammad B. Kassim. "Photocatalytic Properties of Two-Dimensional Graphene and Layered Transition-Metal Dichalcogenides Based Photocatalyst for Photoelectrochemical Hydrogen Generation: An Overview." *International Journal of Hydrogen Energy* 43, no. 41 (October 11, 2018): 18925–45. https://doi.org/10.1016/j.ijhydene.2018.08.126.

[128] Walter, Michael G., Emily L. Warren, James R. McKone, Shannon W. Boettcher, Qixi Mi, Elizabeth A. Santori, and Nathan S. Lewis. "Solar Water Splitting Cells." *Chemical Reviews* 110, no. 11 (November 10, 2010): 6446–73. https://doi.org/10.1021/cr1002326.

[129] Li, Zhao-Qian, Xue-Tai Chen, and Zi-Ling Xue. "Microwave-Assisted Synthesis and Photocatalytic Properties of Flower-like Bi2WO6 and Bi2O3–Bi2WO6 Composite." *Journal of Colloid and Interface Science* 394 (March 15, 2013): 69–77. https://doi.org/10.1016/j.jcis.2012.12.002.

[130] Rosman, Nurul Nabila, Rozan Mohamad Yunus, Lorna Jeffery Minggu, Khuzaimah Rowley-Neale, Samuel J., Graham C. Smith, and Craig E. Banks.

"Mass-Producible 2D-MoS2-Impregnated Screen-Printed Electrodes That Demonstrate Efficient Electrocatalysis toward the Oxygen Reduction Reaction." *ACS Applied Materials & Interfaces* 9, no. 27 (July 12, 2017): 22539–48. https://doi.org/10.1021/acsami.7b05104.

[131] Zhuang, Houlong L., and Richard G. Hennig. "Computational Search for Single-Layer Transition-Metal Dichalcogenide Photocatalysts." *The Journal of Physical Chemistry C* 117, no. 40 (October 10, 2013): 20440–45. https://doi.org/10.1021/jp405808a.

[132] Barczuk, Piotr J., Adam Lewera, Krzysztof Miecznikowski, Artur Zurowski, and Pawel J. Kulesza. "Enhancement of Catalytic Activity of Platinum-Based Nanoparticles towards Electrooxidation of Ethanol through Interfacial Modification with Heteropolymolybdates." *Journal of Power Sources* 195, no. 9 (May 1, 2010): 2507–13. https://doi.org/10.1016/j.jpowsour.2009.11.066.

[133] Bai, Song, Limin Wang, Xiaoyi Chen, Junteng Du, and Yujie Xiong. "Chemically Exfoliated Metallic MoS2 Nanosheets: A Promising Supporting Co-Catalyst for Enhancing the Photocatalytic Performance of TiO_2 Nanocrystals." *Nano Research* 8, no. 1 (January 1, 2015): 175–83. https://doi.org/10.1007/s12274-014-0606-9.

[134] Kang, Yimin, Sina Najmaei, Zheng Liu, Yanjun Bao, Yumin Wang, Xing Zhu, Naomi J. Halas, et al. "Plasmonic Hot Electron Induced Structural Phase Transition in a MoS2 Monolayer." *Advanced Materials* 26, no. 37 (2014): 6467–71. https://doi.org/10.1002/adma.201401802.

[135] Zhang, Peng, Mamoru Fujitsuka, and Tetsuro Majima. "Hot Electron-Driven Hydrogen Evolution Using Anisotropic Gold Nanostructure Assembled Monolayer MoS2." *Nanoscale* 9, no. 4 (January 26, 2017): 1520–26. https://doi.org/10.1039/C6NR07740D.

[136] Kumar, Ritesh, Deya Das, and Abhishek Kumar Singh. "C2N/WS2 van Der Waals Type-II Heterostructure as a Promising Water Splitting Photocatalyst." *Journal of Catalysis* 359 (March 1, 2018): 143–50. https://doi.org/10.1016/j.jcat.2018.01.005.

[137] Luo, Yi, Sake Wang, Kai Ren, Jyh-Pin Chou, Jin Yu, Zhengming Sun, and Minglei Sun. "Transition-Metal Dichalcogenides/Mg(OH)2 van der Waals Heterostructures as Promising Water-Splitting Photocatalysts: A First-Principles Study." *Physical Chemistry Chemical Physics* 21, no. 4 (January 23, 2019): 1791–96. https://doi.org/10.1039/C8CP06960C.

[138] Chen, Jihan, Connor S. Bailey, Yilun Hong, Li Wang, Zhi Cai, Lang Shen, Bingya Hou, et al. "Plasmon-Resonant Enhancement of Photocatalysis on Monolayer WSe2." *ACS Photonics* 6, no. 3 (March 20, 2019): 787–92. https://doi.org/10.1021/acsphotonics.9b00089.

[139] Chang, Kun, Xiao Hai, Hong Pang, Huabin Zhang, Li Shi, Guigao Liu, Huimin Liu, Guixia Zhao, Mu Li, and Jinhua Ye. "Targeted Synthesis of 2H- and 1T-Phase MoS2 Monolayers for Catalytic Hydrogen Evolution." *Advanced Materials* 28, no. 45 (2016): 10033–41. https://doi.org/10.1002/adma.201603765.

[140] Han, Qing, Nan Chen, Jing Zhang, and Liangti Qu. "Graphene/Graphitic Carbon Nitride Hybrids for Catalysis." *Materials Horizons* 4, no. 5 (August 29, 2017): 832–50. https://doi.org/10.1039/C7MH00379J.

[141] Lee, Jae Yoon, Jun-Hwan Shin, Gwan-Hyoung Lee, and Chul-Ho Lee. "Two-Dimensional Semiconductor Optoelectronics Based on van der Waals Heterostructures." *Nanomaterials* 6, no. 11 (November 2016): 193. https://doi.org/10.3390/nano6110193.

[142] Yuan, Ye, Pengfei Shen, Quanjun Li, Gong Chen, Huafang Zhang, Luyao Zhu, Bo Zou, and Bingbing Liu. "Excellent Photocatalytic Performance of Few-Layer MoS2/Graphene Hybrids." *Journal of Alloys and Compounds* 700 (April 5, 2017): 12–17. https://doi.org/10.1016/j.jallcom.2017.01.027.

[143] Chang, Kun, Zongwei Mei, Tao Wang, Qing Kang, Shuxin Ouyang, and Jinhua Ye. "MoS2/Graphene Cocatalyst for Efficient Photocatalytic H2 Evolution under Visible Light Irradiation." *ACS Nano* 8, no. 7 (July 22, 2014): 7078–87. https://doi.org/10.1021/nn5019945.

[144] Ren, Kai, MingLei Sun, Yi Luo, SaKe Wang, Jin Yu, and WenCheng Tang. "First-Principle Study of Electronic and Optical Properties of Two-Dimensional Materials-Based Heterostructures Based on Transition Metal Dichalcogenides and Boron Phosphide." *Applied Surface Science* 476 (May 15, 2019): 70–75. https://doi.org/10.1016/j.apsusc.2019.01.005.

[145] Cui, Zhen, Kai Ren, Yiming Zhao, Xia Wang, Huabing Shu, Jin Yu, Wencheng Tang, and Minglei Sun. "Electronic and Optical Properties of van der Waals Heterostructures of G-GaN and Transition Metal Dichalcogenides." *Applied Surface Science* 492 (October 30, 2019): 513–19. https://doi.org/10.1016/j.apsusc.2019.06.207.

[146] Chojak, Malgorzata, Aneta Kolary-Zurowska, Renata Wlodarczyk, Krzysztof Miecznikowski, Katarzyna Karnicka, Barbara Palys, Roberto Marassi, and Pawel J. Kulesza. "Modification of Pt Nanoparticles with Polyoxometallate Monolayers: Competition between Activation and Blocking of Reactive Sites for the Electrocatalytic Oxygen Reduction." *Electrochimica Acta*, Surface Imaging/Spectroscopy at Solid/Liquid Interface (ISSIS), 52, no. 18 (May 10, 2007): 5574–81. https://doi.org/10.1016/j.electacta.2007.01.063.

[147] Liu, Xupo, Yunfeng Zhang, Shaofeng Deng, Cuicui Li, Jiaming Dong, Jiaying Wang, Zehui Yang, Deli Wang, and Hansong Cheng. "Semi-Interpenetrating Polymer Networks toward Sulfonated Poly(Ether Ether Ketone) Membranes for High Concentration Direct Methanol Fuel Cell." *Chinese Chemical Letters* 30, no. 2 (February 1, 2019): 299–304. https://doi.org/10.1016/j.cclet.2018.09.021.

[148] Khan, Abuzar, Ibrahim Khan, Mohd Yusuf Khan, Hatim Dafallah, and Ahsanulhaq Qurashi. "Facile Synthesis of 1T-WS2/Graphite Nanocomposite for Efficient Solar-Driven Oxygen Evolution Reaction." *International Journal of Hydrogen Energy, Proceedings of 5th International Conference on Nanotechnology, Nanomaterials & Thin Films for Energy Applications – 2018*, 45, no. 45 (September 14, 2020): 24045–53. https://doi.org/10.1016/j.ijhydene.2019.07.170.

[149] Lee, Chuan-Pei, Chun-Ting Li, and Kuo-Chuan Ho. "Use of Organic Materials in Dye-Sensitized Solar Cells." *Materials Today* 20, no. 5 (June 1, 2017): 267–83. https://doi.org/10.1016/j.mattod.2017.01.012.

[150] Singh, Eric, and Hari Singh Nalwa. "Graphene-Based Dye-Sensitized Solar Cells: A Review." *Science of Advanced Materials* 7, no. 10 (October 1, 2015): 1863–1912. https://doi.org/10.1166/sam.2015.2438.

[151] Wei, Wei, Kai Sun, and Yun Hang Hu. "An Efficient Counter Electrode Material for Dye-Sensitized Solar Cells—Flower-Structured 1T Metallic Phase MoS2." *Journal of Materials Chemistry A* 4, no. 32 (August 9, 2016): 12398–401. https://doi.org/10.1039/C6TA04743B.

[152] Mohammad Al-Mamun, Haimin Zhang, Porun Liu, Yun Wang, Jun Cao, and Huijun Zhao. "Directly Hydrothermal Growth of Ultrathin MoS_2 Nanostructured Films as High Performance Counter Electrodes for Dye-Sensitised Solar Cells." *RSC Advances* 4, no. 41 (May 15, 2014): 21277–83. https://doi.org/10.1039/c4ra00583j.

[153] Ahn, Sung Hoon, and Arumugam Manthiram. "Edge-Oriented Tungsten Disulfide Catalyst Produced from Mesoporous WO3 for Highly Efficient Dye-Sensitized Solar Cells." *Advanced Energy Materials* 6, no. 3 (2016): 1501814. https://doi.org/10.1002/aenm.201501814.

[154] Meng, Xiangtong, Chang Yu, Bing Lu, Juan Yang, and Jieshan Qiu. "Dual Integration System Endowing Two-Dimensional Titanium Disulfide with Enhanced Triiodide Reduction Performance in Dye-Sensitized Solar Cells." *Nano Energy* 22 (April 1, 2016): 59–69. https://doi.org/10.1016/j.nanoen.2016.02.010.

[155] Tiwari, Jitendra N., Rajanish N. Tiwari, Gyan Singh, and Kwang S. Kim. "Recent Progress in the Development of Anode and Cathode Catalysts for Direct Methanol Fuel Cells." *Nano Energy* 2, no. 5 (September 1, 2013): 553–78. https://doi.org/10.1016/j.nanoen.2013.06.009.

[156] Zoladek, Sylwia, Iwona A. Rutkowska, Katarzyna Skorupska, Barbara Palys, and Pawel J. Kulesza. "Fabrication of Polyoxometallate-Modified Gold Nanoparticles and Their Utilization as Supports for Dispersed Platinum in Electrocatalysis." *Electrochimica Acta*, Selected Papers from the 61st *ISE Meeting*, Nice, France, 2010, 56, no. 28 (December 1, 2011): 10744–50. https://doi.org/10.1016/j.electacta.2011.04.020.

[157] Kannan, Ramanujam, Change Dae Lim, Ae Rhan Kim, Min Ho Lee, and Dong Jin Yoo. "Bifunctional Electrocatalyst of Pd-C@MoS2-RGO Hybrid Nanostructures—Size Confined Green Synthesis for Direct Alcohol Fuel Cells." *Journal of Nanoscience and Nanotechnology* 19, no. 8 (August 1, 2019): 4520–28. https://doi.org/10.1166/jnn.2019.16352.

[158] Kulesza, Pawel J., Izabela S. Pieta, Iwona A. Rutkowska, Anna Wadas, Diana Marks, Karolina Klak, Leszek Stobinski, and James A. Cox. "Electrocatalytic Oxidation of Small Organic Molecules in Acid Medium: Enhancement of Activity of Noble Metal Nanoparticles and Their Alloys by Supporting or Modifying Them with Metal Oxides." *Electrochimica Acta, Electrochemistry for Advanced Materials, Technologies and Instrumentation*, 110 (November 1, 2013): 474–83. https://doi.org/10.1016/j.electacta.2013.06.052.

[159] Zhang, Hongmei, Fengxing Jiang, Rong Zhou, Yukou Du, Ping Yang, Chuanyi Wang, and Jingkun Xu. "Effect of Deposition Potential on the Structure and Electrocatalytic Behavior of Pt Micro/Nanoparticles." *International Journal of*

Hydrogen Energy 36, no. 23 (November 1, 2011): 15052–59. https://doi.org/10.1016/j.ijhydene.2011.08.072.

[160] Li, Yongjie, Wei Gao, Lijie Ci, Chunming Wang, and Pulickel M. Ajayan. "Catalytic Performance of Pt Nanoparticles on Reduced Graphene Oxide for Methanol Electro-Oxidation." *Carbon* 48, no. 4 (April 1, 2010): 1124–30. https://doi.org/10.1016/j.carbon.2009.11.034.

[161] Zeradjanin, Aleksandar R., Jan-Philipp Grote, George Polymeros, and Karl J. J. Zhai, Chunyang, Mingshan Zhu, Duan Bin, Fangfang Ren, Caiqin Wang, Ping Yang, and Yukou Du. "Two Dimensional MoS2/Graphene Composites as Promising Supports for Pt Electrocatalysts towards Methanol Oxidation." *Journal of Power Sources* 275 (February 1, 2015): 483–88. https://doi.org/10.1016/j.jpowsour.2014.11.030.

[162] Zhang, Hongmin, Jie He, Chunyang Zhai, and Mingshan Zhu. "2D Bi2WO6/MoS2 as a New Photo-Activated Carrier for Boosting Electrocatalytic Methanol Oxidation with Visible Light Illumination." *Chinese Chemical Letters* 30, no. 12 (December 1, 2019): 2338–42. https://doi.org/10.1016/j.cclet.2019.07.021.

[163] Liu, Yuanyue, Jingjie Wu, Ken P. Hackenberg, Jing Zhang, Y. Morris Wang, Yingchao Liu, Yumin, Yafei Cao, Hua Lv, Shuang Li, and Han Zhang. "Construction of a 2D Nanoplate-like MoS2/Bi2WO6 Heterojunction with Enhanced Visible-Light Photocatalytic Activities." *Materials Letters* 188 (February 1, 2017): 99–102. https://doi.org/10.1016/j.matlet.2016.11.060.

[164] Schmidt, T. J., U. A. Paulus, H. A. Gasteiger, and R. J. Behm. "The Oxygen Reduction Reaction on a Pt/Carbon Fuel Cell Catalyst in the Presence of Chloride Anions." *Journal of Electroanalytical Chemistry* 508, no. 1 (July 27, 2001): 41–47. https://doi.org/10.1016/S0022-0728(01)00499-5.

[165] Suresh, Chinnathambi, Sankararao Mutyala, and Jayaraman Mathiyarasu. "Support Interactive Synthesis of Nanostructured MoS2 Electrocatalyst for Oxygen Reduction Reaction." *Materials Letters* 164 (February 1, 2016): 417–20. https://doi.org/10.1016/j.matlet.2015.11.052.

[166] J. Rowley-Neale, Samuel, Jamie M. Fearn, Dale A. C. Brownson, Graham C. Smith, Xiaobo Ji, and Craig E. Banks. "2D Molybdenum Disulphide (2D-MoS 2) Modified Electrodes Explored towards the Oxygen Reduction Reaction." *Nanoscale* 8, no. 31 (2016): 14767–77. https://doi.org/10.1039/C6NR04073J.

[167] Kang, Yimin, Yongji Gong, Zhijian Hu, Ziwei Li, Ziwei Qiu, Xing Zhu, Pulickel M. Ajayan, and Zheyu Fang. "Plasmonic Hot Electron Enhanced MoS2 Photocatalysis in Hydrogen Evolution." *Nanoscale* 7, no. 10 (February 26, 2015): 4482–88. https://doi.org/10.1039/C4NR07303G.

[168] Balanta, Angelica, Cyril Godard, and Carmen Claver. "Pd Nanoparticles for C–C Coupling Reactions." *Chemical Society Reviews* 40, no. 10 (September 20, 2011): 4973–85. https://doi.org/10.1039/C1CS15195A.

[169] Huang, Huodi, Xiaofeng Zhang, Yi Zhang, Baohua Huang, Jiannan Cai, and Shen Lin. "Facile Synthesis of Laminated Porous WS2/C Composite and Its Electrocatalysis for Oxygen Reduction Reaction." *International Journal of Hydrogen Energy* 43, no. 17 (April 26, 2018): 8290–97. https://doi.org/10.1016/j.ijhydene.2018.03.059.

[170] Lukowski, Mark A., Andrew S. Daniel, Fei Meng, Audrey Forticaux, Linsen Li, and Song Jin. "Enhanced Hydrogen Evolution Catalysis from Chemically Exfoliated Metallic MoS2 Nanosheets." *Journal of the American Chemical Society* 135, no. 28 (July 17, 2013): 10274–77. https://doi.org/10.1021/ja404523s.

Chapter 8

Field Emission Applications of Metallic 2D TMDs

Dattatray J. Late[1] and Chandra Sekhar Rout[2,*]

[1]Centre for Nanoscience and Nanotechnology Amity University Maharashtra, Mumbai-Pune Expressway, Bhatan, Post, Somathne, Panvel, Mumbai, Maharashtra, India

[2]Centre for Nano and Material Sciences, Jain University, Jain Global Campus, Ramanagaram, Bangalore, India

Abstract

This chapter starts with a broad introduction to the field emission technique along with basic principle and applications using various two-dimensional metallic emitter. Field electron emission is emission of electrons from condensed phase into vacuum phase under the influence of high electrostatic field (typically 10^6-10^7 V/cm). Such high electrostatic fields are normally produced by employing very sharp tip/edge of nanosheets (apex diameter < 500 nm). One-dimensional carbon nanotubes possess tip diameter < 50 nm, hence exhibit promising field emission properties. Whereas, in case of two-dimensional (2D) semimetal materials such as graphene, semiconducting 2D material such as MoS_2 and metallic 2D materials such as VS_2, VSe_2, ReS_2, VSe_2 and VSe_2-RGO hybrid structure the high electrostatic field is created at sharp/nanometric edges which are mainly responsible for enhancing the electric field. In this chapter, we believe that the electrons are emitted from the nanometric protrusions of 2D metallic sheets due to presence of sharp edges in the form of nanometric protrusions responsible for the enhanced field emission properties.

* Corresponding Author's Email: r.chandrasekhar@jainuniversity.ac.in.
* Corresponding Author's Email: datta099@gmail.com.

In: 2D Metallic Transition Metal Dichalcogenides
Editors: Chandra Sekhar Rout and Brahmananda Chakraborty
ISBN: 978-1-68507-965-9

Keywords: 2D materials, field emission, current stability

8.1. Introduction

Electrons are emitted from a metal or semiconductor on applying a strong electric field by a process called "Field Emission". The phenomenon was first reported by R. W. Wood in 1897; subsequent field emission studies have made significant contribution to the technological advances and basic science similar to the other fundamental electron emission processes from solids like thermionic emission, photoelectric emission and secondary electron emission. A very good review of early work is found in an article by R. O. Jenkins [1]. Classical theories failed to explain the experimental results that field emission was independent of temperature in the region well below the onset of thermionic emission. The basic process was then explained, using the wave mechanical formulation by Fowler and Nordheim in 1928 [2].

The practical difficulty in producing very high fields required for field emission (10^6-10^7V/cm) put a serious limitation in obtaining reliable experimental results on this effect. E. W. Müller placed a cathode in the form of a sharply etched tungsten wire in front of a screen, in vacuum, in order to see the way in which the electrons were ejected from the tip. Field emission from well cleaned, rounded tip was obtained after heating the tip at high temperatures. Reproducible current-voltage (I-V) characteristics could be measured with this two dimensional "Projection Microscope", the now called field emission microscope (FEM), projecting the tip geometry of the surface onto the screen. Müller obtained beautiful patterns of tungsten, molybdenum, nickel and copper. These patterns illustrated the emission anisotropy of the various planes of the single crystal emitters. Müllers measurement of the field emission occurred at a field strength of 3.5×10^7 V/cm, was in full agreement with the Fowler-Nordheim (F-N) theory.

Dyke and co-workers contributed substantially to the development of field emission cathodes by employing both the continuous and the pulsed emission modes [3]. The work showed the potential of the field emitters in obtaining high current density cathodes. There have been many theoretical contributions together with the development of modern techniques, for measurements of energy distribution, work function, noise characteristics and current stability [4].

8.1.1. Field Electron Emission Process

The usual theoretical treatment of electron emission from metal into vacuum on the application of high electric field is based on the F-N model which assumes

1. The temperature of the metal is 0°K.
2. The metal has a free electron band structure.
3. The emitter surface is taken to be smooth and plane so that the electric field applied is uniform.
4. The potential barrier close to the surface in vacuum region consists of an image force potential and a potential due to the applied electric field.
5. The local work function is uniform across the emitting surface and is not affected by external electric field.

The potential energy V(x) of an electron as a function of its distance x to the surface of the metal under the influence of an external applied electric field is shown schematically in Figure 8.1.

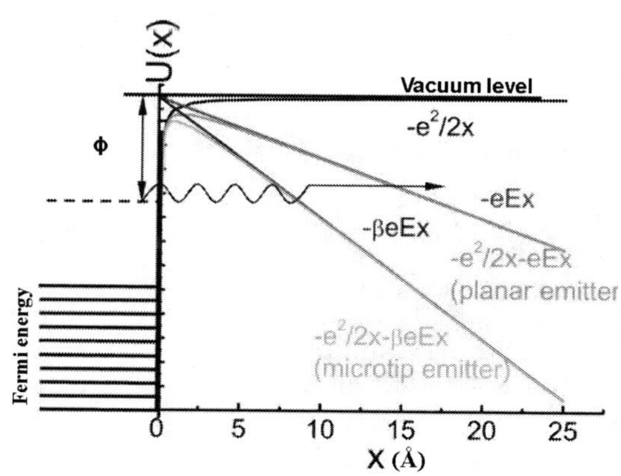

Figure 8.1. Potential energy of an electron V(x) in eV as a function of its distance X to the surface in Å. (Reproduced with permission from [55]).

8.1.2. Fowler Nordheim (F-N) Equation

According to the Summerfield free electron model, the potential energy of the electrons inside the metal V_p, is constant. Thus,

$$V(x) = V_p \; X \langle \; 0 \tag{8.1}$$

Outside the metal,

$$V(x) = -\frac{e^2}{4X} - e\,Fx \; X \rangle \; 0 \tag{8.2}$$

Where e is the charge of the electron and F is the electric field strength. The first term on right hand side in equation (1.2) is potential due to the image force on the electron due to the induced charge in the metal and the second term is the contribution due to the external electric field.

The F-N model of electron emission is based upon tunneling of electrons through the one-dimensional barrier. If the E is the total energy and W the x part of this energy, then

$$W = \frac{P_X^2}{2m} + V(x) \tag{8.3}$$

Where P_X is the momentum of the electrons in the x direction and m is its mass. It is assumed that only the kinetic energy normal to the surface determines the penetration probability of the potential barrier by the electron. The probability of the barrier penetration D (W) is found by solving the time independent Schrödinger equation. Applying the WKB approximation valid for W $\langle\langle$ $V_{max}(x)$, where $V_{max}(x)$ is the maximum potential at the Schottky hump and considering that the electrons penetrating the barrier, originate in the neighborhood of the Fermi level, one obtains:

$$D(W) = \exp\left[-c + \frac{W-\phi}{d}\right] \tag{8.4}$$

where

$$c = \frac{4\left(2mj^{3}\right)^{1/2}}{3\hbar e F} \times f(y) \tag{8.5}$$

$$d = \frac{\hbar e F}{2\left(2\, m j\ \right)^{1/2}\, t(y)} \tag{8.6}$$

and

$$t\left(y\right) = f\left(y\right) - \frac{2}{3}\, y \frac{d\, f\left(y\right)}{d\ y} \tag{8.7}$$

where

$$y = \frac{\left(e^{3}\, F\right)^{1/2}}{j} \tag{8.8}$$

The f (y) and t (y) are the functions of y [5]. Now, P (W) dW, the number of electrons in the range of X component of energy between W and W+dW tunneling through the barrier, can be calculated by multiplying D (W) by the supply function N(W) dW, the number of electrons with the X component of the energy in the range W to W+ dW incident normal to the surface per unit area per unit time. Thus, the normal energy distribution is given by

$$P\ (W) = N\ (W)\ D\ (W)\ dW \tag{8.9}$$

The current density J is found by multiplying P(W)dW by e, the electronic charge and integrating over all values of W.

$$J = e \int_{V_P}^{\infty} P\left(W\right) dW \tag{8.10}$$

The supply function N (W) dW is obtained from the Fermi-Dirac energy distribution. Equilibrium distribution of electrons inside the metal is assumed, although electrons escape from the metal.

$$N(W)dW = \frac{4\pi m k_B T}{h^3} \ln\left[1 + \exp\left(-\frac{W-\phi}{K_B T}\right)\right] dW \tag{8.11}$$

where K_B is the Boltzman constant, T the absolute temperature and h the Plank's constant. For low temperatures,

$$\left|\left(\frac{W-\phi}{k_B T}\right)\right| \rangle\rangle 1 \text{ and if } W \leq \phi,$$

$$k_B T \ln\left[1+\exp\left(-\frac{W-\phi}{k_B T}\right)\right]=\phi-W \tag{8.12}$$

Combining (1.9), (1.11) and (1.12), and substituting in (1.10), integrating over all W, yields

$$J=\frac{e^3 F^2}{8\pi\hbar\phi t^2(y)} \exp\left[-\frac{4(2m)^{1/2}}{3\hbar e}\frac{\phi^{3/2}}{F} f(y)\right] \tag{8.13}$$

Expressing J in A/cm^2, F in V/cm, ϕ in eV and substituting the numerical values of other constants, the Fowler-Nordheim equation is obtained and is given by the expression,

$$J=1.54\times 10^{-6}\frac{F^2}{\phi t^2(y)} \exp\left[-6.8\times 10^7 \frac{\phi^{3/2}}{F} f(y)\right] \tag{8.14}$$

The F-N equation has been found adequate for describing field electron emission. This equation as discussed earlier has been arrived at by following the normal energy distribution and assuming the free electron model inside the metal. The normal energy distribution is experimentally difficult to observe in view of the stringent geometrical conditions of the emitter and the collector, although the total current-voltage (I-V) measurements generally agree with the F-N equation.

8.1.3. Fowler Nordheim (F-N) Plot

In a field emission experiment the measurable quantities are the field emission current (I) and the electric potential (V) applied between the cathode and the anode. J and E are related to these quantities as

$$I = JA \tag{8.15}$$

$$F = \beta V \tag{8.16}$$

where A is the emitter area and β is the field enhancement factor determined by the geometry of the emitter. The F-N equation takes the following form when the above substitutions are made

$$J = \frac{1.54\times10^{-6}\ \beta^2\ V^2 A}{\phi\, t^2(y)} \exp\left[\frac{-6.8\times10^7\,\phi^{3/2}\,f(y)}{\beta V}\right] \tag{8.17}$$

Plot of ln (I/V^2) Vs $(10^4/V)$ from this equation results in a straight line for metallic emitter called Fowler Nordheim plot. The slope of this straight line is given by

$$m = \frac{-6.8\times10^3\,\phi^{3/2}\ s(y)}{\beta} \tag{8.18}$$

where

$$s(y) = f(y) - \frac{y}{2}\left(\frac{df(y)}{dy}\right) = 1 \text{ (unity)}$$

It is assumed that ϕ, β and A are independent of V and the function s(y) has a value close to unity so that m is fairly constant resulting in straight line F-N plots. The change in the slope of the F-N plots upon adsorption of atoms is a measure of the change in work function. However, the β is not altered by adsorption of atoms at low coverage, which is expected to be valid in this case. The change in the β factor and the emitting area A, if any, will be reflected in the pre-exponent term for the complete understanding of the work function variations.

From the assumption of F-N model, it is clear that the theory was developed for only metals and it must be told that the field emission theory for semiconductors cannot be straightforward as metal emitters, because of field penetration into the semiconductor which leads to band bending at the surface [3].

8.1.4. Calculation of Effective Emission Area

The good linearity in the F-N plot suggests that the field emission from the metallic emitter agrees with a metal–vacuum field emission model as described by the F-N equation. In addition, by knowing the work function of the emitter, the slope m of the F–N plot, can be used to estimate the lateral dimensions of the emitter with field enhancement factor $\beta \sim \frac{1}{\alpha r}$, where $\alpha \sim 5$ for the hemispherical geometry of the emitter. The effective emission area A is derived after substituting β and ϕ into the y-intercept b of the F-N plot as

$$b = \ln\left(\frac{1.5\times10^{-6} A\beta^2}{\phi}\right) + \frac{10.4}{\phi^{1/2}} \tag{8.19}$$

Although the reason why only a small spot of the relatively large area of the whole covered tip is activated to emit electrons is still unclear, the small emitting area is a desired feature for high-resolution electron optical applications where point electron sources are needed. The emission current density is estimated by dividing the total emission current to the effective emission area.

8.1.5. Field Emission Microscope (FEM)

Field emission microscope (FEM) allows the observation of a single crystal of material with a large magnification (10^5-10^6) and a high resolution (20-30 Å). It has an advantage of observing various crystallographic planes simultaneously in clean and UHV conditions. Figure 8.2 shows photograph of a typical all glass field emission microscope tube.

Field emission microscope is basically a projection microscope. It consists of a sharp needle of a refractory metal like tungsten or molybdenum of about 1-2 mm length, spot welded onto a heating loop. The free end of the wire assumes a hemispherical shape of about 500-2000 Å radius, after flashing at high temperatures in vacuum. There is almost invariably a single crystal at this end of the needle, with a preferred orientation. Emission takes place at low applied voltages as few kilovolts. The emitted electrons are made to fall on the phosphor coated conducting screen to produce visible pattern representative of structural features of emitting area. The field at the surface of such an emitter is given by an expression

$$F = \frac{V}{kr} \tag{8.20}$$

where r is the tip radius and k is a geometrical factor arising because of the presence of conical shank of the actual tip. It has, in many cases, a value of about 5 near the tip apex. The magnification is given by

$$M = \frac{X}{\gamma . r} \tag{8.21}$$

where C is the tip to screen distance (about 5 -7 cm) and γ is a factor of value about 1.5 reducing the magnification. This arises from the fact that the emitter shank decreases the field and compresses the lines of force towards the axis so that the image is almost uniformly compressed with axial symmetry over the displayed portion of the emitter. Magnification of the order of 10^5 to 10^6 is usually obtainable.

Figure 8.2. Photograph of conventional field emission microscope tube. (Reproduced with permission from [55]).

Resolution of a field emission microscope can be defined as the minimum distance between two objects on the tip surface that will produce separable images on the screen.

The three factors governing the resolution are

1. The divergence of the electron beam.
2. Statistical distribution of the transverse velocities.
3. Diffraction of an electron wave passing through the emitting aperture.

The resolution varies with the square root of the tip radius and is independent of tip to screen distance. Ordinarily the resolution of a field emission microscope ranges from 20 to 30 Å.

Field emission microscope is an ideal tool for surface analysis; the crystal symmetry is reflected in the field emission pattern. The average work function determination is also possible. Field emission has been extensively used in the characterization of the surface structure and electronic properties. The analysis of field emission current fluctuation can yield quantitive information regarding the surface phenomena occurring on the emitter surface.

The linear dependence of *I* on the voltage *V* means that a F-N plot of $\ln(I/V^2)$ versus 1/V gives a straight line, and is a good check on the field emission mechanism. Assuming that the calibration of *F* can be made accurately, the slope gives the work function and the pre-exponential gives the emitting area. Field emission requires a very good vacuum, and often, even in UHV, emission is not due to the clean surface. A typical field emitter needs to be 'flashed' to clean it, usually by passing a current through a loop on which it is mounted. After flashing the emission current is high, but rather unstable; the current decays with time, and becomes more stable as it does so. This is due to contamination of the tip, either from the vacuum, or more often from diffusion of adsorbed surface species to the tip.

8.1.6. Total Energy Distribution (TED)

The supply function N(W,E) dWdE is equal to the number of electrons with their total energy in the range E to E + dE and the component normal to the surface in the range W to W+dW, incident on the surface per unit area, per unit time. The probability for barrier penetration D (W) is identical with that used in the normal energy distribution. The TED is written as a product of a supply function N (W, E) and the barrier penetration D (W) and then integrating over all normal energies consistent with a total energy E. Thus,

$$P(E)=\int_{0}^{E} N(E,W)\,dW \tag{8.22}$$

Therefore the total energy distribution is given by equation,

$$P(E)dE=\frac{4\pi m d}{h^{3}}\left[\exp\left(-c-\frac{\phi}{d}\right)\right]\left[\frac{\exp\left(\frac{E}{d}\right)}{\exp\left(\frac{E-\phi}{K_{B}T}\right)+1}\right]dE \tag{8.23}$$

The component c and d are given by equations (8.5 and 8.6) respectively. The field emitted current is computed by using the equilibrium Fermi-Dirac distribution and integrating over the spherical constant energy surface in the electron approximation. It has been shown that the F-N equation can be obtained via the total energy distribution. A consequence of the free electron model for the metal in the derivation outlined above is that the field emitted current is isotropic and independent of the crystal indices of the emitting surfaces, ignoring the work function differences. However, anomalous humps in the total energy distribution of field emitted electrons in certain crystallographic directions have been observed, contrary to the free electron theoretical curves. These results imply a complicated topology of the energy surfaces and have been explained in various theoretical treatments. Attempts have been made to include band structure effects. Band structure consideration in the calculation of the tunneling probability made particularly for the d electrons in the transitions metals; predict fine structure in the TED, in full agreement with the experimental results.

Thus, the energy analysis of field emitted electrons can be a useful probe in band structure and other electronic properties of solid surfaces. More details regarding TED and FEM are elaborated in an excellent article by Gadzuk and Plummer [14].

8.2. Importance of Field Enhancement Factor in Nanometric Field Emitter

The field enhancement factor determines the size of the emitter. The nanometric dimensions of materials are excellent field emitters due to their favorable aspect ratio and size. The field enhancement of an individual nanowire is approximately proportional to the aspect ratio $\frac{l}{r}$, where l and r are the length and radius of the nanowire. The high current densities and low threshold field are a result of high field enhancement factor.

The enhancement factor can also be estimated from the Filip model [6] for the planar geometry of the emitter using

$$F_{local} = \beta F = \frac{\beta V}{d} \tag{8.24}$$

where F_{local} is the local field at the emitter tip, d is the interelectrode spacing, V is the applied voltage and β quantifies the ability of the emitter to amplify the electric field (F) and is defined as the field enhancement factor. Under this frame, the F-N law is expressed as [13]

$$J = \eta a \left(\frac{\beta^2 F^2}{\phi} \right) \exp \left(\frac{-b\phi^{3/2}}{\beta F} \right) \tag{8.25}$$

where a = 1.54310^{-6} $A.V^{-2}$ eV, b = 6.83×10^9 Vm^{-1} $eV^{-3/2}$, and ϕ is the work function, The factor η describes the geometrical efficiency of electron field emission, i.e., it is equal to the ratio of an actual emitting surface area to an overall surface area. To determine β, it is easy to trace the F-N plot through ln (J/F^2) Vs 1/F, which follows a linear relationship with the slope dependent on ϕ and β. The field enhancement factor β can be thus determined from the slope and the ϕ values.

8.3. Applications of Field Emission Microscope

8.3.1. Field Emission Cathode as a Current Source

Field emission cathode can be an attractive electron source for fine point applications because of its two main features viz. high electron optical brightness and extremely small virtual source size, making it an attractive electron source for fine focus applications in comparison with thermionic cathodes [5].

1. The electron optical brightness 'B' is given by

$$B \, \alpha \, \frac{J}{E_d}$$

   where, J is current density and E_d is energy spread of the emitted electrons. Compared to a thermionic cathode, the field emission cathode has a comparable E_d (0.2-0.3 eV) but up to 10^7 times higher J.
2. The maximum beam current 'I', which can be focused into an extremely small spot, is proportional to the optical brightness of the electron source.

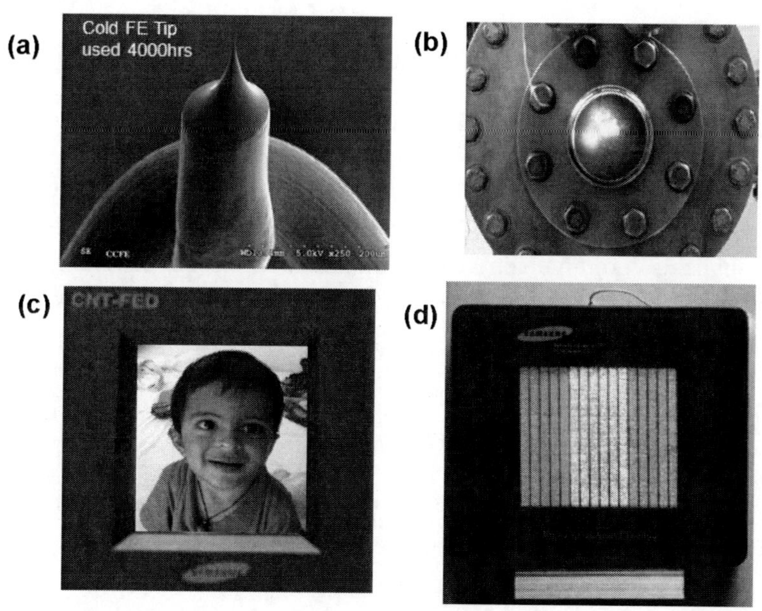

Figure 8.3. (a) Small source size –fine beam small energy spread: 0.2~0.3eV, (b) field emission pattern with high brightness, (c) typical CNT-FED and (d) 30 inch CNT based field emission display demonstrated by Samsung in 1999. (Reproduced with permission from [56]).

For very small focal spots

$$I \alpha B d^{8/3},$$

where, B is electron optical brightness and d is diameter of focal spot.

Properties of point electron sources include:

(i) Low turn on field
(ii) Low beam energy spread
(iii) Small virtual size ($< 10^{-6}$ cm)
(iv) High emission current density
(v) Stable electron emission
(vi) Operable in modest background pressure
(vii) Longer lifetime of the emitter
(viii) High mechanical strength to withstand intense electric field

8.3.2. Field Emission Displays

A field emission display (FED) is a low power, flat cathode ray tube type display that uses a matrix-addressed cold cathode to produce light from a screen coated with phosphor materials. A field emission display typically includes an electron emission tip configured for emitting a flux of electrons upon application of an electric field to the field emission device. An array of miniaturized field emission devices can be arranged on a plate and used for forming a visual display on a display panel. The emission tip is specifically shaped to facilitate effective emission of electrons, and may for example be conical, pyramidal, or ridge-shaped in surface profile, or alternatively the tip may comprise a flat emitter surface of low work function material. In the initial FEDs, the emitters were formed of what are referred to as Spindt-type metal tips (or microtips). A main material of the metal tips is molybdenum. In a field emission display, several hundreds to thousands of micro tips or carbon nanotubes (CNTs) per pixel are provided as an electron emission source on a back plate of FED. Figure 8.3 (a) shows the typical small source size –fine beam small energy spread ~ 0.2 eV, (b) show the higher brightness and resolution. Figure 8.3 (c, d) shows the typical CNT-Field Emitter Display marked by Samsung.

8.3.3. Field Emitter Arrays

Active investigation on field-emitter arrays (FEAs) began in the early 1970s with a view to developing vacuum microelectronic devices including flat panel displays. Historically, microstructured tips have been at the heart of the field emission technology, produced from molybdenum and subsequently other materials. The advantages of electron field emitters compared to thermal and Schottky emitters are a high brightness, small energy spread, good coherence and particularly the possibility of miniaturization. The latter allows for reducing the operating voltage significantly and creating arrays of gated field emitters by micro and nanostructure. Most important applications so far are flat panel displays and electron sources with high current densities for high power microwave tubes. In this area the emission characteristic of an individual emitter is not a critical issue, since a redundancy of 100 or even 1000 emitters per functional unit is high enough to compensate for instability and failure. On the other hand, an electron source providing several independent beams may be applied to fast direct write electron-beam lithography, electron-beam test systems or new types of memory devices. For this purpose, every single emitter has to meet much more severe demands on emission stability, lifetime and uniformity. Thus, examination of many emitters in parallel but independent from each other would be very helpful for research and development of miniaturized field electron emitters. Moreover, an emitter tip with a relatively large and smooth surface could behave more stable due to a large emission area. Therefore, tips that are blunt compared to sharp silicon emitters or carbon nanotubes may be good candidates for individual stable field emitters. Therefore, this kind of addressable field emitter array (FEA) is a very useful tool for experimentally designing field emitters with improved emission characteristics. The field emission tip itself is fabricated by electron beam induced deposition and subsequent physical vapor deposition of a thin metal layer. Thus, it is possible to vary the geometrical setup of tip and extraction electrode as well as the tip shape easily and also to define exactly material properties of the emitter, i.e., work function and sticking coefficient.

With the growth of high demand market for relatively large area domestic high-definition television and space saving desktop monitors, the requirement for flat panel displays (FPD) with similar image specification to the cathode ray tube (CRT) has become urgent. With regard to the physical construction, among all types of FPDs, the Field Emission Display (FED) shares many similarities with the CRT. Instead of using three hot electron guns (as in the case of a colour CRT) that reside some distance behind the faceplate to scan the electrons across the display screen to produce an image, FEDs feature millions of micro-field emission electron guns placed directly behind the display screen, which are used to fluoresce the display. This allows FEDs to be scaled down to few millimeters in depth [7]. There are still some technological difficulties, which need to be overcome before putting the FED techniques into mass production, although a small volume of mono-colour FED products have been available in the market for the last few decades.

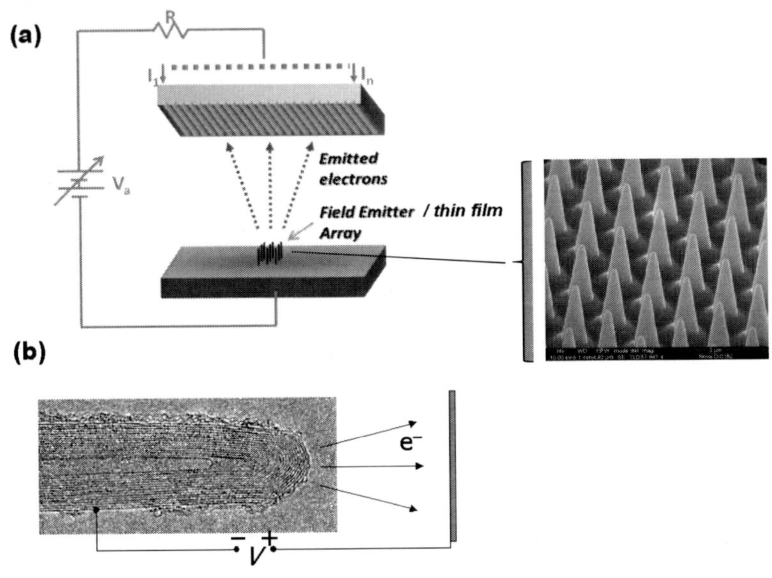

Figure 8.4. (a) Typical Close- proximity/planer diode configuration for field emission [57a]). (b) A potential difference is applied between a nanotube (or an assembly of nanotubes) or arrays of emitter and a counter electrode. (Reproduced with permission from [57b, c]).

A FED structure usually comprises of a cathode back plane, a narrow vacuum gap in the middle, and an anode front phosphor screen on which the image is formed [8, 9]. Among these, the cathode back plane is the most complicated structure on which micro-field emission cathodes reside. One of the successfully developed FED structures is the Spindt type with molybdenum micro-cold field emitters [7, 9-11]. Field emission studies show that electron emission current from this micro-field emitter is unstable and may trigger irreversible catastrophic breakdown of the device [12, 13]. To overcome this problem, a ballast resistive layer was introduced to be incorporated at the base of the micro-cone shaped tips [15–18]. The ballast resistive layer can effectively reduce the potential difference between the cathode tip and the gate electrode as more electron current flows out of the emitter. Meyer of LETI demonstrated the world's first FED and showed stable field emission from Spindt type field emitters by utilizing a resistive layer [19]. This stimulated the development of field emitters as electron sources for FEDs by many researchers in the 1990s.

Currently, FEDs are fabricated mostly on soda-lime glass substrates, which is compatible to the technology of Liquid Crystal Display (LCD). Soda-lime glass has demonstrated some advantages over other candidates. For instance, it is vacuum compatible, is inexpensive, it can be made very flat over a relatively large area and can be coated with a variety of materials using very large scale integrated (VLSI) thin film technology [20]. However, the fabrication of a Spindt cathode, using micro-fabrication techniques, has difficulties to produce large area field emitters with acceptable uniformity [8]. So far, FEDs with Spindt type field emitters have only been fabricated in small and medium size structures. Figure 8.4 shows the (a) typical Close-proximity/planer diode configuration for field emission. (b) A potential difference is applied between a nanotube (or an assembly of nanotubes) or arrays of emitter and a counter electrode.

8.4. Field Emission from Nanomaterials

Field emission cathodes development research has a long history, which starts with W. P. Dyke's field emission work in the 1940s. He devoted his research life to develop field emission cathodes to achieve controlled, stable, electron emission, at high intensity levels, which have the applications in a variety of devices. He and his co-workers at Linfield College studied tungsten cathodes extensively. They could find some applications to microwaves, fine focus electron optical devices and flash X-ray systems. But some big challenges like field emission current stability and life of the field emission cathodes at industrial vacuum conditions remained unanswered. However, Dyke made comb-structure field emitter arrays and established a solid foundation for field emission cathode development research while working at Linfield College, USA. A good account of an earlier research done by Dyke and his co-workers can be found in a review article by Charbonnier [5].

To have adequate conductance for microwave devices, compact multitip arrays were needed instead of single tip emitter. In 1958, Shoulders [21] introduced the concept of field emitter arrays. After ten years, in 1968, using thin film technology and molybdenum tips, Spindt [10] reported the fabrication and operation of gated field emitter arrays. In 1972, Thomas and his co-workers [22] demonstrated silicon based ungated microfabricated field emitter arrays. Gray [23-26] introduced the concept of an integrated micro fabricated vacuum transistor based on gated silicon, FEAs in 1980s and afterwards put the concept into practice. Till the end of last decade, there were several attempts reported by researchers to develop different kinds of field emission cathodes by cathode surface coating as well as without coating, which yield stable current at high intensity level and has long life in the industrial vacuum conditions for making practical devices [27]. Dyke with his co-workers performed extensive research on tungsten field emitter cathodes [5]. He came to the conclusion that the long-term field emission current stability is possible from tungsten field emitter at the residual pressure well below 10^{-10} mbar during field emission cathode operation, which is not feasible to maintain in practical devices. Therefore, it is necessary to heat the tungsten field emitter periodically, e.g., every hour at 10^{-9} mbar, to desorbs contaminants and to repair sputtering damage. Also, he concluded that a temperature of 1000K is sufficient to repair the sputtering damage and at least 1800K is required to desorb all contaminants on tungsten surface. In this way, his conclusions about tungsten field emission cathodes were not very good to use them in practical devices [5]. The story of field emission from LaB_6 coated field emitters is not very old. The reports tell us that the low onset voltage, very high current density and long-term current stability are obtained from coated LaB_6 field emitters [27].

Field emitter arrays (FEAs) have been emerging as micron size electron source for a number of advanced and new generation devices such as flat panel displays, ultrahigh-frequency devices, and so on [11, 28]. However, there are limitations of micro-fabricated FEAs in their reliability and efficiency [29]. Highly efficient and reliable FEAs can be realized first, by emitter tip sharpening, second, by reducing emitter-to-gate distance precisely, third, by using lower work function emitter materials, finally, and by improving uniformity and reproducibility. The above four necessary conditions give the low operation voltage preventing deterioration of FEAs caused by bombardment of ionized residual gas and breakdown effects as well as providing highly efficient field emission. However, though emitter materials exert a great influence on field emission characteristics because of its direct contribution to emission efficiency according to the F-N theory [2], they have not been investigated to a great extent. One of the reasons is that comparison and evaluation of emitter materials in the usage of FEAs is markedly difficult due to non-uniformity and non-reproducibility of emitter shapes. For example, popular technique of coating low work function emitter materials on the conventional emitters such as Mo and Si emitters causes reduction of emitter tip radii and non-uniformity of emitter shapes and has difficulty in emitter material evaluation. Nakamota and Fukuda et al. used the transfer mold technique to fabricate LaB_6 field emitter arrays [30-33] to solve these problems and realize sharp, uniform and low operation voltage FEAs using LaB_6. Transition metal borides and transition metal nitrides, for example, LaB_6 and titanium nitride (TiN) are well known for their low work function of 2.6 and 2.9 eV, respectively, compared to 4.3 – 4.5 eV of conventional emitter material [34-44]. Nakamota and Fukuda et al. [33] reported that extremely sharp, uniform, and low operation voltage FEAs using LaB_6 prepared by transfer mold technique by combining anisotropic Si wet etching, physical vapor deposition of emitter materials, and electrostatic bonding is possible. Their FEA structures and electrical characteristics are compared with Mo FEAs. Because of their emitter tip sharpness and low work function, they exhibit high performance of low operation voltage at the low vacuum level of 4×10^{-7} mbar were shown in Table 8.1.

Table 8.1. Comparative performance of thermionic and field emission energy spread

Emitter	Thermionic emitter Energy Spread	Field emitter Energy Spread
Tungsten	3 eV	0.3 eV
LaB_6	1.5 eV	0.7 eV
Carbon fiber/CNT	-	≈ 0.3 eV *

Recently, nanostructures of various materials have also been reported to show low turn-on voltage, high current density, high field enhancement factor and stable electron emission. These field-emission phenomena are believed to originate from their nanometric dimensions of the emitters such as CNTs [45-46], ZnO [47], SnO_2 [48], GaN[49], MoO_3 [50], MoS_2 nanoflower [51], Mo nanowires [52] and Graphite nanofiber [53]. However, these materials have a large work function, including carbon nanotubes, and require stringent working conditions for their utilization. The materials with good electrical, optical and mechanical properties found to be best emitter due to the fact that a suitable contribution of mechanical and electronic properties makes the material a good field emitter. The atomically sharp edges create intense electric fields, which leads to excellent field emission properties. Further, the metal emitters are known for their inherent potential for delivering high current density, whereas semiconducting emitters are more suitable for device applications.

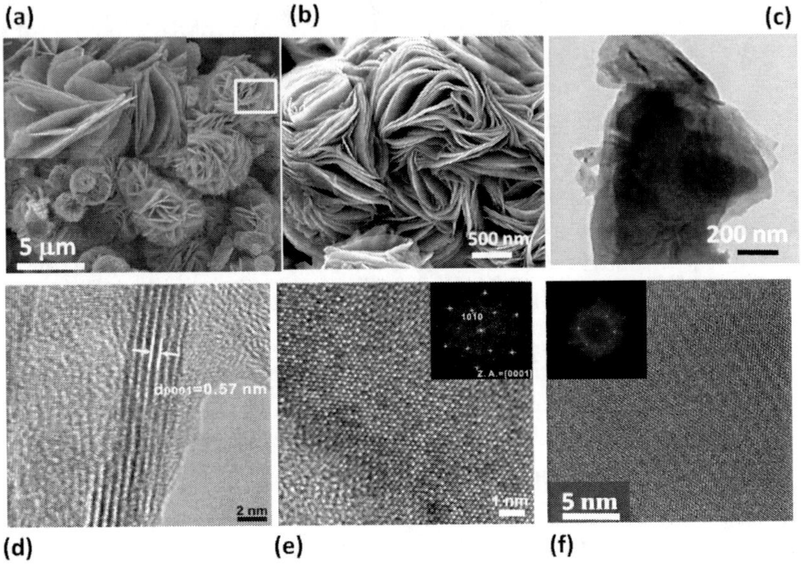

Figure 8.5. (a, b) FESEM images of single- and few-layre VS_2 nanosheets. TEM analysis of VS_2 sheets (c) low magnfication image, (d-f) high resutlon TEM image and inset of (e) selected area diffraction pattern and inset of (f) fast Fourier transform of the electron diffraction pattern of a few layers of VS_2. (Reproduced with permission from [60] *Eur. J. Inorg. Chem.* 2014, 5331–5336).

8.5. Importance of 2D Metallic Materials for Field Emission Based Devices

It's well known that one dimensional CNT's are best field emitters owing to high aspect ratio. In recent times revealed 2D graphene possesses better electrical and mechanical properties in comparison with the CNT's. Hence, it is being widely explored for field emission properties. One of the important causes of exploring the two-dimensional materials is its planar structure which could be useful for planar device technology. For example, the layered MoS_2

[58] is an inorganic analogue of graphene. It possesses a direct band gap of 1.8 eV when thinned to monolayer and has promising electronic properties. It also possesses sharp/nanometeric edges like graphene which can enhance the field emission behavior. The nonlinearity is a direct consequence of semiconducting nature of a field emitter due to the presence of surface states and inherently limited number of charge carriers in the conduction band. Hence, in the semiconducting emitter such as MoS_2 the observed nonlinearity is attributable to the semiconducting nature of emitter. The nonlinearity of F-N plot for semiconductors does indicate involvement of surface states and band-bending effects [58]. Modifications to the original Fowler-Nordheim theory for different semiconducting materials have been worked out. However, these are difficult to apply to the practical field emitter because of insufficient knowledge of the surface states density, etc. [59]. Graphene like metallic layered materials have emerged as perspective field emitter materials for the next generation devices due to useful and enhanced properties. Also, the observed turns on values are very well comparable with the existing and known field emitters.

Further, Chandra Sekhar Rout et al. [60] reported for the first time the enhanced and superior field emission performance of metallic few-layered VS_2 nanosheets synthesized using simple and cost-effective hydrothermal method. Figure 8.5 (a, b)shows the typical FESEM images of single- and few-layre VS_2 nanosheets. The TEM analysis of VS_2 sheets, Figure 8.5 (c) shows the low magnfication image, (d-f) high resutlon TEM image and inset of (e) selected area electron diffraction pattern and inset of (f) show the typical fast Fourier transform of the electron diffraction pattern of a few layers of VS_2. The turn on and threshold field required to draw an emission current density of 1 μA/cm^2 and 10 μA/cm^2 is found to be 4 and 5.01 V/μm respectively. Field emission performance from few-layered VS_2 nanosheets, Figure 8.6 (a) shows the typical plot of applied electrical field as a function of emission current density. Figure 8.6 (b) show the F-N plot showing linear behaviour indicating emission current from the metalic emitter. Figure 1.6 (c) shows the typical long term field emission current stabilty measurement analysis indiacting the good emission current stability of the VS_2 emitter. Figure 8.6 (d) shows the field emission pattern taken during the long term current stability experimental measurement of the emitter. The enhanced performance in the field emission such as low turn on and threshold field values are attributed to the high field enhancement factor (~2484) which further attributed to the fact that the presence of sharp protrusions/edges of VS_2 nanosheets. The field emission current stability of the VS_2 emitter was tested over a longer duration and emitter current observed to be stable over a period of 12h. The results are also supported by of first-principles density functional calculations carried out for VS_2 monolayer systems and which clearly show that the work function of the 2D metallic VS_2 is of the order of 6.01 eV. Figure 8.7 shows (a) the unit cell orientations of VS_2 and (b, c) shows the out-of-plane electrostatic potential and work function for VS_2.

Sachin R. Suryawanshi et al. [61] reported the photosensitive field emission and low frequency noise analysis of few-layer $PbBi_2Se_4$ nanosheets emitter synthesized using hydrothermal method at the base pressure of ~1x10^8 mbar. Figure 8.8 (a) shows the Crystal structure of $PbBi_2Se_4$ showing the layered intergrowth structure. Each single layer constitutes seven covalently bonded atomic planes. One unit cell (c ~3.92 nm) of $PbBi_2Se_4$ is made of 3 single layers. Figure 8.8 (b) shows the typical Powder XRD pattern of as synthesized $PbBi_2Se_4$ nanosheets. Figure 8.9 (a, b) depicts the typical AFM images of $PbBi_2Se_4$ nanosheets. Figure 8.9 (c, d) shows the typical FESEM images of the few layers of $PbBi_2Se_4$ nanosheets. Figure 8.10 (a) shows the field emission current density (J) versus applied electric field (E) plots before and after visible photon illumination for $PbBi_2Se_4$ nanosheets emitter. The J-E characteristic is further analyzed by plotting the graph of ln (J/E^2) versus (1/E), known as a (F-N) plot in Figure 8.10 (b). The turn on field required to draw an emission current density of ~1 $\mu A/cm^2$ is found to be ~3.90 and 4.05 V/μm respectively with and without visible photon illumination respectively. The decrease in turn on field by illumination of photon light is credited to the fact that photoemission of electron from $PbBi_2Se_4$ nanosheets emitter. Furthermore, the visible photo illumination on $PbBi_2Se_4$ nanosheets emitter shows almost two times improvement in the field emission current. Owing to the low turn on field and planar (sheet like) morphology, along with photosensitive nature of the $PbBi_2Se_4$ nanosheets emitter has great potential for its utilization for the next invention of photo-switch based device application as well. In addition, the power spectral density clearly depicts $1/f^{\alpha}$ type noise with power of α ~0.99 for few layers $PbBi_2Se_4$ nanosheets emitter. The observed current fluctuations in the field emission current attributed to the fact that during emission current measurements the adsorption/desorption of the residual gas atoms/molecules and ion bombardment on emitter surface takes place which finally leads to the temporal change in the 'local' work function at the field emission sites.

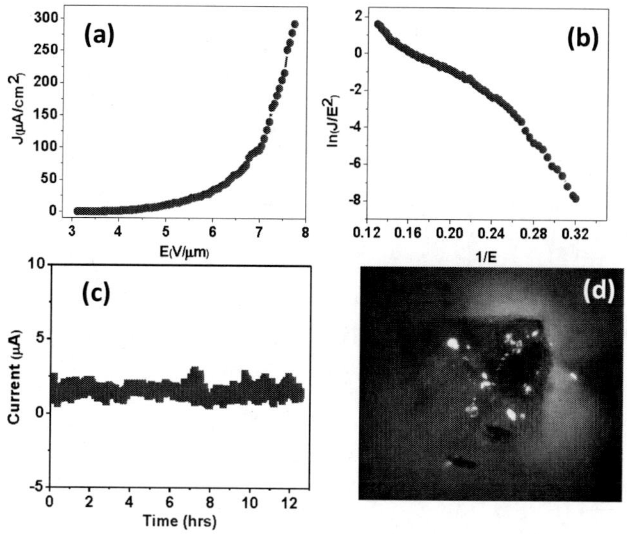

Figure 8.6. Field emission from few-layered VS_2 nanosheets. (a) Applied electrical field as a function of emission current density. (b) F-N plot showing linear behaviour indicating emission current from the metalic emitter. (c) Long term field emission current stabilty indiacting good emission current stability. (d) Field emission pattern taken during the long term stability study of the emitter. (Reproduced with permission from [60], *Eur. J. Inorg. Chem.* 2014, 5331–5336).

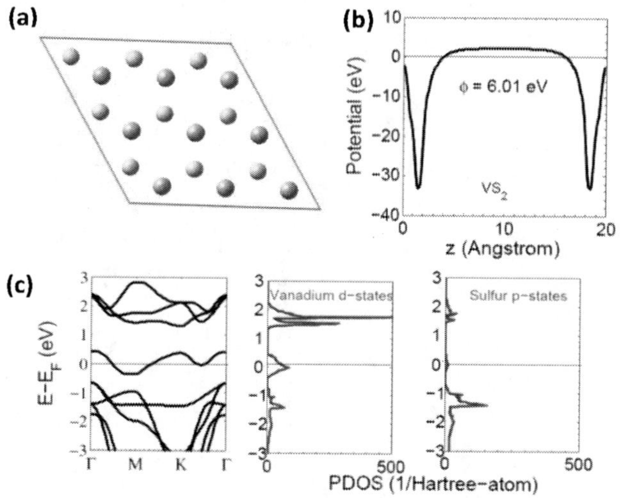

Figure 8.7. (a) The unit cell orientations of VS_2. (b, c) The out-of-plane electrostatic potential and work function for VS_2..(Reproduced with permission from [60], *Eur. J. Inorg. Chem.* 2014, 5331–5336).

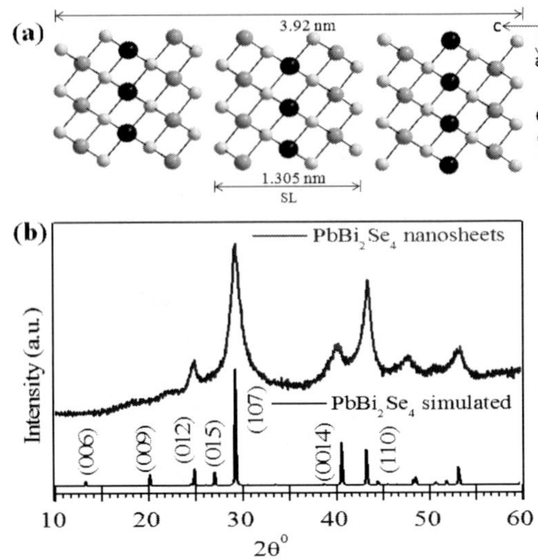

Figure 8.8. (a) Crystal structure of $PbBi_2Se_4$ showing the layered intergrowth structure. Each SL constitutes seven covalently bonded atomic planes. One unit cell (c ~3.92 nm) of $PbBi_2Se_4$ is made of 3 SLs. (b) Typical Powder XRD pattern of as synthesized $PbBi_2Se_4$ nanosheets (Reproduced with permission from [61], *J. Mater. Chem.* C, 2016, 4, 1096-1103).

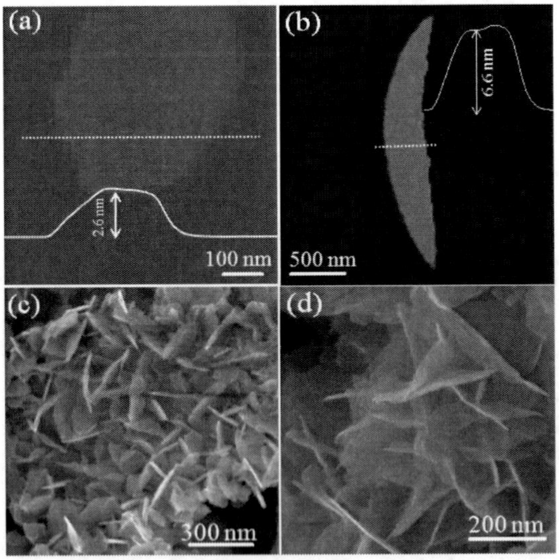

Figure 8.9. (a, b) depicts the AFM images $PbBi_2Se_4$ sheets. (c, d) shows FESEM images of few layers of $PbBi_2Se_4$ nanosheets. (Reproduced with permission from [61], *J. Mater. Chem.* C, 2016, 4, 1096--1103).

Satyajit Ratha et al. [62] reported the field emission performance of pristine VSe_2 nanosheets and its hybrid structure with reduced graphene oxide (RGO) in detail. The Synthesis VSe_2 has been carried out by hydrothermal method with 40.6 mg of sodium metavanadate ($NaVO_3$,) and 74 mg of selenium dioxide (SeO_2) were mixed together and added to 5 ml of DI water under constant stirring. Further, it was transferred to a Teflon-lined stainless-steel autoclave inside a hot air oven at 200°C for 24 h. The VSe_2/RGO hybrid was prepared using the exactly similar method only difference that the vanadium and selenium precursors were put directly into a 0.15 wt% uniform aqueous dispersion of GO [63]. The reduction of GO to RGO occurred during the process. Figure 8.11 shows the typical FESEM images showing (a) low and (b) high magnification images of VSe_2, (c) and (d) respectively show low and high magnified images of the VSe_2/RGO hybrid. Further, the Promising field emission performance were noted with lower values of turn-on and threshold fields of ~1.3 V μm^{-1} and 2.12 V μm^{-1} respectively for the VSe_2/RGO hybrid which is much more superior than that of pristine VSe_2 emitter and other 2D materials. Field emission results obtained from VSe_2 were shown in figure 8.12 (a) the J-E response plot, (b) F-N plot and (c) emission stability test plot. Figure 8.13 shows the field emission results obtained from VSe_2/RGO hybrid showing (a) typical J-E plot, (b) F-N plot and (c) emission current stability test plot. The density functional theory (DFT) calculations were carried out to understand the enhanced field emission performance of VSe_2/RGO hybrid field emitter. It was interesting to note that the calculations qualitatively support experimental observations through exhibition of Density of States and calculation of work function for pristine VSe_2 and hybrid VSe_2/RGO. Figure 8.14 shows the DFT optimized structures of (a) (001) VSe_2 surface (b) graphene with (001) VSe_2 surface; the golden, purple and blue spheres indicate V, Se and C atoms respectively. The lowering of work function 4.79 eV and appearance of additional states near Fermi level from carbon *p* orbital of RGO is responsible for superior field emission characteristics in hybrid VSe_2/RGO compared to pristine VSe_2 5.33 eV. The experimental data and theoretical insight from electronic structure simulations strongly infer that the hybrid VSe_2/RGO has great potential to be tailored as efficient and high-performance cathode material in field emission-based devices. It is interesting to note the fact that, as VSe_2 and graphene (RGO) have different lattice parameters, there will be lattice mismatch when a hybrid of VSe_2/RGO is formed. Due to this lattice mismatch, diverse types of defects may be introduced, especially in the boundary region. A few types of defects (not all) may reduce the work-function, e.g., the graphene with Stone–

Thrower–Wales defects have smaller work functions than that of perfect one [64]. The enhancement in Field emission in RGO-VSe_2 hybrid structure field emitter due to the fact of lowering in the work function of hybrid materials, which further boost the increase in electrical conductivity of the material and appearance of additional states near Fermi level from p orbital of RGO. Table 8.2 show the comparative field emission performance of various 2D metallic emitters with other well-known materials in the literature.

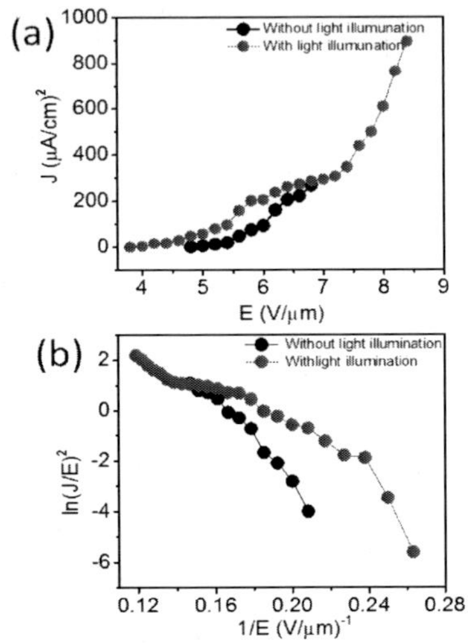

Figure 8.10. (**a**) The field emission current density (J) versus applied electric field (E) plots before and after visible photon illumination for $PbBi_2Se_4$ nanosheets emitter. The J-E characteristic is further analyzed by plotting the graph of ln (J/E^2) versus (1/E), known as a (F-N) plot in (**b**). (Reproduced with permission from [61], *J. Mater. Chem.* C, 2016, 4, 1096--1103).

Mohit Kumar et al. reported [73] the Wafer Scale growth of ReS_2 with "Tunable" Geometry for Field Emission Application. The large area ReS_2 nanostructures were grown using a radio–frequency sputter deposition technique over wafer-scale. Notwithstanding high potential, the guarantee of 2D materials has not yet been in realized practical device application due to the fact that it's limitation of miniature size and in tricky for manipulation of dynamic spot. The exploitation of 2D layers is the crucial approach, which can be approached by large-scale fabrication. Further, the controllability of ReS_2 structure has been carried out to form representative thin films or vertically associated layers which were further utilized for the field emission application.

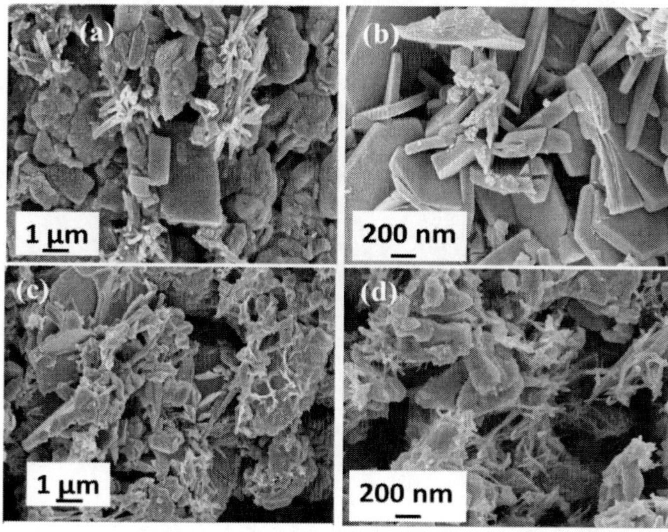

Figure 8.11. FESEM data showing (a) low and (b) high magnification images of VSe_2, (c) and (d) respectively show low and high magnified images of the VSe_2/RGO hybrid. (Reproduced with permission from [62]).

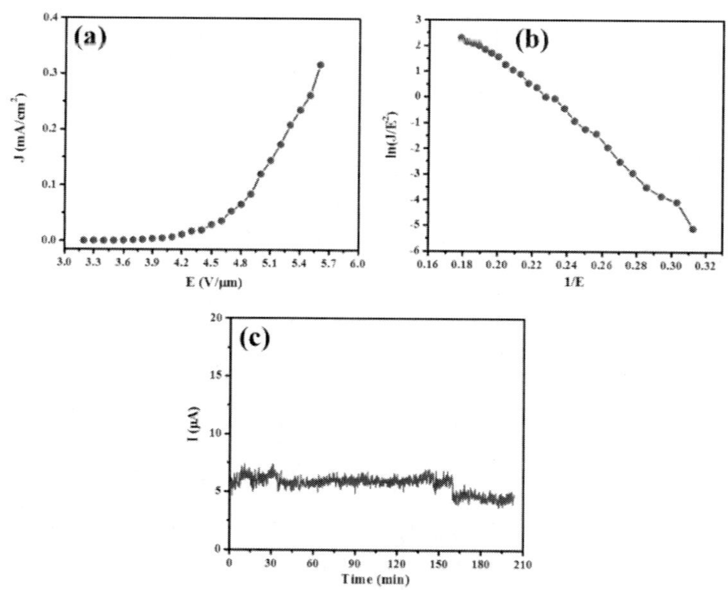

Figure 8.12. Field emission results obtained from VSe_2 showing (a) the J-E response plot, (b) F-N plot and (c) emission stability test plot. (Reproduced with permission from [62]).

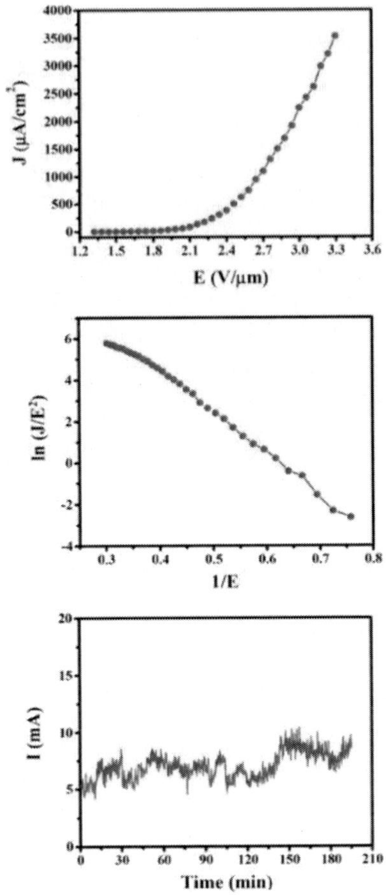

Figure 8.13. Field emission results obtained from VSe_2/RGO hybrid showing (a) J-E response plot, (b) F-N plot and (c) emission stability test plot. (Reproduced with permission from [62]).

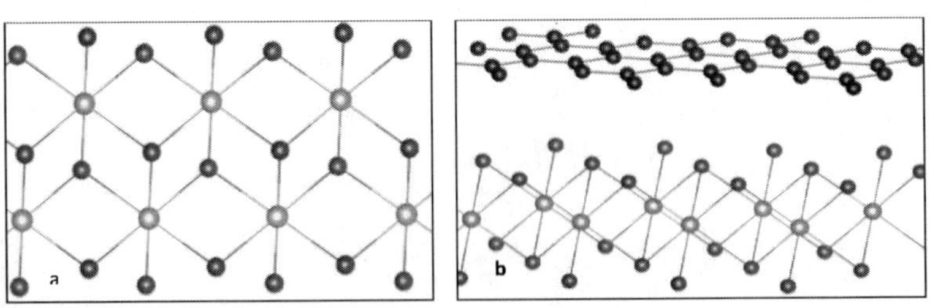

Figure 8.14. DFT optimized structures of (a) (001) VSe_2 surface (b) graphene with (001) VSe_2 surface; the golden, purple and blue spheres indicate V, Se and C atoms respectively. (Reproduced with permission from [62]).

Table 8.2. Comparative field emission performance of various 2D metallic emitters with other well known materials in the literature (reproduced with permission from [62])

Field emitter	Turn on field 1 μA/cm²	Threshold field 10 μA/cm²	Enhancement Factor (β)	Reference
SnS_2-rGO	2.65	-	3700	[65]
RGO-Ag nanowire	2.40	-	1985	[66]
Graphene-SnO_2	3.86	8.6	1787	[67]
WS_2-RGO	2	2.5	2978	[59]
VS_2	4	5.01	2500	[60]
WS_2/HμSi	1.1	1.5	3580	[68]
MoS_2	-	3.05	1138	[58]
FLG/μSi	-	2.3	-	[69]
rGO:HμSi	1.3	-	4819	[70]
rGO/P3HT	0.6		8900	[71]
rGO:P3HT	0.9	-	1900	[72]
VSe_2	3.7	4.2	1428	[62]
VSe_2-RGO	1.5	1.8	4097	[62]

Figure 8.15 depicts the growth and characterization of 2D ReS_2 material with various growth paramcters. Figure 8.15 (a) shows the typical optical photograph of ReS_2 directly deposited on entire Si wafer using radio–frequency sputter deposition technique. Figure 8.15 (b-d) shows the typical SEM images of the ReS_2 samples grown at T_{RT}, T_{200} and T_{400}, t_d =30 min respectively. Figure 8.15 (e) shows the typical TEM image of the vertically grown ReS_2 layers and the inset shows the 5 stacked layers with spacing ~ 0.61 nm. Figure 8.15 (f) shows the HAADFSTEM image with corresponding EDS maps of element (g) Re and (h) S. Figure 8.15 (i-k) shows the tilted-view SEM images of ReS_2 grown for 2, 10 and 30 minutes (Ts = 400°C) respectively which clearly depicting the layer growth evolution. Figure 8.15 (l) show the parametric phase diagram of ReS_2 growth at a fixed RF deposition power, the arrows given indicate that a transition from thin film to vertical layers is achievable *in situ* and it can be applied to the nanoscale engineering by varying the various parameter including substrate temperature. Figure 8.15 (m) shows the Raman spectra for Ts = RT, 200 and 400°C, td = 30 min grown ReS_2. It is important to note that the 2D ReS_2 shows more intricate Raman spectrum than the conventional other 2D transition metal dichalcogenides due to the reduced crystal symmetry. The typical Raman of ReS_2 show two prominent mode first order in-plane (E_g) at 165 cm^{-1} and out-of-plane (A_{1g}-

like) ~ 229 cm^{-1} respectively. It is noted that with varying *T*s these two peak positions do not change much due to weak interlayer coupling between nearby layers and is consistent with earlier reports in the literature [74, 75]. The chemical composition and bonding atmosphere of the elements for as–grown ReS_2 samples were investigated by XPS. The Re 4*f* and S 2*p* core level spectra corresponding to ReS_2 were depicted in Figure 8.15 (n) and (o) respectively.

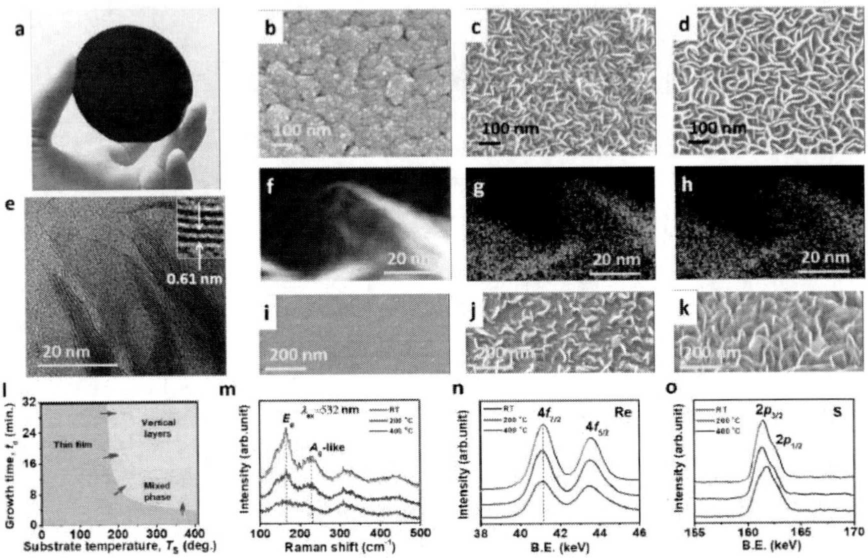

Figure 8.15. Growth and characterization of 2D ReS_2. (a) Optical photograph of ReS_2 deposited on Si wafer. (b-d) SEM images for T_{RT}, T_{200} and T_{400}, t_d =30 min ReS_2 samples. (e) TEM image of vertically aligned ReS_2 layers, inset shows the 5 stacked layers with 0.61 nm width. (f) HAADFSTEM image with corresponding EDS maps of (g) Re and (h) S. (i-k) Tilted-view SEM images of ReS_2 grown for 2, 10 and 30 minutes (Ts = 400°C) respectively depicting the layer growth evolution. (l) Parametric phase diagram of ReS_2 growth at a fixed RF deposition power, arrows indicate that a transition from thin film to vertical layers is possible *in situ* and it can be applied to nanoscale engineering by varying the substrate temperature. (m) Raman spectra for Ts = RT, 200 and 400°C, td = 30 min grown ReS_2. (n) and (o) XPS spectra of Re 4f and S2p respectively. (Reproduced with permission from [73]).

Figure 8.16 shows the comparative field emission performance of different ReS_2 samples. Figure 8.16 (a) shows the typical band alignment diagram and also showing the presence of potential barrier between cathode and anode for field electron emission. Figure 8.16 (b) shows the schematic diagram for field electron emission from ReS_2 sample. Figure 8.16 (c) shows the typical plot of emission current density Vs applied electric field (*J-E* for the different ReS_2 samples. The vertically aligned ReS_2 layers exhibit ultralow turn-on field (0.6 V μm^{-1}) along with the high emission current density (0.6 mA cm^{-2}) and considerably lower threshold electric field (0.8 V μm−1) along with exceptional emission current stability. Figure 8.16 (d) shows the corresponding Fowler–Nordheim plots for the different ReS_2 samples.

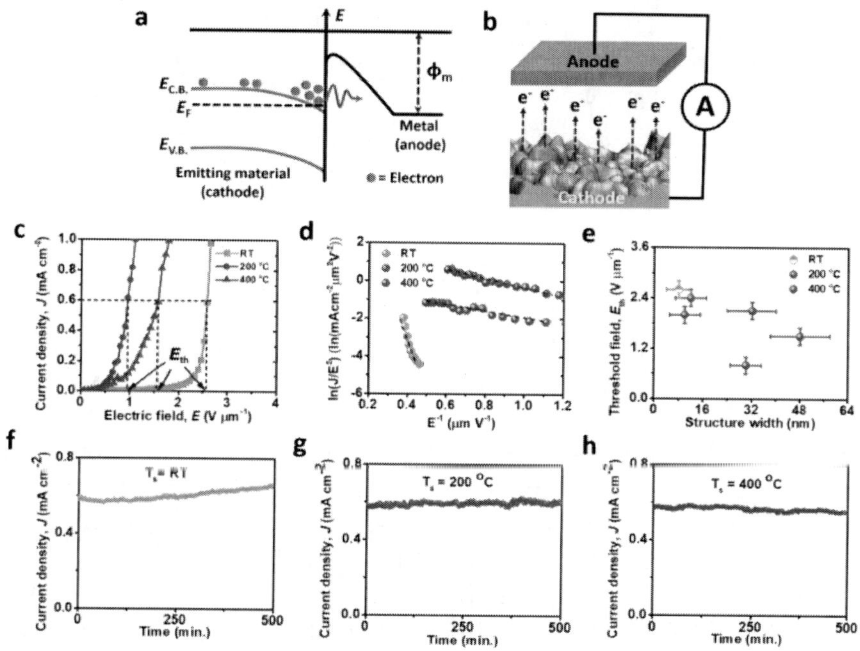

Figure 8.16. Field emission performance of ReS_2. (a) Band alignment and presence of potential barrier between cathode and anode for field electron emission, (b) Schematic diagram for FE measurement setup. (c) Plot of *J-E* characteristics for the different ReS_2 samples: T_{RT} (yellow square); T_{200} (red spheres); T_{400} (green triangles). (d) The Fowler–Nordheim plots. The dotted lines represent linear fit to the F-N equation. (e) Variation of E_{th} as a function of structural width of nanostructures (e.g., grain size and vertical layer width) grown at different *T*s and *t*d. (f-h) Field emission current stability over 500 min at a fixed E_{th} for T_{RT}, T_{200} and T_{400} samples respectively. (Reproduced with permission from [73]).

The dotted lines in the plot are an indicative of linear fit to the F-N equation. Figure 8.16 (e) shows the typical observation of Variation of threshold field E_{th} as a function of structural width of nanostructures (e.g., grain size and vertical layer width) grown at different *T*s and *t*d. Figure 8.16 (f-h) shows the emission current stability over 500 min at a fixed E_{th} for various 2D ReS_2 samples. The field emission results are attributed to the weakly attached ReS_2 layers and high geometrical field enhancement factor (~1.08×10^5). Further, the Kelvin probe force microscopy measurements were carried out to confirm the decrease in the work function of the $ReS_2$is not exclusively responsible for achieving the low operating threshold field. Furthermore, predetermined element simulation clearly suggests that not only the length, width, and partition of the nanostructures but also the local slope plays a crucial role to suppress the screening effects. The obtained results clearly suggest the materials can be engineered for potential application in field emission-based devices.

Conclusion

In conclusion, field emission performance of various two-dimensional transition metal dichalcogenides of the type MX_2, where M is a transition metal atom (such as Mo or W) and X is a chalcogen atom (such as S, Se or Te), has been analyzed. The results provide a promising alternative for field emission-based devices because of their robustness, high emission current density, low turn on and threshold field and longer emission stability. The turn-on and threshold voltages observed greatly suggest that the large area uniform atomically thin films of transition metal dichalcogenides are better suited for large-area emitter applications such as flat panel display devices whereas vertically grown films or wall like morphology of these metallic transition metal dichalcogenides would be ideal candidates for high current density electron beam applications.

References

[1] R. O. Jenkins, "Field emission of electrons", *Report. Prog. Phys.*, Vol 9, page 77, 1942.

[2] T. E. Starn, B. S. Gossling, R. H. Fowler, Further Studies in the Emission of Electrons from Cold Metals, *Proc. Roy. Soc*, (London), Vol 124, No. 795 699-723, 1929.

[3] W. P. Dyke and W. W. Dolan, Field emission, *Advan. Electron. Electron Phys.*, 8 (1956) 89-185.

[4] L. W. Swanson and A. E. Bell, *Advan. Electron. Electron Phys.*, 32 (1973)194.

[5] F. Charbonnier, Developing and using the field emitter as a high intensity electron source, *Appl. Surf. Sci.*, Vol 94/95, page, 26-43, 1996.

[6] V. Filip, D. Nicolaescu, M. Tanemura and F. Okuyama, Modeling the electron field emission from carbon nanotube films, *Ultramicroscopy*, Vol 89, page 39-49, 2001.

[7] S. Itoh, M. Tanaka, T. Tonegawa, Development of field emission displays, *J. Vac. Sci. Technol. B*, Vol 22 (3) (2004) 1362.

[8] I. Shah, *Phys. World Proceeding*, page 45., 1997

[9] B.R. Chalamala, Y. Wei, B.E. Gnade, FED up with fat tubes, *IEEE Spectr.*, Vol 35 (No. 4), Page 42-51, 1998.

[10] C.A. Spindt, A ThinFilm Field Emission Cathode, *J. Appl. Phys.*, Vol 39 No. (7), 3504, 1968.

[11] C.A. Spindt, I. Brodie, L. Humphrey, E.R. Westerberg, Physical properties of thin-film field emission cathodes with molybdenum cones, *J. Appl. Phys.*, Vol 47 (12) 5248, 1976.

[12] I. Brodie, C.A. Spindt, Vacuum Microelectronics, *Adv. Electronics Electron Phys.*, Vol 83 1-106, 1992.

[13] P.R. Schwoebel, I. Brodie, Surface-science aspects of vacuum microelectronics *J. Vac. Sci. Technol. B*, Vol 13 (4) 1391, 1995.

[14] J. W. Gadzuk and E.W. Plummer, Field Emission Energy Distribution (FEED) *Rev of Modern Phys*, Vol 45, 1, 1973.

[15] C. Py, R. Baptist, Stability of the emission of a microtip, *J. Vac. Sci. Technol. B*, Vol 12 (2) 685, 1994.

[16] A. Ghis, R. Mayer, P. Ramboud, F. Levy, T. Leroux, Sealed vacuum devices: fluorescent microtip displays, *IEEE Trans. Electron Dev.*, 38 (10), 2320-2322, 1991.

[17] S. Itoh, T. Niiyama, M. Taniguchi, T. Watanabe, A new structure of field emitter arrays, *J. Vac. Sci. Technol. B* Vol 14 (3), 1996, 1977.

[18] R. Baptist, F. Bachelet, C. Constancias, Microtips and resistive sheet: A theoretical description of the emissive properties of this system, *J. Vac. Sci. Technol. B*, Vol 15 (2), 385, 1997.

[19] R. Meyer, A. Ghis, P. Rambaud, F. Muller, Technic. *Digest Jpn. Display*, Vol 86, 513, 1986.

[20] H. Ohsaki, Y. Kokubu, Global market and technology trends on coated glass for architectural, automotive and display applications, *Thin Solid Films*, Vol 351, Issue 1–2, page 1-7, 1999.

[21] K.R. Shoulders, Microelectronics Using Electron-Beam-Activated Machining Techniques, *Adv. Computers*, Vol, 2, page 135-293, 1961.

[22] R. N. Thomas and H. C. Nathanson, Photosensitive field emission from silicon point arrays, *Appl. Phys. Lett.*, Vol 21, 384, 1972.

[23] H. F. Gray, *Proc. Colloquium on Advances in Electron Emission*, page 24, 1982.

[24] H. F. Gray, *Proc. The 29th Int. Field Emission Symposium*, page 111, 1982.

[25] H. F. Gray, G. J. Campasi, R. F. Greene, *Technical Digest of the Int. Electron devices Meeting IEDM*, Vol, 86, 776, 1986.

[26] H. F. Gray, G. J. Campasi, *Mat. Res. Soc. Symp. Proc.*, Vol 76, 25, 1987.

[27] D. S. Joag, PhD Thesis, Pune University, India Dec. (1978).

[28] P. Vaudaine, R. Meyer, *Technical Digest of International Electron Devices Meeting*, page 197, 1991.

[29] H. H. Busuta, J.E. Pogemiller, B.J. Zimmerman, Scaling of emission currents and of current fluctuations of gated silicon emitter ensembles, *J. Vac. Sci. Technol. B*, Vol 12, 689, 1994.

[30] M. Nakamoto, T. Ono, Y. Nakamura, in: *Proceedings of the International Vacuum Microelectronics Conference*, Late News # 28, 1993.

[31] M. Nakamoto, K. Ichimura, T. Ono, Y. Nakamura, in: *Proceedings of the International Vacuum Microelectronics Conference*, page 186, 1995.

[32] M. Nakamoto, T. Ono, T. Sakai, US Patent no. 5, no. 499, no. 938 (1996).

[33] M. Nakamoto and K. Fukuda, Field electron emission from LaB_6 and TiN emitter arrays fabricated by transfer mold technique, *Appl. Surf. Sci.*, Vol 202, 289-294, 2002.

[34] J.M. Lafferty, Boride Cathodes, *J. Appl. Phys.*, Vol 22, 299, 1951.

[35] R. Shimizu, Y. Kataoka, T. Tanaka, S. Kawai, Field Emission Pattern of LaB6-Single Crystal Tip, Jpn. *J. Appl. Phys.*, Vol 14, 1089, 1975.

[36] P.H. Schmidt, D.C. Joy, Low work function electron emitter hexaborides, *J. Vac. Sci. Technol.*, Vol 15, 1809, 1978.

[37] L.W. Swanson, D.R. McNeely, Work functions of the (001) face of the hexaborides of Ba, La, Ce and Sm, *Surf. Sci.*, Vol 83, page 11-28, 1979.

[38] M. Futamoto, M. Nakazawa, K. Usami, S. Hosoki, U. Kawabe, Thermionic emission properties of a single-crystal LaB_6 cathode, *J. Appl. Phys.*, Vol 51, page 3869, 1980.

[39] M. Futamoto, M. Nakazawa, U. Kawabe, Thermionic emission properties of hexaborides, *Surf. Sci.*, Vol 100, 470-480, 1980.

[40] M. Futamoto, M. Nakazawa, U. Kawabe, High temperature surface composition of hexaboride thermionic electron emitters, *Vacuum*, vol 33, page 727-732, 1983.

[41] D.M. Goebel, Y. Hirooka, T.A. Sketchley, Large-area lanthanum hexaboride electron emitter, *Rev. Sci. Instruments*, Vol 56, page 1717, 1985.

[42] P.R. Davis, L.W. Swanson, J. J. Hutta, D. L. Jones, Fabrication and characterization of rare earth hexaboride single-crystal materials, *J. Mater. Sci.*, Vol 21, page 825-836, 1986.

[43] L.T. Toth, *Transition Metal Carbides and Nitrides*, Academic Press, New York, 1971.

[44] J. Rivor, J.M. Behaghel, S. Berthier, J. Lafait, Optical properties of substoichiometric TiN_x, *Thin Solid Films*, Vol 78, page 161-165, 1981.

[45] J. M. Bonard, H. Kind, T. Stockli, L. O. Nilsson, Field emission from carbon nanotubes: the first five years, *Solid-State Electron.*, Vol 45, page 893-914, 2001.

[46] N. de Jonge, Y. Lamy, K. Schoots, T. H. Oosterkamp, High Brightness Electron Beam From a Multi-Walled Carbon Nanotube, *Nature*, Vol 420, page 393, 2002.

[47] N. S. Ramgir, I. S. Mulla, K. Vijaymohanan, D. J. Late, A. B. Bhise, M. A. More and D. S. Joag, Ultralow threshold field emission from a single multipod structure of ZnO, *Appl. Phys. Lett*, Vol 88, page 42107, 2006.

[48] A. B .Bhise, D. J. Late, N. S. Ramgir, M. A. More, I. S. Mulla, K. Vijaymohanan, D. S. Joag, Field emission investigations of RuO_2-doped SnO_2 wires, *Appl. Surf. Sci.*, 253 (2007) 9159-9163.

[49] H. M. Kim, T. W. Kang, K. S. Chung, J. P. Hong, W. B. Choi, Field emission displays of wide-bandgap gallium nitride nanorod arrays grown by hydride vapor phase epitaxy, *Chem. Phys. Lett.*, Vol 377, pages 491-494, 2003.

[50] Y. B. Li, Y. Bando, D. Golberg, K. Kurashima, Field emission from MoO3MoO3 nanobelts, *Appl. Phys. Lett.*, Vol 81, 5048, 2002.

[51] Y. B. Li, Y. Bando, D. Golberg, MoS_2 nanoflowers and their field-emission properties, *Appl. Phys. Lett.*, Vol 83, 1962, 2003.

[52] J. Zhou, N. S. Xu, S. Z. Deng, J. Chen, J.C. She, Z. L. Wang, Large-Area Nanowire Arrays of Molybdenum and Molybdenum Oxides: Synthesis and Field Emission Properties, *Ad. Mat.*, Vol 15, page 1835, 2003.

[53] W. J. Zhang, Y. Wu, W. K. Wong, X. M. Meng, C. Y. Chan, I. Bello, Y. Lifeshitz, S. T. Lee, Structuring nanodiamond cone arrays for improved field emission, *Appl. Phys. Lett.*, Vol 83, page 3365, 2003.

[54] M. Gesley, L. W. Swanson, A determination of the low work function planes of LaB_6, *Surf. Sci.*, Vol page 146 583-599, 1984.

[55] D. J. Late, *Synthesis and Characterization of LaB_6 thin films on tungsten, rhenium and silicon substrates and their investigations as Field Emitters*, PhD thesis, submitted to SPPU, formally University of Pune, India (Feb. 2008).

[56] A. A. Talin, K. A. Dean, J. E. Jaskie, Field emission displays: a critical review, *Solid State Electronics*, Vol 43, page 963-976, 2001.

[57] (a) https://www.nist.gov/image/fieldemissiontoprt1png (reproduced with permission from NIST, Accessed on 9 April 2020). (b) Review on "*Field Emission of Carbon Nanotubes* by Tharanga Kariyawasam, (c) J. M Bonard, H.Kind, ockli T. St and L.O. Nilsson, Field emission from carbon nanotubes: the first five years, *Solid-State Electronics* Vol 45, 893-914, 2001.

[58] R. V. Kashid, D. J. Late, S. S. Chou, Y. Huang, M. De, D. S. Joag, M. A. More, and V. P. Dravid, Enhanced field-emission behavior of layered MoS_2 sheets, *Small*, Vol 9, page 2730-2734, 2013.

[59] C. S. Rout, P. D. Joshi, R. V. Kashid, D. S. Joag, M. A. More, A. J. Simbeck, M. Washington, S. K. Nayak, and D. J. Late, Superior field emission properties of layered WS_2-RGO nanocomposites, *Sci. Rep.*, Vol 3, page 3282, 2013.

[60] C. S. Rout, R. Khare, R. V. Kashid, D. S. Joag, M. A. More, N. A. Lanzillo, M. Washington, S. K. Nayak, D. J. Late, Metallic Few-Layer Flowerlike VS_2 Nanosheets as Field Emitters, *Eur. J. Inorg. Chem.*, 2014, page 5331–5336.

[61] S. R. Suryawanshi, S. N. Guin, A. Chatterjee, V. Kashid, M. A. More, D. J. Late and K. Biswas, Low frequency noise and photo-enhanced field emission from ultrathin PbBi2Se4 nanosheets, *J. Mater. Chem. C*, Vol 4, page 1096—1103, 2016.

[62] S. Ratha, J. N. Behera, P. Bankar, M.A. More, D.J. Late, A. Gangan, B. Chakraborty, C.S. Rout, VSe_2-reduced graphene oxide as efficient cathode material for field emission, *Journal of Physics and Chemistry of Solids*, Vol 128, page 384-390, 2019.

[63] S. R. Marri, S. Ratha, C. S. Rout, J. N. Behera, 3D cuboidal vanadium diselenide embedded reduced graphene oxide hybrid structures with enhanced supercapacitor properties, *Chem. Commun.*, Vol 53, page 228–231, 2017.

[64] G. Bae, J. Cha, H. Lee, W. Park, N. Park, Effects of defects and non-coordinating molecular overlayers on the work function of graphene and energy-level alignment with organic molecules, *Carbon*, Vol 50, page 851–856, 2012 .

[65] C.S. Rout, P.D. Joshi, R. V. Kashid, D.S. Joag, M.A. More, A.J. Simbeck, M. Washington, S.K. Nayak, D.J. Late, Enhanced field emission properties of doped graphene nanosheets with layered SnS_2, *Appl. Phys. Lett.*, Vol 105, page 43109, 2014.

[66] A.K. Samantara, D.K. Mishra, S.R. Suryawanshi, M.A. More, R. Thapa, D.J. Late, B.K. Jena, C.S. Rout, Facile synthesis of Ag nanowire–rGO composites and their promising field emission performance, *RSC Adv.*, Vol 5, pp 41887–41893, 2015.

[67] J. Ding, X. Yan, J. Li, B. Shen, J. Yang, J. Chen, Q. Xue, Enhancement of Field Emission and Photoluminescence Properties of Graphene-SnO_2 Composite Nanostructures, *ACS Appl. Mater. Interfaces.*, Vol 3, 4299–4305, 2011.

[68] G. Viskadouros, A. Zak, M. Stylianakis, E. Kymakis, R. Tenne, E. Stratakis, Enhanced Field Emission of WS_2 Nanotubes, *Small.*, Vol 10, pp 2398–2403, 2014.

[69] E. Stratakis, G. Eda, H. Yamaguchi, E. Kymakis, C. Fotakis, M. Chhowalla, Free-standing graphene on microstructured silicon vertices for enhanced field emission properties, *Nanoscale.*, Vol 4, pp 3069–3074, 2012.

[70] G. Viskadouros, D. Konios, E. Kymakis, E. Stratakis, Electron field emission from graphene oxide wrinkles, *RSC Adv.*, Vol 6, pp 2768–2773, 2016.

[71] G. Viskadouros, D. Konios, E. Kymakis, E. Stratakis, Direct laser writing of flexible graphene field emitters, *Appl. Phys. Lett.*, Vol 105, pp 203104, 2014.

[72] G.M. Viskadouros, M.M. Stylianakis, E. Kymakis, E. Stratakis, Enhanced Field Emission from Reduced Graphene Oxide Polymer Composites, *ACS Appl. Mater. Interfaces.*, Vol 6 pp 388–393, 2014.

[73] M. Kumar, D. Y. Park, R. Singh, M. S. Jeong, T. Som and J. Kim, Growth of Wafer-Scale ReS_2 with "Tunable" Geometry toward Electron Field-Emission Application, *ACS Appl. Mater. Interfaces*, Vol 11 issue 39, pp 35845–35852, (2019)

[74] M. Rahman, K. Davey, S.Qiao, Advent of 2D Rhenium Disulfide (ReS_2): Fundamentals to Applications, *Adv. Funct. Mater.*, Vol 27, pp 1606129, 2017.

[75] J. Qin, W.Shao, Y. Li, C. Xu, X. Song, L. Zhen, D. Ren, van der Waals epitaxy of large-area continuous ReS_2 films on mica substrate, *RSC Adv.*, Vol 7, pp 24188–24194, 2017.

Chapter 9

Conclusion and Future Directions

Chandra Sekhar Rout[1,*]
and Brahmananda Chakraborty[2,3]
[1]Centre for Nano and Material Science, Jain University, Jain Global Campus, Jakkasandra, Ramanagaram, Banglore, India
[2]High Pressure and Synchrotron Radiation Physics Division, Bhabha Atomic Research Centre, Trombay, Mumbai, India
[3]Homi Bhabha National Institute, Mumbai, India

Introduction

Post the advent of graphene and its wondrous applications in numerous fields, there is much interest in other similar materials that can mimic the structural properties of graphene while showing improvements in other areas, such as electronic properties, availability, and invariably cost [1]. Transition metal dichalcogenides (TMDCs) have been investigated exhaustively to this end, and continue to gain attention towards various investigations and applications [2-4]. Through this book, we have attempted to bring the vista of information available on TMDCs in a systematic format, focusing first on their properties, with particular interest in the properties of metallic 2D TMDCs. Among the metallic and semi-conducting versions of 2D TMDCs, the former seems to be less studied [5]. It is essential to be aware of all the relevant properties of this material, which could shed some light on its possible applications.

Chapter 2 provides an overview of the various properties of TMDCs, starting from structural properties like phases and their orientations, electronic properties including band structure of the different phases, optical, and

* Corresponding Author's Email: r.chandrasekhar@jainuniversity.ac.in.

In: 2D Metallic Transition Metal Dichalcogenides
Editors: Chandra Sekhar Rout and Brahmananda Chakraborty
ISBN: 978-1-68507-965-9

magnetic properties [6-8] Practical applications of metallic 2D TMDCs are hampered by its high reactivity and readiness to undergo oxidation [9]. Strategies to modify these materials by functionalization, intentional introduction of dopants or defects, or hybridizing these structures with other materials can contribute significant changes in properties [10-12]. The investigation of material properties can be increasingly material-, time-, and cost-intensive, considering the enormous number of permutations that exist to functionalize these materials, and further research work is needed to assess whether the synthesized material will be suitable for a given application. Theoretical simulations and computational modelling can greatly reduce the strain on experimentalists by evaluating the properties and applications of the materials and assessing their potency for various applications [13, 14]. The intricacies of quantum simulations and their potential has been elaborated at length in Chapter 3.

We have summarized the recent advances in synthesizing metallic 2D TMDCs in Chapter 4, going beyond the popular methods of chemical exfoliations to more novel and facile techniques as reported in recent literature [15-17]. Maintaining the TMDCs in their metastable metallic phase can be accomplished by phase engineering, regarding which several exciting strategies have been designed [18-20]. Further, the success of a synthesis method invariably requires exhaustive characterizations, which have been expanded on [21, 22]. With the in-depth understanding of the properties, synthesis strategies, and characterization techniques of metallic 2D TMDCs, and the insight about how theoretical calculations can contribute to discovering new materials and their possible applications, we have then shifted our focus towards the applications of these materials.

Chapter 5 summarizes the chemical and biosensing applications of metallic 2D TMDCs. Sensors are important tools in diagnostics, environmental monitoring, routine testing, and so on. The need for non-invasive, inexpensive, and easily operable sensors for various applications is pronounced [23-25], and in these characteristics, 2D metallic TMDCs have shown great potential to be fabricated into sensor devices [26-28]. Other important societal problems of today include the energy crises, with depleting reserves of fossil fuels putting pressure on economies around the globe, fueling the development of alternate energy sources. Besides energy sources, energy storage devices can help facilitate the changeover from fossil fuels to alternate fuels, in both stationery and transport applications [29]. Chapter 6 gives an outline of energy conversion and storage strategies using metallic 2D TMDCs.

Catalytic applications of 2D metallic TMDCs, specifically in the catalysis of the water-splitting reaction, are upcoming and extremely important as a method for efficient hydrogen production [30-33]. Current methods include noble catalysts in the platinum group, which elevate costs. The produced hydrogen can fuel hydrogen-based vehicles making use of fuel cells for energy conversion. Further, electrocatalysts within fuel cells are also expensive Pt group metals and their variations, where again, metallic 2D TMDCs show ample promise as catalyst replacements [34, 35]. The electrocatalytic, photocatalytic, and fuel cell applications of these materials are expanded on in Chapter 7. Metallic 2D TMDCs also show promise in field emission, wherein high electrostatic fields can induce emission of electron streams [36, 37].

This property of TMDCs can be effectively utilized for the various applications of field emission such as microscopy, spectroscopy, electron emission guns, etc. The theoretical background and applications of TMDCs towards field emission have been explained in Chapter 8.

In the reporting of all these applications, an effort has been made to bring to the reader a comparison of the performance of different materials for the benefit of the reader. The merits and demerits of existing and conventional materials are juxtaposed with the performance of novel materials, with concerns of practical feasibility reasonably addressed. Our comprehensive study of the properties and applications of 2D metallic TMDCs have led us to believe that for bringing these materials into the mainstream commercially is a long way off with abundant scope available for improvement in reproducibility, facile synthetic methods, and application-specific device performance.

While semiconducting 2D TMDCs are comparatively well-researched, metallic 2D TMDCs are still an upcoming field. Considering the significant differences in properties between the two, the specific applications for which metallic TMDCs are more suitable are bound to motivate more research into development of these materials. Further, improvement in properties and application-based performance is noted for many cases of functionalized materials as compared to their pristine counterparts. This opens up infinite opportunities to tune materials and optimize their performance for a given application. Functionalization here includes doping, decorating, number of layers, distance between layers, vacancy defects, and/or generating hybrid structures using two different materials, and/or any combination of these, or more. Focus of upcoming research needs to be on stability of the as-prepared materials, reproducibility in synthesis as well as application, stability in device performance for assessing longevity, facile production methods, overall cost-effectiveness, etc. In conclusion, we hope this book provides the reader with a summative account of metallic TMDCs with reference to their properties and applications in a single platform, effectively.

References

[1] Fan, Feng Ru, Ruoxing Wang, Hua Zhang, and Wenzhuo Wu. "Emerging Beyond-Graphene Elemental 2D Materials for Energy and Catalysis Applications." *Chemical Society Reviews* 50, no. 19 (October 4, 2021): 10983-31. https://doi.org/10.1039/C9CS00821G.

[2] Manzeli, Sajedeh, Dmitry Ovchinnikov, Diego Pasquier, Oleg V. Yazyev, and Andras Kis. "2D Transition Metal Dichalcogenides." *Nature Reviews Materials* 2, no. 8 (June 13, 2017): 1-15. https://doi.org/10.1038/natrevmats.2017.33.

[3] Wei, Zhongming, Bo Li, Congxin Xia, Yu Cui, Jun He, Jian-Bai Xia, and Jingbo Li. "Various Structures of 2D Transition-Metal Dichalcogenides and Their Applications." *Small Methods* 2, no. 11 (2018): 1800094. https://doi.org/10.1002/smtd.201800094.

[4] Zhang, Yu, Yuyu Yao, Marshet Getaye Sendeku, Lei Yin, Xueying Zhan, Feng Wang, Zhenxing Wang, and Jun He. "Recent Progress in CVD Growth of 2D Transition Metal Dichalcogenides and Related Heterostructures." *Advanced Materials* 31, no. 41 (2019): 1901694. https://doi.org/10.1002/adma.201901694.

[5] Zhao, Bei, Dingyi Shen, Zucheng Zhang, Ping Lu, Mongur Hossain, Jia Li, Bo Li, and Xidong Duan. "2D Metallic Transition-Metal Dichalcogenides: Structures, Synthesis, Properties, and Applications." *Advanced Functional Materials* 31, no. 48 (2021): 2105132. https://doi.org/10.1002/adfm.202105132.

[6] Castro Neto, A. H. "Charge Density Wave, Superconductivity, and Anomalous Metallic Behavior in 2D Transition Metal Dichalcogenides." *Physical Review Letters* 86, no. 19 (May 7, 2001): 4382-85. https://doi.org/10.1103/PhysRevLett.86.4382.

[7] Yin, Xinmao, Chi Sin Tang, Yue Zheng, Jing Gao, Jing Wu, Hua Zhang, Manish Chhowalla, Wei Chen, and Andrew T. S. Wee. "Recent Developments in 2D Transition Metal Dichalcogenides: Phase Transition and Applications of the (Quasi-)Metallic Phases." *Chemical Society Reviews* 50, no. 18 (September 20, 2021): 10087-115. https://doi.org/10.1039/D1CS00236H.

[8] Shi, Jianping, Min Hong, Zhepeng Zhang, Qingqing Ji, and Yanfeng Zhang. "Physical Properties and Potential Applications of Two-Dimensional Metallic Transition Metal Dichalcogenides." *Coordination Chemistry Reviews* 376 (December 1, 2018): 1-19. https://doi.org/10.1016/j.ccr.2018.07.019.

[9] Chhowalla, Manish, Damien Voiry, Jieun Yang, Hyeon Suk Shin, and Kian Ping Loh. "Phase-Engineered Transition-Metal Dichalcogenides for Energy and Electronics." *MRS Bulletin* 40, no. 7 (July 1, 2015): 585-91. https://doi.org/10.1557/mrs.2015.142.

[10] Tong, Yun, Yuqiao Guo, Kejun Mu, Huan Shan, Jun Dai, Yi Liu, Zhe Sun, et al. "Half-Metallic Behavior in 2D Transition Metal Dichalcogenides Nanosheets by Dual-Native-Defects Engineering." *Advanced Materials* 29, no. 40 (October 2017): 1703123. https://doi.org/10.1002/adma.201703123.

[11] Sarkar, Deblina, Xuejun Xie, Jiahao Kang, Haojun Zhang, Wei Liu, Jose Navarrete, Martin Moskovits, and Kaustav Banerjee. "Functionalization of Transition Metal Dichalcogenides with Metallic Nanoparticles: Implications for Doping and Gas-Sensing." *Nano Letters* 15, no. 5 (May 13, 2015): 2852-62. https://doi.org/10.1021/nl504454u.

[12] Li, Shisheng, Jinhua Hong, Bo Gao, Yung-Chang Lin, Hong En Lim, Xueyi Lu, Jing Wu, et al. "Tunable Doping of Rhenium and Vanadium into Transition Metal Dichalcogenides for Two-Dimensional Electronics." *Advanced Science* 8, no. 11 (2021): 2004438. https://doi.org/10.1002/advs.202004438.

[13] Guan, Jie, Hsun-Jen Chuang, Zhixian Zhou, and David Tománek. "Optimizing Charge Injection across Transition Metal Dichalcogenide Heterojunctions: Theory and Experiment." *ACS Nano* 11, no. 4 (April 25, 2017): 3904-10. https://doi.org/10.1021/acsnano.7b00285.

[14] Fan, Shaoxun, Xiaolong Zou, Hongda Du, Lin Gan, Chengjun Xu, Wei Lv, Yan-Bing He, Quan-Hong Yang, Feiyu Kang, and Jia Li. "Theoretical Investigation of the Intercalation Chemistry of Lithium/Sodium Ions in Transition Metal Dichalcogenides." *The Journal of Physical Chemistry C* 121, no. 25 (June 29, 2017): 13599-605. https://doi.org/10.1021/acs.jpcc.7b05303.

[15] Sokolikova, Maria S., and Cecilia Mattevi. "Direct Synthesis of Metastable Phases of 2D Transition Metal Dichalcogenides." *Chemical Society Reviews* 49, no. 12 (June 22, 2020): 3952-80. https://doi.org/10.1039/D0CS00143K.

[16] Zhang, Zhepeng, Pengfei Yang, Min Hong, Shaolong Jiang, Guanchao Zhao, Jianping Shi, Qin Xie, and Yanfeng Zhang. "Recent Progress in the Controlled Synthesis of 2D Metallic Transition Metal Dichalcogenides." *Nanotechnology* 30, no. 18 (May 3, 2019): 182002. https://doi.org/10.1088/1361-6528/aaff19.

[17] Wang, Qichen, Yongpeng Lei, Yuchao Wang, Yi Liu, Chengye Song, Jian Zeng, Yaohao Song, Xidong Duan, Dingsheng Wang, and Yadong Li. "Atomic-Scale Engineering of Chemical-Vapor-Deposition-Grown 2D Transition Metal Dichalcogenides for Electrocatalysis." *Energy & Environmental Science* 13, no. 6 (June 19, 2020): 1593-1616. https://doi.org/10.1039/D0EE00450B.

[18] Ma, Rui, Huairuo Zhang, Youngdong Yoo, Zachary Patrick Degregorio, Lun Jin, Prafful Golani, Javad Ghasemi Azadani, et al. "MoTe2 Lateral Homojunction Field-Effect Transistors Fabricated Using Flux-Controlled Phase Engineering." *ACS Nano* 13, no. 7 (July 23, 2019): 8035-46. https://doi.org/10.1021/acsnano.9b02785.

[19] Wang, Mengmeng, Dawei Li, Kun Liu, Qitong Guo, Sumei Wang, and Xin Li. "Nonlinear Optical Imaging, Precise Layer Thinning, and Phase Engineering in MoTe2 with Femtosecond Laser." *ACS Nano* 14, no. 9 (September 22, 2020): 11169-77. https://doi.org/10.1021/acsnano.0c02649.

[20] Wang, Shuxi, John Cavin, Zahra Hemmat, Khagesh Kumar, Alexander Ruckel, Leily Majidi, Hamed Gholivand, et al. "Phase-Dependent Band Gap Engineering in Alloys of Metal-Semiconductor Transition Metal Dichalcogenides." *Advanced*

Functional Materials 30, no. 51 (2020): 2004912. https://doi.org/10.1002/adfm.202004912.

[21] Özdemir, Onur, Iñigo Ramiro, Shuchi Gupta, and Gerasimos Konstantatos. "High Sensitivity Hybrid PbS CQD-TMDC Photodetectors up to 2 Mm." *ACS Photonics* 6, no. 10 (October 16, 2019): 2381-86. https://doi.org/10.1021/acsphotonics.9b00870.

[22] Leonhardt, Alessandra, Daniele Chiappe, Valeri V. Afanas'ev, Salim El Kazzi, Ilya Shlyakhov, Thierry Conard, Alexis Franquet, Cedric Huyghebaert, and Stefan de Gendt. "Material-Selective Doping of 2D TMDC through AlxOy Encapsulation." *ACS Applied Materials & Interfaces* 11, no. 45 (November 13, 2019): 42697-707. https://doi.org/10.1021/acsami.9b11550.

[23] Javaid, Mohd, Abid Haleem, Shanay Rab, Ravi Pratap Singh, and Rajiv Suman. "Sensors for Daily Life: A Review." *Sensors International* 2 (January 1, 2021): 100121. https://doi.org/10.1016/j.sintl.2021.100121.

[24] Mathew, Minu, Sithara Radhakrishnan, Antara Vaidyanathan, Brahmananda Chakraborty, and Chandra Sekhar Rout. "Flexible and Wearable Electrochemical Biosensors Based on Two-Dimensional Materials: Recent Developments." *Analytical and Bioanalytical Chemistry* 413, no. 3 (January 1, 2021): 727-62. https://doi.org/10.1007/s00216-020-03002-y.

[25] Vaidyanathan, Antara, Minu Mathew, Sithara Radhakrishnan, Chandra Sekhar Rout, and Brahmananda Chakraborty. "Theoretical Insight on the Biosensing Applications of 2D Materials." *The Journal of Physical Chemistry B* 124, no. 49 (December 10, 2020): 11098-122. https://doi.org/10.1021/acs.jpcb.0c08539.

[26] Liao, Jianfei, Jianfei Liao, Lei Han, and Chaoyu Xu. "Comparison of the Sensitivity by SPR in a Metal-ITO-BlueP/TMDC Structure." *Applied Optics* 60, no. 17 (June 10, 2021): 5161-68. https://doi.org/10.1364/AO.425903.

[27] Rahman, M. Saifur, Md. Shamim anower, and Lway Faisal Abdulrazak. "Utilization of a Phosphorene-Graphene/TMDC Heterostructure in a Surface Plasmon Resonance-Based Fiber Optic Biosensor." *Photonics and Nanostructures - Fundamentals and Applications* 35 (July 1, 2019): 100711. https://doi.org/10.1016/j.photonics.2019.100711.

[28] Jiang, Li, Shuwen Zeng, Qingling Ouyang, Xuan-Quyen Dinh, Philippe Coquet, Junle Qu, Sailing He, and Ken-Tye Yong. "Graphene-TMDC-Graphene Hybrid Plasmonic Metasurface for Enhanced Biosensing: A Theoretical Analysis." *Physica Status Solidi (a)* 214, no. 12 (2017): 1700563. https://doi.org/10.1002/pssa.201700563.

[29] Singh, Sandeepa. "Energy Crisis and Climate Change." In: *Energy*, 1-17. John Wiley & Sons, Ltd, 2021. https://doi.org/10.1002/9781119741503.ch1.

[30] Frydendal, Rasmus, Elisa A. Paoli, Brian P. Knudsen, Björn Wickman, Paolo Malacrida, Ifan E. L. Stephens, and Ib Chorkendorff. "Benchmarking the Stability of Oxygen Evolution Reaction Catalysts: The Importance of Monitoring Mass Losses." *ChemElectroChem* 1, no. 12 (2014): 2075-81. https://doi.org/10.1002/celc.201402262.

[31] Ahmad, H., S. K. Kamarudin, L. J. Minggu, and M. Kassim. "Hydrogen from Photo-Catalytic Water Splitting Process: A Review." *Renewable and Sustainable Energy Reviews* 43 (March 1, 2015): 599-610. https://doi.org/10.1016/j.rser.2014.10.101.

[32] Chen, Shuai, Sapanbir S. Thind, and Aicheng Chen. "Nanostructured Materials for Water Splitting - State of the Art and Future Needs: A Mini-Review." *Electrochemistry Communications* 63 (February 1, 2016): 10-17. https://doi.org/10.1016/j.elecom.2015.12.003.

[33] Jamesh, Mohammed-Ibrahim, and Xiaoming Sun. "Recent Progress on Earth Abundant Electrocatalysts for Oxygen Evolution Reaction (OER) in Alkaline Medium to Achieve Efficient Water Splitting - A Review." *Journal of Power Sources* 400 (October 1, 2018): 31-68. https://doi.org/10.1016/j.jpowsour.2018.07.125.

[34] Sumesh, C. K., and Sebastian C. Peter. "Two-Dimensional Semiconductor Transition Metal Based Chalcogenide Based Heterostructures for Water Splitting Applications." *Dalton Transactions* 48, no. 34 (August 27, 2019): 12772-802. https://doi.org/10.1039/C9DT01581G.

[35] Gan, Xiaorong, Dangyuan Lei, and Kwok-Yin Wong. "Two-Dimensional Layered Nanomaterials for Visible-Light-Driven Photocatalytic Water Splitting." *Materials Today Energy* 10 (December 1, 2018): 352-67. https://doi.org/10.1016/j.mtener.2018.10.015.

[36] Patra, Abhinandan, Mahendra A. More, Dattatray J. Late, and Chandra Sekhar Rout. "Field Emission Applications of Graphene-Analogous Two-Dimensional Materials: Recent Developments and Future Perspectives." *Journal of Materials Chemistry C* 9, no. 34 (September 2, 2021): 11059 78. https://doi.org/10.1039/D1TC02054D.

[37] Giubileo, Filippo, Antonio Di Bartolomeo, Laura Iemmo, Giuseppe Luongo, and Francesca Urban. "Field Emission from Carbon Nanostructures." *Applied Sciences* 8, no. 4 (April 2018): 526. https://doi.org/10.3390/app8040526.

About the Editors

Chandra Sekhar Rout

Centre for Nano and Material Science, Jain University, Jain Global Campus, Jakkasandra, Ramanagaram, Banglore 562112, India

Dr. Chandra Sekhar Rout is a full Professor at Centre for Nano & Material Sciences, Jain University. Before joining CNMS, he was a DST-Ramanujan Fellow at Indian Institute of Technology Bhubaneswar, India (2013-2017). He received his BSc (2001) and MSc (2003) degrees from Utkal University and his PhD from Jawaharlal Nehru Centre for Advanced Scientific Research, India (2008) under the supervision of Prof. C.N.R. Rao, Bharat Ratna. He did his postdoctoral research at National University of Singapore (2008-2009), Purdue University, USA (2010-2012) and UNIST, South Korea (2012-2013). His research interests include preparation and characterization of two-dimensional layered nanomaterials and their nanocarbon hybrids for chemical sensors and biosensors, supercapacitors and energy storage devices, field emitters and electronic devices. He has authored or co-authored more than 150 research papers in international journals and edited 4 books. He has been awarded prestigious Ramanujan Fellowship and Young Scientist award of Department of Science and Technology, Govt. of India in 2013, Emerging investigator award 2017 from Elsevier, IAAM medal 2017 from International Association of Advanced Materials 2017, Young researcher award from Venus International Foundation, India in 2015. His current google scholar h-index is 44 with total citations ~9000. He was ranked top 2% scientists in India by the Stanford study in 2020. He is associate editor of "*RSC Advances*" a journal of Royal Society of Chemistry and "*American Journal of Engineering and Applied Sciences*" of Science Publications. He also reviews for various top scientific journals published by ACS, Wiley, RSC, Springer, IOP, AIP, Nature and ScienceDirect etc.

Brahmananda Chakraborty
High Pressure and Synchrotron Radiation Physics Division,
Bhabha Atomic Research Centre, Trombay, Mumbai 400085, India
Homi Bhabha National Institute, Mumbai 400094, India

Dr. Brahmananda Chakraborty is a scientist in High Pressure & Synchrotron Radiation Physics Division of Bhabha Atomic Research Centre, Mumbai and Associate Professor in Homi Bhabha National institute. His current research interest includes Density Functional Theory (DFT) & Molecular Dynamics (MD) simulations of materials for energy materials, biosensors, reactor fuel and structural materials, nanomaterials and behavior of materials under high pressure. Dr. Chakraborty got his PhD degree from IIT, Bombay for his research topic on 'Irradiation Effects and Hydrogen Storage on Carbon Nanotubes'. He did two years post-doctoral research in North Carolina State University, USA. He received International Association of Advanced Material (IAAM) Scientist Medal in 2017. He has published more than 120 papers in reputed international journals.

He has developed several software packages for material simulations using Molecular Dynamics and Reactor Simulations using Monte Carlo Techniques. He has experience teaching Density Functional Theory at North

Carolina State University, USA and Computational Physics at Homi Bhabha National Institute, Mumbai. Dr. Chakraborty is also a regular reviewer in reputed international journals like Journal of Physical Chemistry, CARBON, Physical Chemistry Chemical Physics etc. Dr. Chakraborty has contributed a lot in Department of Atomic Energy's Nuclear Programme in designing various reactors.

List of Contributors

Aaryashree - Shibaura Institute of Technology, Toyosu, Koto City, Tokyo 1358548, Japan.

K. A., Sree Raj - Centre for Nano and Material Science, Jain University, Jain global campus, Jakkasandra, Ramanagaram, Banglore - 562112, India.

Kandasamy, Manikandan - Nonlinear Optical Materials Laboratory, Department of Physics, Bharathidasan University, Tiruchirappalli 620 024, Tamil Nadu, India.

Late, Dattatray J. - Centre for Nanoscience & Nanotechnology Amity University Maharashtra, Mumbai-Pune Expressway, Bhatan, Post – Somathne, Panvel, Mumbai, Maharashtra 410206, India.

Nair, Heera T. - School of Pure and Applied Physics, Mahatma Gandhi University, Kerala, India.

Pardhi, Avinash Macchindra - Department of Chemistry, Ramnarain Ruia Autonomous College, Mumbai, India 400019.

Sanyal, Gopal - Mechanical Metallurgy Division, Bhabha Atomic Research Centre, Trombay, Mumbai 400085, India.

Seetharaman, Amreetha - Nonlinear Optical Materials Laboratory, Department of Physics, Bharathidasan University, Tiruchirappalli 620 024, Tamil Nadu, India.

Vaidyanathan, Antara - Department of Chemistry, Ramnarain Ruia Autonomous College, Mumbai, India 400019.

Wagh, Vaibhav - Department of Chemistry, Ramnarain Ruia Autonomous College, Mumbai, India 400019.

Index

D

E

F

G

H

I

J

K

L

M

S

T

V

X